THE ANTHROPOCENE

T0075506

THE ANTHROPOCENE

101 Questions and Answers for Understanding the Human Impact on the Global Environment

B. L. TURNER II

agenda
publishing

© B. L. Turner II 2023

This book is copyrighted under the Berne Convention.
No reproduction without permission.
All rights reserved.

First published in 2023 by Agenda Publishing

Agenda Publishing Limited
The Core
Bath Lane
Newcastle Helix
Newcastle upon Tyne
NE4 5TF
www.agendapub.com

ISBN 978-1-78821-511-4 (paper)
ISBN 978-1-78821-512-1 (ePdf)
ISBN 978-1-78821-513-8 (ePub)

British Library Cataloguing-in-Publication Data
A catalogue record for this book is available from the British Library

Typeset by Newgen Publishing UK
Printed and bound in the UK by CPI Group (UK) Ltd, Croydon, CR0 4YY

CONTENTS

CONTENTS

CONTENTS

PREFACE

The initial inspiration for this book evolved from casual conversations with individuals who ultimately became an explicit cohort of sorts: those who engage in small cruise ship adventures, a cohort I did not encounter until later in my life when my wife insisted that we engage in such adventures. She especially enjoys the ad hoc seating arrangements at dinner involving four to six other people, usually couples, and the conversations that ensue. At the beginning of a voyage, when the cast of cruisers is unfamiliar, each seating begins with introductions of the place called home and a rendition of past cruise experiences, before questions pop up about one's profession, usually past professions given the number of retirees present. Once I am identified as a researcher of the geographical, environmental and sustainability sciences, queries from the more inquisitive or provocative at the table invariably arise about climate change and, in rare cases, environmental change more broadly. These questions range from informed individuals seeking to glean a better understanding of some facet of the state of the Earth system, such as sea-level rise, to politically inspired, rhetorical questions challenging environmental change and the anthropogenic role.

I recall vividly a highly educated couple who denied that climate warming was taking place, human-induced or not, and that the evidence on which that science is based was, well, not really scientific evidence. That couple actively engaged in a lengthy conversation as we walked through the "evidence" of the changes and inability of natural causes to explain the scale and pace of those changes. We eventually made it to the uncertainties for economies and the environment at large, the possible responses – mitigation or adaptation – that might be taken to address warming, and the role of international agreements given those uncertainties. It was at this point in the discussion that one of the couple conceded that human-induced climate warming might be real. This concession, however, was followed with the observation that international agreements in-the-making (i.e. what would become the Paris climate agreement) would privilege parts of the world, China especially, relative to the United States, in turn affecting the "bottom line" of the company owned by the individual speaking.

This discussion had moved from a denial of a physical reality and the disbelief in climate change science to the underlying reason for the denial and critique in the first place. This conversation, from my first cruise, was subsequently followed, if sporadically, by others grounded in problems that I characterize as those of the Anthropocene. These experiences generated the idea for a question-and-answer book for those holding a mild interest in those problems but, for whatever reason, did not undertake the labour of extensive internet searches to engage their queries.

Another inspiration followed directly from my profession. I have instructed in universities for going on 50 years. For 40 of them, I have not used textbooks because I have never found one that fits the full content of the courses I teach. The textbook dilemma follows partly from my research expertise, what today we label the human–environmental sciences and its subset field of land systems and sustainability. These sciences are only a few of the increasing number that are integrative across research fields; they seek to understand the dynamics and consequences of the interactions between the human/social and biogeophysical worlds, drawing on the past and present and projecting possible futures. The extant texts for these sciences tend to be anchored in the concerns and approaches of specific disciplines. More so, none of the texts fully captured all of the core dimensions and topics that I consider important for my students (or the public at large) to become "literate" about the human–environmental science: to understand the nature of the Anthropocene, the Earth system processes that maintain the biosphere, the history of human impacts on the biosphere, and the role of sustainability science to provide understanding of the social and environmental conditions of the Anthropocene. Developed over a quarter of a century, my course content influenced the structure and organization of the questions and answers that follow. As such, this book may serve as companion volume for a multitude of human–environmental and sustainability courses that now dot the academic landscape.

The range of the 101 questions and answers found in this work crosses many environmental and social sciences, and a few humanistic fields. No individual can claim research expertise across this range. In some cases, I have spent my professional life researching and writing on the question; in many cases, however, I synthesize from the literature, even moving into topics that are relatively new to me. As such, some answers provide more depth and complexity than others, although in all cases I have attempted to provide succinct syntheses. I focus on what I interpret to be the consensus of the expert community, but endeavour to identify challenges or contested interpretations of the answers.

This book has been too long in the making, impeded by my research and that of my graduate students, and the multiple obligations of all active faculty members. A nine-month sabbatical in 2020, for which I thank Arizona State University (ASU), provided the time to complete a rough draft based on starts and stops over the previous several years. Completion proved more complicated than I had imagined, in part owing to my own "catching up" on various advances across the various research fields on which I draw, the constant new discoveries that appear in various research fields touched upon, and marshalling the permission to use the graphics or their redrafting from other publications. In regard to the last, I am neither sufficiently patient nor interested in the systematic tracking of permissions or the details of graphic design and preparation. I am, therefore, indebted to Garima Jain for her diligence in sourcing the graphics and tables and gaining the permission for their use and seeking alternative graphics, in some cases, and to Barbara Trapido-Lurie who took over the graphics production, keeping a detailed eye on the products, finding alternative figures as needed, and managing all activities within the two units at ASU handling the production of the graphics. Without their assistance, this book would not have moved forward.

I also thank Shea Lemar (director), Yueling Li (research specialist) and James Ruberto (research aide) of the Geospatial Research Solutions (School of Geographical Sciences and Urban Planning–SGSUP) and Megan Joyce of VisLabs (School of Life Sciences), both units at ASU, for their graphic products. Heidi McGowan and Angelene Capello assisted with the various payments required for the production of figures. Appreciation is given to all of the entities permitting the use of their graphics or our simulations of them. A special thanks is given to T. Beach and S. Luzzadder-Beach for the use of their aerial and lidar data and the images from them produced by B. A. Smith of the Beach Labs in Geography and Environment and the Programme for Belize Archaeology Project, at the University of Texas at Austin.

Throughout the preparation of this work I called on my fellow faculty members at ASU, foremost in the SGSUP, School of Sustainability and School of Life Sciences, especially Randall Cerveny, Ronald Dorn, Matei Georgescu, Osvaldo Sala, Nancy Grimm and Susanne Neuer. For Question and Answer 80, I turned to my brother, Dr Matt W. Turner, a humanist with a strong botanical background, to formulate the "elaboration". He has always been an inspiration to me. My current and former doctoral advisees have long been a wealth of knowledge to be thanked. Finally, I recognize my formal mentor, William M. Denevan, and subsequent mentor-colleagues, the late Robert W. Kates and my former colleagues at Clark University, and William C. Clark, who influenced my formulation of human–environmental science, from historical to contemporary sustainability themes.

Two final appreciations must be given. Camilla Erskine of Agenda Publishing walked me through the entire production effort, offering helpful insights throughout the process, keeping my prose on track as much as anyone can, and editing the graphics. Carol Snider, my wife, gave me great latitude during my sabbatical time and subsequent summer, providing the time to complete this work while reducing the activities that she wished to undertake. Absent this time, perhaps I would not have completed the effort.

B. L. Turner II
Tempe, Arizona

METRICS AND MEASURES

Elements and compounds

CO_2	carbon dioxide
CH_4	methane
CFC	chlorofluorocarbon
H_2O	water
HCFC	hydrochlorofluorocarbon
NH_3	ammonia
NO_x	nitrogen oxides
NO_3	nitrate
N_2O	nitrous oxide
O_3	ozone
Pu	plutonium
SO_2	sulphur dioxide

Groups of elements and compounds

GHG	greenhouse gases
ODS	ozone-depleting substances

Area

gha	global hectares
ha	hectare = 1,000 square metres or 2.47 acres
ha^{-1}	per hectare
km^2	square kilometre = 100 ha or 0.38 square miles (mi^2)
Mkm^2	million square kilometres
m^2	square metre or 10.7 square feet
MH	million hectares

Distance/length/size

cm	centimetre = 0.30 inch
km	kilometre = 1,000 m or 0.62 miles
m	metre = 1.09 yards
micron (μm)	micrometre = one-millionth of a metre
mm	millimetre = 0.10 cm or 1,000th m or 0.039 inch
nm	nanometre = one-billionth of a metre

Concentration

mg/l	milligrams per litre
$mg\ l^{-1}\ O_2$	milligrams per litre of oxygen
$nmol\ mol^{-1}$	nanomole per mole or one-billionth of a mole; a mole is $6.02214076 \times 10^{23}$ particles
ppm	parts per million
ppb	part per billion
Ω_{ar}	aragonite saturation level

Energy

W	watt or power of one joule of work per second or 1/746 horsepower
W/m^2	watt per square metre (energy/area)

Rate and speed

$gC/m^2/y$	grams of carbon per square metre per year
ha min^{-1}	hectares per minute
km/hr	kilometres per hour (0.62 miles/hr)
$mg/m^3\ yr^{-1}$	milligram per cubic metre per year
$mtCO_2e^{yr-1}$	metric tonnes of carbon dioxide equivalents (all greenhouse gas emissions converted to CO_2 warming power) per year
MSY	million species year or number of extinctions per 10,000 species per 100 years
PBq	petabecquerel or 10^{15} becquerels where Bq is radioactive material decay of one nucleus per second
yr^{-1}	per year (rate); some graphics use $^{yr-1}$

Sound

dB	decibel or relative unit of measurement of the amplitude of sound

Salinity

g/l	gram per litre = 0.13 ounces per gallon

Acidity

pH	acidity or basicity of water-based solutions (1–14 in which <7 = acid; >7 = basic)

Temperature

°C	degrees Celsius or Centigrade (0 = freezing point of water)
°F	degrees Fahrenheit (32 = freezing point of water)
°K	degrees Kelvin (273.15 = freezing point of water)

Volume

km^3	cubic kilometre = 0.24 cubic miles
litre	1,000 cubic centimetres = 1.74 pints or 0.26 gallons

Weight

CO_2e	carbon dioxide equivalents of greenhouse gases, in tonnes
g	gram or 0.035 ounces
gC	grams of carbon
Gg	gigagram in which 1 Gg equates to 1,000 tonnes
Gt	gigaton or 1 billion metric tonnes
GtC/y	gigaton of carbon per year
Kt	kiloton or 1,000 tonnes
Mg	megagram or 1 tonne
mg	milligram = 1/1000 grams
Mt	megatonne = one million metric tonnes
mt	metric tonne (see tonne)
Pg	petagram = 1 quadrillion or 1 billion metric tons
tC ha^{-1}	tonnes of carbon per hectare
Tg	teragram = one trillion grams
ton	imperial measure = 1,016.047 kg or 2,240 lbs
tonne	metric measure = 1,000 kg or 2,204.6 lbs

Time

BCE	before common era replaces BC (before Christ)
CE	common era replaces AD (Anno Domini) or after Christ (colloquial)
kya	thousand years ago
Mya	million years ago
ya	years ago

INTRODUCTION

Our species, *Homo sapiens* (or wise human!), evolved during the last two seconds of Earth's geological clock (Fig. 0.1), which, in our case, started ticking about 4.6 billion years ago. In those two seconds we have risen to become the dominant species on Earth, reaching the capacity during the last one-tenth of a second to rival the forces of nature in our influence on the Earth system. This ascendancy is a chronicle of evolving human–environment relationships. For our purposes, it begins during the Late Palaeolithic (or Late Stone Age, about 100,000 to 70,000 years ago) as migrations of modern humans out of Africa ultimately reached the far corners of all the continents, apart from Antarctica. That chronicle continues today through rapidly advancing technological capacities that supply the demands of nearly eight billion people whose material consumption, on average, is at an unprecedented per capita level. In the process, we have changed – more often than not degraded – the natural capital and environmental services of the Earth system (nature) that have supported our rise to dominance. For most of this ascent, our environmental impacts were local to regional in scale and our responses to declines in environmental services involved moving to new locations, importing resources from afar, or innovating management and technology to maintain or improve those resources. The human transformation of the planet in the last one-tenth of a second of the clock, however, not only challenges these responses but has become global in scope and threatens the functioning of the Earth system.

This transformation raises serious concerns about the capacity of the Earth system to sustain the biosphere – the very part of our planet that our species occupies. The human–environmental conditions that create these concerns are called the Anthropocene. The implications of the Anthropocene for society and the environment are variously interpreted, however, be they the lessons of human-induced environmental changes in the past to forecasts of the well-being of society and the environment in the future. The differences in these interpretations reside partially in the evidence and significantly in the worldviews and values of the interpreter. Society at large navigates among the interpretations through its day-to-day behaviour and the signals sent (e.g. through voting) to its decision-makers. The signals sent could be informed by the state of the science – both the evidence and interpretations – relevant to understanding past, present and future human–environment relationships.

This book provides a baseline for understanding these relationships. Through a series of interconnected questions and answers, it addresses: (1) the concept of the Anthropocene

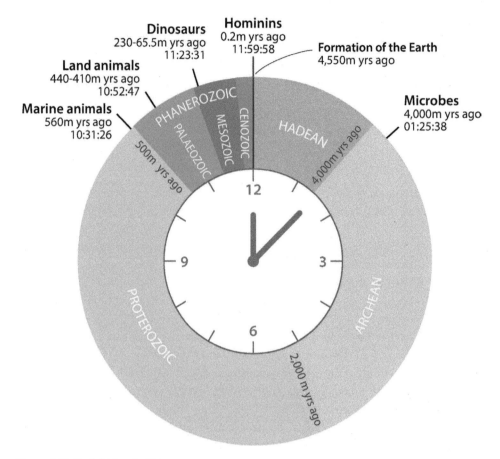

Figure 0.1: Earth history in 12 hours

Condensing our planet's history into a 12-hour period, our species would appear only in the last two seconds, and the Anthropocene perhaps one-tenth of the last second. Different depictions of Earth history clocks as well as different assessments of Earth history apply somewhat different time and date ranges for the appearances of different forms of life. *Source*: adapted from Wikimedia Commons (commons.wikimedia.org/wiki/File:Geologic_Clock_with_events_and_periods. svg; accessed 12 February 2022).

and the broad-stroke history of human-induced changes in the environmental conditions of the planet that provide different indicators of the emergence of this period; (2) the major human impacts currently inscribed on the Earth system and their consequences; (3) the different causes and rationales applied to understanding the changes made; and (4) various elements of sustainability science, the field of study that informs society about sustainable development in the Anthropocene.

The connections among the themes, phenomena and processes in this book are expansive and complex, requiring multiple encyclopedias to cover their entirety. Indeed, various collections exist for the Anthropocene, Earth system and sustainability (DellaSala & Goldstein 2017; Meyers 2012; Nierenberg 1992). This work differs from these compendia in several ways.

First, it takes a broad, historical sweep of human impacts on the Earth system, providing insights on the past and present activities and consequences as they affect interpretations of the Anthropocene and sustainability. The topics addressed are those this author finds important to that history, the current human–environment condition, and the science of sustainability. The empirical evidence applied to the topics and the interpretations of this evidence is presented through a science-based lens focused on what was and is, not what ought to be. Moral and philosophical considerations of the topics, as important as they may be, are underemphasized for the most part.

Second, a question-and-answer (Q&A) format attempts to balance the breadth and depth of knowledge of the topics included. A brief answer, typically a sentence or two, follows each question. An elaboration of the brief answer deliberately reduces the complexities of the issues explored, but provides a more complete answer and justifications for it as well as references to support the claims and metrics presented. These references include new evidence or claims in the literature as they appeared up to July 2022. Those that are especially new and novel must be weighed in light of the fact that they have yet to stand the test of subsequent research attention that support or not or alter the observations of the new work (e.g. Clements *et al.* 2022).

Third, in most cases, the Q&As are expected to be read as needed and not necessarily in their numerical order. As such, the questions are cross-listed in the text based on their relevance with one another. For example, **Question 94** asks if it is possible to place a monetary value on the Earth system. Part of the answer resides in the economic value of environmental services, a core concept used throughout the book. In **Question 94** and elsewhere among other answers, these services are coded to the original question addressing them, in this case **Q7**, guiding the reader to the elaboration of the phenomenon which is not found in **Q94**.

Fourth, the Q&As are organized in sections, as noted below, to provide coherence. Each section introduces the rationale for the Q&As within it and provides insights of relevance to the section theme not otherwise found in the Q&As. Such insights are especially important for those sections addressing the human impacts on the different spheres of the Earth system (Sections III–VI) relative to natural drivers of change. These sections, therefore, briefly introduce the natural drivers, identifying those dimensions that differ from human-induced changes.

The framing of the Q&As could be arranged in various ways. The approach adopted takes us from the meaning of the Anthropocene, through the changes in the Earth system created by societies past and present, the broader, science-based means of understanding these changes, and the emerging field seeking to address a more sustainable human–environment relationship.

Section I. identifies the meaning of the human–environmental relationships of the Anthropocene, the Earth system and global environmental change. Attention is given to the roles of the Anthropocene as a concept and, possibly, a geological time unit or event.

Section II. The exact timing of the start of the Anthropocene is variously interpreted. Arguments and evidence about its antiquity and recency exist. A broad-stroke history identifies the commonalities and distinctions of past and current human-induced environmental change. This history provides a basis from which to interpret the emergence of the Anthropocene, including those dimensions involved in considerations of the Anthropocene as a formal geological unit. Major human impacts on the stocks and flows of the Earth system are identified for different periods of human history, including the evidence, often contested, that the impacts in question were the outcome of natural forcing or drivers in the Earth system or some combination of them and human activities.

Sections III–VI. What are the major human impacts on environments globally and what Earth system consequences have followed from these impacts? These four sections explore the nature and scale of current human changes through the different "spheres" comprising the Earth system: land cover of the lithosphere, hydrosphere, atmosphere and biosphere – or the land surface, water, air and life, respectively. Each sphere is essential to functioning of the Earth system and has been changed variously by human activities. The spheres differ, however, in the discreteness of their geographic boundaries and, as may be expected in a system, the environmental processes interact among the spheres, many flowing across them. Notably, the land surface per se is not considered a sphere, although it covers the lithosphere. The atmosphere is distinctive in its position relative to the lithosphere. In contrast, the hydrosphere and biosphere physically cross-cut one another and the other two spheres. In addition, biogeochemical cycles, a critical dimension of the Earth system, flow through the spheres.

These dimensions require decisions about which phenomena and processes to treat in which section. With a few exceptions, the decision lies in the sphere space in which the phenomenon physically exists or the process takes place, or in which the phenomenon or process receive the most research attention. For example, carbon cycling is the centrepiece of climate warming; it and other greenhouse gas Q&As are addressed primarily in the atmosphere section, despite their cross-sphere flows. The most abundant human-induced change to the terrestrial surface of the Earth – part of the lithosphere – is that of land cover, captured largely by vegetation changes in terms of spatial dimensions. As such, land-cover/vegetation change is placed in the lithosphere/land cover section, despite biota constituting life and, hence, the biosphere. The hydrosphere flows through the other spheres in different forms (i.e. liquid, solid, gas), but all dimensions of water are treated within the context of the hydrological cycle, regardless of their spatial location. Given the vegetative link to land cover, the biosphere section focuses on the biological kingdom of Animalia, both terrestrial or marine, and microorganisms.

Section VII. How do we explain the underlying human drivers that generate the environmental outcomes? Human innovations to overcome the vagaries of the biophysical environment to improve the provisioning of food, fibre, fuel and shelter and to reduce environmental hazards, such as floods and droughts, mark the history of human–environment relationships. In some cases, the technologies and strategies employed in these efforts enhance environmental services and attenuate hazards. In others, however, they degrade the environment and/or lead to unintentional Earth system consequences. This section examines the driving forces that have been advanced for the changes in question, ranging from the demands on nature from the growth in the global population of our species to differing values and norms of societies. The various roles and supporting evidence proposed for each cause and the challenges to it are addressed.

Section VIII. Experts differ in their interpretations of the dynamics of human–environment relationships and the future consequences of dynamics now under way. These differences have as much to do with worldviews of the interpreter as they do with the evidence. This section addresses those worldviews as entertained by experts committed to perspectives grounded in mainstream science. It begins with Q&As about the character of the integrative science addressing the Anthropocene, followed by considerations of opposing worldviews, with examples.

Section IX. How do we deal with the conditions of the Anthropocene? Is it possible to create a sustainable human–environment relationship and what would that relationship entail? These are the questions of sustainability science, the emergence of which is examined. The goals and content of sustainability science are identified, followed by considerations of its various dimensions, including issues such as the vulnerability and resilience of human–environmental systems and the capacity to measure sustainable development.

Strategies to address sustainable human–environment conditions are limited to the broader themes for which consensus exists among the expert community, such as the need for adaptive management. Specific strategies and the technologies associated with them are not discussed. There are several reasons for this omission. Many of the strategies proposed to mitigate climate warming, for example, involve controversial approaches, such as placing aerosols in the atmosphere to block solar radiation. Furthermore, the means of assessing the costs of mitigating strategies versus subsequent adaptations to the changes as they occur involve significant disagreements among the expert communities, such as the appropriate discount rates to apply in assessments of the costs. These two examples alone would require expansive text to treat adequately, let alone address the other issues at play, such as the means to calculate environmental services that are not part of the market system.

The text includes a large range of metrics and measures dealing with environmental changes and associated drivers of those changes. The International System of Units (SI) is used in most cases. Some literature, however, uses the term "ton" in United States-based sources without reference to a US spelling of tonne or to the use of the imperial system. If ton is used in context of all other measures in the metric system or if the cited reference for the measure comes from a source that invariably employs metric measures, such as the United Nations, this text assumes that the US spelling of tonne has been employed and the metric spelling used. If neither of these options proves viable, the imperial/US figures are provided as reported in the source from which they were taken. In addition, scientific notations of metrics and measures are used throughout. As an example, 10 tonnes per hectare is noted as $10 \, t^{ha-1}$, not 10 t/ha. In some complex cases, however, measures use the per sign (/) as in $gC/m^2/y$. The notations are listed in the section on metrics and measures.

As noted, the sources of the data and claims for each answer elaboration are referenced, providing a starting point for those readers seeking more in-depth discussion of the topic. In almost all cases, references involve research articles appearing in English language journals and available online through their DOIs and Google Scholar or other scholarly search engines. Books, encyclopedias and non-journal internet sources are sparingly referenced, with the exception of major reports, largely from international agencies, such as the Intergovernmental Panel on Climate Change and the United Nations. Numerous figures are drawn from digital sources, however. Note that figure and table references are treated as reference material and placed in the reference section of the individual Questions and Answers unless inappropriate to do so. In addition, research articles overwhelmingly focus on those that have appeared after 2000. Older articles are sparingly cited, but are referenced if they were seminal to a particular issue and a more recent article does not fully cover the original base theme. For each Q&A, references that provide reviews and syntheses or constitute a starting point for exploration are identified in **bold** print in the reference list.

The text uses a large number of abbreviations and acronyms, be they for elements and compounds in the Earth system or various metrics, time periods, and so forth. These are listed in the Metric and Measures section at the beginning of the book. Definitions of the phenomena, processes and concepts referred to in the book are listed in the Glossary at the end of the book. With a few exceptions, the text does not refer to the international science programmes or activities associated with changes in the Earth system or sustainability interests. A list and brief description of those that are mentioned can be found in the Appendix.

References

Clements, J. C. *et al.* (2022). "Meta-analysis reveals an extreme 'decline effect' in the impacts of ocean acidification on fish behavior". *PLoS Biol* 20 (2): e3001511. doi.org/10.1371/journal.pbio.3001511.

DellaSala, D. & M. Goldstein (eds) (2017). *Encyclopedia of the Anthropocene*. Amsterdam: Elsevier.

Meyers, R. (ed.) (2012). *Encyclopedia of Sustainability Science and Technology*. Berlin: Springer.

Nierenberg, W. (ed.) (1992). *Encyclopedia of Earth System Science*. San Diego, CA: Academic Press.

SECTION I

THE ANTHROPOCENE AND THE EARTH SYSTEM: FOUNDATIONAL CONCEPTS

Never before has humankind dominated the planet Earth as it does today, provisioning, if inequitably, 7.8 billion people, at the historically highest level of material life on a per capita average (Fig. I.1). This achievement has relied on advances in technologies and strategies, accompanied by changing political economies, that, in general, have increasingly stressed natural resources, local to regional environments, and the functioning of the Earth system. The Anthropocene emerges from the resulting human–environment conditions **(Q1)**, a period challenging the capacity of the Earth system to maintain itself suitable for life. Securing a more equitable provisioning of material life for the nearly 10 billion people expected to exist by the middle of the twenty-first century raises concerns (Clark & Harley 2019). This provisioning is desired without reducing the capacity of the Earth system to provide the resources and environmental conditions which humankind expects (Griggs *et al.* 2013). As such, the conditions of the Anthropocene and Earth system, linked to ways in which we seek to understand them, provide a means by which questions of more sustainable human–environment relationships (Section IX) may be addressed. This section introduces foundational concepts and the key phenomena and process critical to the concepts of the Anthropocene and Earth systems, including several important dimensions of the Earth system that cross multiple thematic sections of the book.

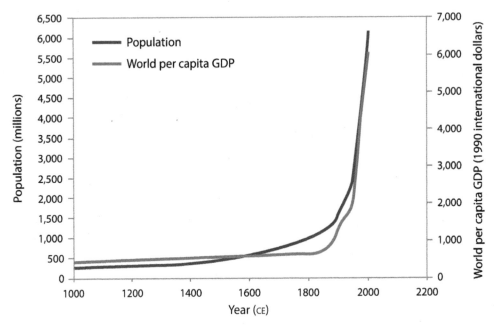

Figure I.1 Global population and per capita gross domestic product (GDP) over 1,000 years
1 international dollar would purchase an amount of goods and services comparable to that of 1 US$ in the United States.
Source: Šlaus & Jacobs (2011).

References

Clark, W. & A. Harley (2019). "Sustainability science: towards a synthesis". *Annual Reviews in Environment and Resources* 45: 331–86. doi.org/10,1146/annurev-environ-012420-043621.

Griggs, D. *et al.* (2013). "Sustainable development goals for people and planet". *Nature* 495(7441): 305–307. doi.org/10.1038/495305a.

Šlaus, I. & G. Jacobs (2011). "Human capital and sustainability". *Sustainability* 3(1): 97–154. doi.org/10.3390/su3010097.

Q1

WHAT IS THE ANTHROPOCENE?

Short answer: the current human–environment condition in which human activity matches and, in some cases, exceeds nature as a force of change in the Earth system.

While not the first to use the term, the late Nobel laureate, Paul Crutzen, advanced the concept of the Anthropocene through a series of publications beginning in 2002 (Crutzen 2002). The term refers to a new human–environmental relationship in which humankind has become a force that equals or exceeds the natural forcings shaping the Earth system **(Q3)**, such as human-induced climate change. This force derives from our actions that change the states or stocks of phenomena comprising the Earth system and the biogeo-chemical cycles or flows **(Q4)** that maintain those conditions. The subsequent changes, in turn, improve or degrade environmental services **(Q7)** and the functioning of the biosphere, which support humankind and life in general (Crutzen & Steffen 2003; Steffen, Crutzen & McNeill 2007).

Each new technological phase in the advancement of our species, associated with increasing global population, has amplified existing demands and generated new pressures on the states and flows of the Earth system. The human footprint of today **(Q96)** is sufficiently large to warrant the Anthropocene identity **(Q9)**, although debate exists over how distant in the past the concept is applicable **(Q15)**. Especially since the "Great Acceleration" (Monastersky 2015; Steffen *et al.* 2015) in population and affluence arising in the mid-twentieth century **(Q62 , Q63, Q64)**, the complexity and pace of changes are such that there are few, if any, analogues in our planet's history to guide the anticipation of the consequences (Steffen, Crutzen & McNeill 2007; Zalasiewicz *et al.* 2015). Owing to the conditions prevailing in this new relationship, various arguments declare that perhaps the Anthropocene constitutes a new geological time unit, an epoch or stage **(Q2)**. Regardless of this proposal, the Anthropocene has become the moniker for contemporary human–environmental relationships (Zalasiewicz *et al.* 2021). It is conceptually powerful and a useful heuristic to examine those relationships.

References

Crutzen, P. (2002). "The 'anthropocene'". *Journal de Physique IV (Proceedings)* 12(10): 1–5. doi.org/10.1051/jp4:20020447.

Crutzen P. & W. Steffen (2003). "How long have we been in the Anthropocene?" *Climatic Change* 62(3): 251–7. doi:10.1023/B:CLIM.0000004708.74871.62.

Monastersky, R. (2015). "Anthropocene: the human age". *Nature News* 519(7542): 144–7. doi.org/10.1038/519144a.

Steffen, W., P. Crutzen, & J. McNeill (2007). "The Anthropocene: are humans now overwhelming the great forces of nature?" *Ambio* 36(8): 614–21. doi.org/10.1579/0044-7447(2007)36[614:TAAHNO]2.0.CO;2.

Steffen, W. *et al.* (2015). "The trajectory of the Anthropocene: the great acceleration". *Anthropocene Review* 2(1): 81–98. doi.org/10.1177/2053019614564785.

Zalasiewicz, J. *et al.* (2015). "When did the Anthropocene begin? A mid-twentieth century boundary level is stratigraphically optimal". *Quaternary International* 383: 196–203. doi.org/10.1016/j.quaint.2014.11.045.

Zalasiewicz, J. *et al.* (2021). "The Anthropocene: comparing its meaning in geology (chronostratigraphy) with conceptual approaches arising in other disciplines". *Earth's Future* 9(3): e2020EF001896. doi.org/10.1002/2016EF000379.

Q2

IS THE ANTHROPOCENE A GEOLOGICAL TIME UNIT?

Short answer: the process is under way to determine the formalization or not of the Anthropocene as a geological epoch or stage.

Geological time units refer to events that change conditions of the Earth system as registered in the stratigraphy of the Earth, such as a mass extinction of biota and the emergence of new organisms, identified by distinguishing markers or "golden spikes" found in the geological (or stratigraphic) record. The markers or Global Boundary Stratotype Sections and Points (GSSPs), include, for example, the absence of specific fossil sets, changes in the magnetic polarity of minerals in rock, and various indicators of climatic change. Until recently, the current geological time unit was recognized as part of the Cenozoic era, Quaternary period and the Holocene epoch (Fig. 2.1 A). The Holocene epoch began about 11,650 years ago at the end of the last glaciation and the beginning of an interglacial period. Recognition of the Anthropocene as an epoch was initially proposed as either the second epoch of the Quaternary period (Fig. 2.1 B), rendering the Holocene as the shortest epoch on record, or shifting the Holocene to the last stage of the Pleistocene epoch (Fig. 2.1 C), making the Anthropocene the first epoch beyond the Pleistocene (Lewis & Maslin 2015).

Heretofore, identification of geological time units has begun with notable distinctions found in stratigraphy, largely associated with a GSSP. The rationale for the Anthropocene epoch, however, is based on the profound and rapid changes in the Earth system created by human activity – those that reside beyond the conditions marking the Holocene – with a major case made for human impacts on biogeochemical cycles (**Q4**), such as human-induced carbon emitted to the atmosphere, triggering climate change (**Q46**). In this case, the time unit is in search of a GSSP. A large number of other indicators suggest that currently the Earth system is functionally and stratigraphically different from that of the Holocene (Waters *et al.* 2016), perhaps marked by the appearance of radionuclides (i.e. unstable nuclear atoms) from nuclear activities, providing a date beginning about 1950 (Fig. 2.2). Formal recognition of the Anthropocene, however, rests with the International Commission on Stratigraphy (ICS), which appointed a working group to consider the case and make a recommendation (Monastersky 2015; Zalasiewicz *et al.* 2017). A final vote on the formal recognition of the Anthropocene is projected to take place in 2024.

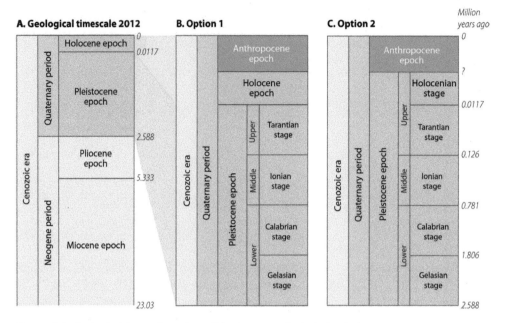

Figure 2.1 Geological time units and possible changes to include the Anthropocene

Figures for boundary start dates are in millions of years. [?] refers to the uncertain date given to the Anthropocene if it is accepted as an epoch. Option 1 would create an extremely abbreviated Holocene, whereas Option 2 moves the Holocene to a stage rather than an epoch.

Source: simplified and altered from Lewis & Maslin (2015). Reprinted by permission from Springer Nature. © (2019).

Various arguments against the new unit exist, however, owing to the identity of its golden spike or the clarity of its dating. For example, epochs and their markers need identification in "deep geologic" time, not that of the short-lived period of the Anthropocene. Humankind's marks on the Earth are, at best, too transient, perhaps constituting a geological event but not yet an epoch (Brannen 2019). Others advance that the Anthropocene emerged over a long period of time with multiple markers but no definitive one **(Q11)** (Ruddiman 2018). Yet others recognize the Anthropocene as an ongoing geological event.

Interestingly, in 2018 the ICS divided the Holocene epoch into three stages, the last beginning 2250 BCE (Walker *et al.* 2018) (Fig. 2.3), generating serious challenges from palaeo-environmentalists (Middleton 2018). How this designation will affect the Anthropocene decision is not yet clear: a new stage or a new epoch starting at 1950 or some other configuration, or an ongoing geological event?

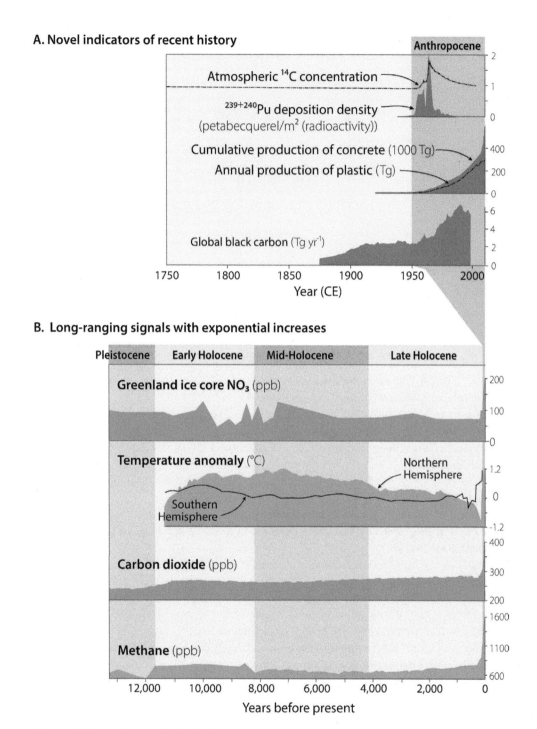

Figure 2.2 Key indicators for the Anthropocene
See Metrics and measures for various symbols.
Source: Waters *et al.* (2016).

Epoch	Stage	Start date (BCE)
Holocene	Meghalayan	2250
	Northgrippian	6236
	Greenlandian	9700

Figure 2.3 2018 Holocene reconfiguration by the International Commission on Stratigraphy
The colour coding matches that of Fig. 2.1 for visual insertion of the 2018 additions to the time units.

References

Brannen, P. (2019). "The Anthropocene is a joke". *The Atlantic*, 13 August. www.theatlantic.com/science/archive/2019/08/arrogance-anthropocene/595795/.

Lewis, S. & M. Maslin (2015). "Defining the Anthropocene". *Nature* 519(7542): 171–80. doi:10.1038/nature14258.

Middleton, G. (2018). "Bang or whimper?" *Science* 361(6408): 1204–05. doi:10.1126/science.aau8834.

Monastersky, R. (2015). "Anthropocene: the human age". *Nature News* 519(7542): 144–7. doi:10.1038/519144a.

Ruddiman, W. (2018). "Three flaws in defining a formal 'Anthropocene'". *Progress in Physical Geography: Earth and Environment* 42(4): 451–61. doi.org/10.1177/0309133318783142.

Walker, M. *et al.* (2018). "Formal ratification of the subdivision of the Holocene Series/Epoch (Quaternary System/Period): two new Global Boundary Stratotype Sections and Points (GSSPs) and three new stages/subseries". *Episodes* 41(4): 213–23. doi.org/10.18814/epiiugs/2018/018016.

Waters, C. *et al.* (2016). "The Anthropocene is functionally and stratigraphically distinct from the Holocene". *Science* 351(6269): 137. doi:10.1126/science.aad2622.

Zalasiewicz, J. *et al.* (2017). "The working group on the Anthropocene: summary of evidence and interim recommendations". *Anthropocene* 19: 55–60. doi.org/10.1016/j.ancene.2017.09.001.

Q3

WHAT IS THE EARTH SYSTEM?

Short answer: the interacting physical, chemical and biological processes operating through the spheres of the planet that generate its biogeophysical conditions.

The Earth system, sometimes referred to as the geosphere (which also includes the interior of the Earth), is composed of interacting subsystems or spheres (Fig. 3.1). The most common set of spheres used to describe the Earth system are the lithosphere (solid mineral earth, including the deep Earth; addressed via land cover in this text as explained in Section III); hydrosphere (water, water vapour, ice; the latter sometimes identified separately as the cryosphere); atmosphere, and biosphere (life) (Sections III–VI). Stocks and states reference the components or phenomena comprising these spheres, such as the amount of fresh surface water and the average temperature of the Earth. These stocks and states affect and are affected by biogeochemical cycles or the flow of physical, chemical and biological elements among the spheres, such as that of water and carbon, including the energy (heat) that elements and compounds carry (**Q4**) (Reid *et al.* 2010). The stocks/states and cycles/flows interact, creating the Earth system. Figure 3.2 illustrates a simplified version of the subsystems and their linkages comprising the Earth system (Rosswall *et al.* 2015). Together, the functioning of the Earth system generates the environmental services (**Q7**) and disservices confronting humankind. Recognition of this system has given rise to the field of Earth system science (Steffen *et al.* 2020).

Biogeophysical drivers or forcings of our planetary and Earth system, all of which continue to operate in the Anthropocene, influence the longer-term conditions of the biosphere. These drivers include: the great tectonic forces within the Earth (i.e. plate tectonics, Section III); variations in incoming solar radiation (i.e. Milankovitch cycles, Section V); waxing and waning of ocean currents (Section IV); occasional large meteorites (Section V); and periods of intensive volcanic activity (Petersen, Dutton & Lohmann 2016; Shen *et al.* 2019) (Section III). The Anthropocene (**Q1**) is marked by the ascendency of human activities as a new driver or forcing (Fig. 3.2, far right) that amplifies and attenuates the stocks/states and cycles/flows within the Earth system and, thus, the dynamics of the subsystems.

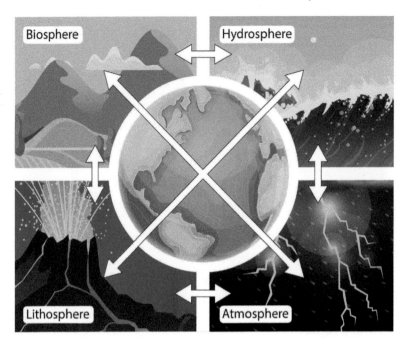

Figure 3.1 The spheres of the Earth system

While other spheres are recognized, the four shown here tend to be the more common set used to identify the Earth system. Each sphere is comprised of many interacting phenomena with changing conditions or states; some phenomena cycle or flow (arrows) through the Earth system.

Source: adapted from Hamilton (2016).

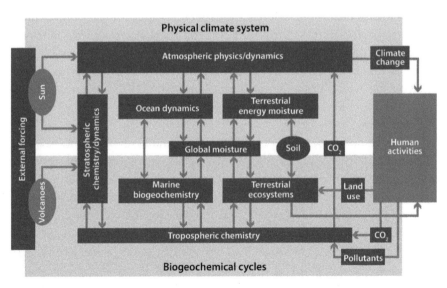

Figure 3.2 Simplified components and linkages comprising the Earth system

The Earth system is a coupling of the subsystems of the physical climate system and biogeochemical cycles (brown) driven by external forcing, such as solar radiation and volcanism, and human activities, such as fossil fuel burning and deforestation (blue).

Source: IGBP (2015).

References

Hamilton, C. (2016). "The Anthropocene belongs to Earth system science". The Conversation. theconversation.com/the-anthropocene-belongs-to-earth-system-science-64105.

IGBP (International Geosphere-Biosphere Programme) (2015). "Reflections on Earth-system science". www.igbp.net/news/features/features/reflectionsonearthsystemscience.5.950c2fa1495db7081ecdc.html.

Petersen, S., A. Dutton & K. Lohmann (2016). "End-Cretaceous extinction in Antarctica linked to both Deccan volcanism and meteorite impact via climate change". *Nature Communications* 7(1): 1–9. doi.org/10.1038/ncomms12079.

Reid, W. *et al.* (2010). "Earth system science for global sustainability: grand challenges". *Science* 330(6006): 916–17. doi:10.1126/science.1196263.

Rosswall, T. *et al.* (2015). "Reflections on Earth-system science". *Global Change* 64: www.igbp.net/download/18.950c2fa1495db7081e754b/1446110005354/NL84-reflections_ES_science.pdf.

Shen, J. *et al.* (2019). "Evidence for a prolonged Permian–Triassic extinction interval from global marine mercury records". *Nature Communications* 10(1): 1–9. doi.org/10.1038/s41467-019-09620-0.

Steffen, W. *et al.* (2020). "The emergence and evolution of Earth System Science". *Nature Reviews Earth & Environment* 1(1): 54–63. doi.org/10.1038/s43017-019-0005-6.

Q4

WHAT ARE BIOGEOCHEMICAL CYCLES?

Short answer: the turnover or movement of chemical substances through the spheres of the Earth system.

Chemical elements and compounds (e.g. natural and synthetic or human-made substances) cycle or "flow" in pathways through the spheres of the Earth system (Galloway *et al.* 2014). A rock cycle also exists but is not considered here. A proportion of chemical substances is

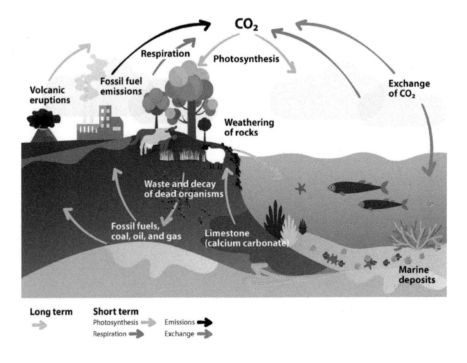

Figure 4.1 The carbon cycle
Long and short term refer to the timescales of the movement of carbon. Volcanic activity is long term in the sense that the carbon it emits has long been stored. Human activity adds to the natural cycle primarily by burning fossil fields and through changes in land use and land cover.
Source: reproduced with permissions from the Dewwool Team.

stored in the spheres for long periods, such as carbon in subterranean coal and oil deposits, or in permafrost and old growth forests. The different long-term storage parts of the various cycles, typically located in the lithosphere and deep oceans, are referred to as "sinks" (Fig. 4.1, yellow arrows flow from sinks). These substances help to create and maintain the Earth system's structure and function. Human-made synthetic compounds cycle as well, typically amplifying the flows in the Earth system. They include industrial-made chemicals that occur in nature (e.g. ammonia) (Fig. 4.2) (Galloway *et al.* 2008) and those that do not, such as chlorofluorocarbons or CFCs **(Q98)** (Montzka *et al.* 2018).

Human impacts on biogeochemical cycling are replete in the Anthropocene given the large range and magnitude of human activities that release elements and compounds from sinks or involve synthetic compounds. For example, the mining and subsequent use of coal, oil and natural gases, stored for multiple millennia in sinks, add significant amounts of carbon dioxide (CO_2) and methane (CH_4) into the reservoirs of the atmosphere and oceans (Fig. 4.1) **(Q44)**. Laboratory-produced (synthetic) ammonia (NH_3) for fertilizers, among other uses, adds nitrogen to its global stock, altering the flux of nitrogen (Fig. 4.2), contributing to global warming and acid rain **(Q43, Q44)**. Likewise, large amounts of methane are human-produced from mining, landfills and livestock rearing (Karakurt, Aydin & Aydiner 2012) or follow from indirect human impacts, such as climate warming reducing permafrost

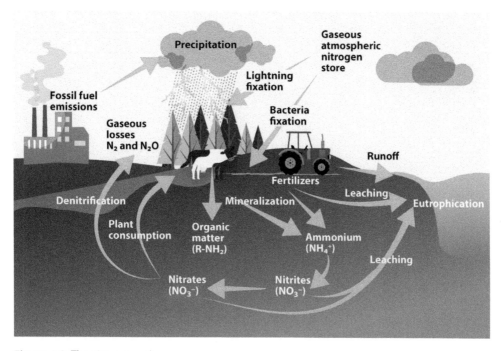

Figure 4.2 The nitrogen cycle
Human activity adds to the cycle through emissions from industries and synthetic fertilizers operating through soils. Emissions constitute a greenhouse gas and create acid rain.
Source: reproduced with permissions from Michael Pidwirny.

in peat and tundra landscapes of the northern latitudes (**Q29**). Human changes in biogeo-chemical cycling loom as a major, if not principal, concern for maintaining a biosphere suitable for humankind. Major human changes to this cycling are a common, but not the only, marker for the existence of the Anthropocene.

References

Galloway, J. *et al.* (2008). "Transformation of the nitrogen cycle: recent trends, questions, and potential solutions". *Science* 320: 889–92. doi:10.1126/science.1136674.

Galloway, J. *et al.* (2014). "Biogeochemical cycles". In J. Melillo, T. Richmond & G. Yohe (eds), *Climate Change Impacts in the United States: The Third National Climate Assessment*, 350–68. US Global Change Research Program. doi:10.7930/J0X63JT0.

Karakurt, I., G. Aydin & K. Aydiner (2012). "Sources and mitigation of methane emissions by sectors: a critical review". *Renewable Energy* 39(1): 40–48. doi:10.1016/j.renene.2011.09.006.

Montzka, S. *et al.* (2018). "An unexpected and persistent increase in global emissions of ozone-depleting CFC-11". *Nature* 557(7705): 413–17. doi.org/10.1038/s41586-018-0106-2.

Q5

WHAT IS ALBEDO?

Short answer: reflectivity of solar radiation off surfaces in the Earth system.

The reflective properties of planetary surfaces (e.g. land, water, vegetation, cities) to solar (shortwave) radiation constitutes albedo. It is measured on a scale of 0 to 1 or as a percentage (or the ratio of shortwave radiation reflected to the total incoming solar radiation), where 0 (or 0 per cent) refers to surfaces absorbing all radiation, usually black in colour, and 1 (or 100 per cent), to surfaces reflecting all such radiation, usually white. Albedo is affected by the spectral and angular distribution of radiation reaching a surface, such as the distinction between the higher angles of the Sun in the summer and the lower angles of the Sun in the winter at high latitudes, illustrated in Figure 5.1 (Gueymard *et al.* 2019). The average albedo of the Earth is about 0.29, in which interannual variability increases towards the poles owing to the seasonal changes noted and the associated seasonal variations in snow and ice cover (Fig. 5.1) (Stephens *et al.* 2015).

The phenomena within the different spheres of our planet (e.g. clouds in the atmosphere and vegetation on the land surface) have different albedos. Common land surface albedos appear in Figure 5.2, ranging from 0.04 from asphalt roadways to 0.8 for fresh snow cover (hydrosphere) (Stephens *et al.* 2015). For the most part, vegetative land cover has relatively low albedo, save when it is covered by snow and ice, which have high albedos. Clouds also have high albedos. Surface water, in contrast, has low albedo.

These reflectivities are critical to the heat balance of the Earth system (Coleman 2013). The greater the reflectivity, the more shortwave radiation is bounced back into the atmosphere and eventually to space. Low albedo, in contrast, absorbs this radiation (i.e. energy or heat) and reradiates it as longwave radiation into the atmosphere where it can be trapped by greenhouse gases, warming the Earth system (**Q44**). Overall, changes in the albedos of the different parts of the Earth system have strong feedbacks involving climate change (Coleman 2013). For instance, global warming reduces Arctic sea ice (high albedo), exposing the ocean's surface (low albedo), which amplifies atmospheric warming (Wunderling *et al.* 2020).

Summer albedo

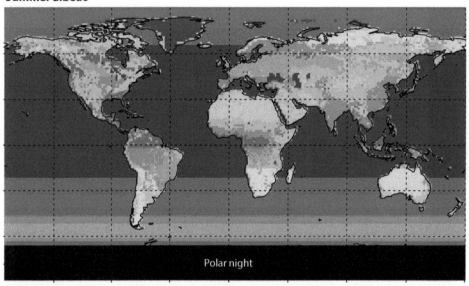

Winter albedo

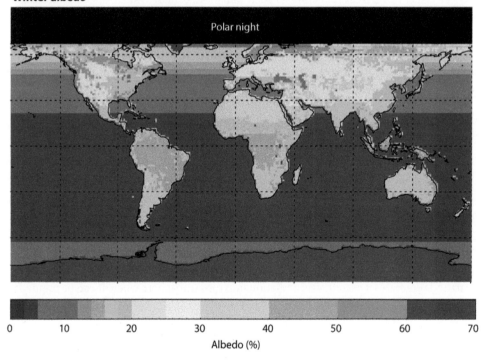

Figure 5.1 Albedo of the Earth: summer and winter

The reflectivity (albedo) of land and water by Northern hemisphere seasons owing to changes in solar radiation is shown. The larger the colour-coded numbers the less albedo. Notice the significant change in albedo in the Northern hemisphere between the summer (top) and winter (bottom). These maps do not include seasonal sea ice and snow cover which would provide more dramatic albedo changes by season.

Source: Coakley (2003).

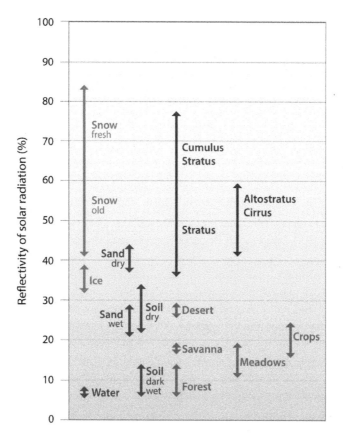

Figure 5.2 Albedo of different surfaces
Stratus, cumulus stratus and altostratus cirrus refer to clouds.
Source: Grobe (2000).

References

Coakley, J. (2003). "Reflectance and albedo, surface". In J. Holton & J. Curry (eds), *Encyclopedia of the Atmosphere*, 1914–23. San Diego, CA: Academic Press.

Colman, R. (2013). "Surface albedo feedbacks from climate variability and change". *Journal of Geophysical Research: Atmospheres* 118(7): 2827–34. doi.org/10.1002/jgrd.50230.

Grobe, H. (2000). "Albedo – percentage of reflected sun light in relation to the various surface conditions of the earth". commons.wikimedia.org/wiki/File:Albedo-e_hg.png.

Gueymard, C. *et al.* (2019). "Surface albedo and reflectance: review of definitions, angular and spectral effects, and intercomparison of major data sources in support of advanced solar irradiance modeling over the Americas". *Solar Energy* 182: 194–212. doi.org/10.1016/j.solener.2019.02.040.

Stephens, G. *et al.* (2015). "The albedo of Earth". *Reviews of Geophysics* 53(1): 141–63. doi. org/10.1002/2014RG000449.

Wunderling, N. *et al.* (2020). "Global warming due to loss of large ice masses and Arctic summer sea ice". *Nature Communications* 11: 51–77. doi.org/10.1038/s41467-020-18934-3.

Q6

WHAT ARE ECOSYSTEMS, LANDSCAPES AND BIOMES?

Short answer: distinctive networks of biota interacting with the abiotic environment (ecosystems); clusters of interacting ecosystems (landscapes); common characteristics of biota associated with climate (biomes).

The land surface and its biota, and associated processes, are categorized throughout this text as biomes, landscapes and ecosystems. A biome refers to the collection of biota with common characteristics, highly associated with prevalent climatic conditions. Biomes may be terrestrial, microbial, marine or anthropogenic in kind. They can be extensive, stretching across continents, such as the vast stretches of the Sahara Desert and the boreal forests crossing the northern Western and Eastern hemispheres. Attention in this text is given to terrestrial biomes that are linked to temperature and precipitation (Fig. 6.1) (Huang *et al.* 2019).

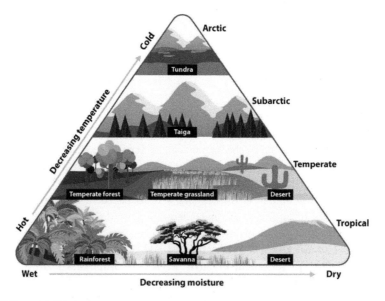

Figure 6.1 Terrestrial biomes
Source: adapted from Kate (2016).

From an ecological perspective, landscapes are composed of heterogeneous clusters of interacting ecosystems or maintain at least one element of interest that is spatially heterogeneous (Wu 2006; Turner 2005). The diversity and complexity following from this heterogeneity generate processes and patterns that distinguish the landscape from an ecosystem (Turner 2005). An ecosystem, in contrast, refers to the material transfer between organisms and the environment (Willis 1997). Coined in 1935, there is no single, accepted definition of ecosystem today, despite the significance of the concept in ecology and environmental science. The interactions among communities of organisms and between those communities and the physical environment (e.g. precipitation, temperature, soil composition) are included in many definitions (Pickett & Cadenasso 2002).

Land units specified as biomes, landscapes and ecosystems are common in research. Nonetheless, these phenomena do not have specified areal dimensions for various reasons. In general, however, a biome tends to be applied to much larger areas than landscapes and ecosystems. Some biomes stretch across continents, such as tundra (Q29) across the northern latitudes of the Eastern and Western hemispheres. Landscapes tend to be applied to biophysical conditions that are much smaller in area than biomes but larger than ecosystems. For the most part, ecosystems refer to biotic–abiotic systems organized and defined well below the global and continental scale, such as a coral reef or a forest, up to a biome, such as tropical forests or savannas. Interestingly, the Earth system itself has been defined as an ecosystem.

Note that landscapes in this text refers to its use and meaning in ecological and environmental science. Cultural landscapes or aesthetic landscapes as used in some social science and humanities are not considered.

References

Huang, M. (2019). "Air temperature optima of vegetation productivity across global biomes". *Nature Ecology & Evolution* 3(5): 772–9. doi.org/10.1038/s41559-019-0838-x.

Kate, M. (2016). "How are biomes distinguished from each other?" Socratic.org. socratic.org/questions/how-are-biomes-distinguished-from-eachother.

Pickett, S. & M. Cadenasso (2002). "The ecosystem as a multidimensional concept: meaning, model, and metaphor". *Ecosystems* **5(1): 1–10. doi:10.1007/s10021-001-0051-y.**

Turner, M. (2005). "Landscape ecology: what is the state of the science?" *Annual Review in Ecology, Evolution, and Systematics* **36: 319–44. doi.org/10.1146/annualrev. ecolsys.36.102003.152614.**

Willis, A. (1997). "The ecosystem: an evolving concept viewed historically". *Functional Ecology* 11(2): 268–71. doi.org/10.1111/j.1365-2435.1997.00081.x.

Wu, J. (2006). "Landscape ecology, cross-disciplinarity, and sustainability science". *Landscape Ecology* 21: 1–4. doi.org/10.1007/s10980-006-7195-2.

WHAT ARE ENVIRONMENTAL (ECOSYSTEM) SERVICES?

Short answer: wanted/expected benefits by people from the Earth system, both natural resources and the environmental conditions and processes that deliver them.

The concept of environmental or ecosystem services gained traction in the first part of the twenty-first century in recognition of the full range of human benefits gained from well-functioning environments or ecosystems; "disservices" refer to unwanted consequences (von Döhren & Haase 2015). Originally labelled as ecosystem services, owing to the goods and services benefitting humankind derived directly from ecosystem functioning, they also include abiotic ones, such as ozone blocking ultraviolet radiation in the stratosphere (**Q53**) or the transfer of heat through the thermohaline circulation (**Q40**) of the oceans that regulates climate. Considering these abiotic services, "environmental services" functions better to identify the full range of Earth system services and is the term used throughout this text.

Natural capital is a complementary term. It refers to the stock of natural resources from which the flow of environmental goods and services are derived (Guerry *et al*. 2015; Mancini *et al*. 2017). Recently, the Intergovernmental Science-Policy Platform on Biodiversity and Ecosystem Services (IPBES) has proposed a new label for environmental services, "nature's contribution to people" (NCP), which recognizes the variations in perceived services by different cultures (Diaz *et al*. 2011, 2015, 2018). These various labels essentially maintain the same meaning: the benefits people obtain from nature.

The broader concept of environmental services emerged partly in an attempt to illuminate the value of well-functioning ecosystems and the Earth system for the maintenance of the biosphere. One initial supposition was that public recognition of the value of the environment increases if its goods and services are linked to a monetary value (Seppelt *et al*. 2011). Not all environmental services are part of the market, however, raising questions about how to provide robust monetary values for them (Chan & Satterfield 2020). Debates exist over valuation methods applied to services not bought and sold (e.g. clean air, biodiversity) and the economic viability of projecting the value of the aggregation of environmental services, including the Earth system at large (**Q95**) (Costanza *et al*. 2014). In addition, the efficacy of placing a monetary value on services incorporated within spiritual and cultural services, including the diversity in phenomena providing those services, has been challenged, leading to the use of the NCP label (Diaz *et al.* 2018). Despite these issues and various other critiques (Schröter *et al.* 2014), the concept of environmental services has emerged as critical to considerations of sustainability of the Anthropocene (Daily & Matson 2008). Environmental services have become a boundary object – information used by different communities for collaborative activities because the term retains legitimacy despite the plasticity of its meaning (Ainscough *et al.* 2019). In this sense, environmental services maintain a range in meanings with various metrics applied to them.

What are these services (MEA 2005)? Human history has relied on and improved the production of resources from nature, foremost water, food, fibre and fuel, or *provisioning* services (Fig. 7.1). Far less attention has been paid, until lately, to *regulating* services, such as the environment's role in cleaning the air, regulating climate, filtering and recharging water into aquifers, or providing pollinators for crops. Provisioning and regulating services

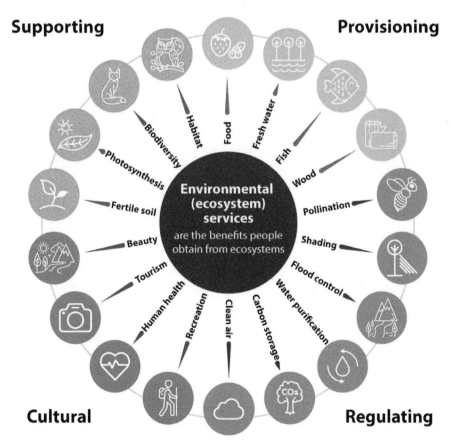

Figure 7.1 Environmental (ecosystem) services
Source: adapted from Metro Vancouver Regional Planning (2018).

depend on *supporting* services (a term that has been dropped in some expert communities), such as nutrient cycling, seed dispersal, primary production and biodiversity. Biotic diversity (or biodiversity) may also be viewed as a provisioning service when it provides a societal want, such as the presence of local wildlife or ecotourism. For the most part, however, specialists tend to treat biodiversity, especially flora, as a factor that generates other services.

Most of the services noted above exist in nature independent of people and commonly benefit biota other than our species. *Cultural* services – the non-material benefits from ecosystems – exist as constructs by individuals and social groups, such as sacred mountains, desirable landscapes, or natural or built environmental features serving as shared cultural heritages (Daniel *et al.* 2012). The significance of these benefits notwithstanding, they are human constructions – even if applied to the material environment – and as such can and have changed through time and across space.

For the most part, society attempts to maintain or improve many environmental services, especially those considered to be natural capital, although numerous examples of their degradation or loss exist throughout human history. For example, the build-up of impervious surfaces in cities and their environmental consequences are increasingly recognized as the reduction of the services provided by the loss of "green space", which in many cases amplifies flood events, such as in Munich, Germany. There, as much as 95 per cent of the precipitation runoff after intense rainstorms is funnelled through the city's impervious surfaces and grey infrastructure, with potential consequences for flooding. Additions of green infrastructure – replacement or supplement of vegetation, such as adding more trees and green roofs to the city – may reduce the runoff by nearly 15 per cent, with increasing percentages as the area of green infrastructure enlarges (Zölch *et al.* 2017). More broadly, changes in ecosystems and landscapes worldwide have reduced environmental capacities to regulate flooding within and beyond cities, perhaps increasing flood impacts to 20–24 per cent of the global population (Tellman *et al.* 2021).

References

Ainscough, J. *et al.* (2019). "Navigating pluralism: understanding perceptions of the ecosystem services concept". *Ecosystem Services* 36: p.100892. doi.org/10.1016/j.ecoser.2019.01.004.

Chan, K. & T. Satterfield (2020). "The maturation of ecosystem services: social and policy research expands, but whither biophysically informed valuation?" *People and Nature* 2: 1021–60. doi:10.1002/pan3.10137.

Costanza, R. *et al.* (2014). "Changes in the global value of ecosystem services". *Global Environmental Change* 26: 152–8. doi.org/10.1016/j.gloenvcha.2014.04.002.

Daily, G. & P. Matson (2008). "Ecosystem services: from theory to implementation". *Proceedings of the National Academy of Sciences* 105(28): 9455–6. doi.org/10.1073/pnas.0804960105.

Daniel T. *et al.* (2012). "Contributions of cultural services to the ecosystem service agenda". *Proceedings of the National Academy of Sciences* 109(23): 8812–19. doi.org/10.1073/ pnas.1114773109.

Díaz, S. *et al.* (2011). "Linking functional diversity and social actor strategies in a framework for interdisciplinary analysis of nature's benefits to society". *Proceedings of the National Academy of Sciences* 108(3): 895–902. doi.org/10.1073/pnas.1017993108.

Díaz, S. *et al.* (2015). "A Rosetta Stone for nature's benefits to people". *PLoS Biol* 13(1): e1002040. doi.org/10.1371/journal.pbio.1002040.

Díaz, S. *et al.* (2018). "Assessing nature's contributions to people". *Science* 359(6373): 270–72. doi:10.1126/science.aap8826.

Guerry, A. *et al.* (2015). "Natural capital and ecosystem services informing decisions: from promise to practice". *Proceedings of the National Academy of Sciences* 112(24): 7348–55. doi.org/10.1073/pnas.1503751112.

Mancini, M. *et al.* (2017). "Stocks and flows of natural capital: implications for ecological footprint". *Ecological Indicators* 77: 123–8. doi.org/10.1016/j.ecolind.2017.01.033.

Metro Vancouver Regional Planning (2018). "Ecosystem services provided by healthy ecosystems". www.metrovancouver.org/services/regional-planning/conserving-connecting/about-ecological-health/Pages/default.aspx (accessed 17 May 2022).

MEA (Millennial Ecosystem Assessment) (2005). *Ecosystems and Human Well-Being: Synthesis.* Washington, DC: Island Press.

Schröter, M. *et al.* (2014). "Ecosystem services as a contested concept: a synthesis of critique and counter-arguments". *Conservation Letters* 7(6): 514–23. doi.org/10.1111/conl.12091.

Seppelt, R. *et al.* (2011). "A quantitative review of ecosystem service studies: approaches, shortcomings and the road ahead". *Journal of Applied Ecology* 48(3): 630–36. doi.org/10.1111/j.1365-2664.2010.01952.x.

Tellman, B. *et al.* (2021). "Satellite imaging reveals increased proportion of population exposed to floods". *Nature* 596(7870): 80–86. doi.org/10.1038/s41586-021-03695-w.

von Döhren, P. & D. Haase (2015). "Ecosystem disservices research: a review of the state of the art with a focus on cities". *Ecological Indicators* 52: 490–97. doi.org/10.1016/j.ecolind.2014.12.027.

Zölch, T. *et al.* (2017). "Regulating urban surface runoff through nature-based solutions – an assessment at the micro-scale". *Environmental Research* 157: 135–44. doi.org/10.1016/j.envres.2017.05.023.

ARE GLOBAL ENVIRONMENTAL CHANGE AND CLIMATE CHANGE DIFFERENT?

Short answer: yes. Global environmental change refers to any changes in the Earth system at large, whereas climate change identifies long-lasting, average changes in global weather. Contemporary attention is given to "human-induced changes" in either category.

Global concern about the consequences of human-induced climate warming, such as that registered by the existence of the Intergovernmental Panel on Climate Change (IPCC; see Appendix) (Pachauri *et al.* 2014) may blur the distinction between this change and that labelled global environmental change (GEC). The two terms are linked but should not be conflated (Turner *et al.* 1990). GEC refers to cumulative or systemic changes (Text box 8.1) in the states/stocks and flows/cycles of the Earth system (**Q3, Q4**), be they the global loss of permafrost, extinction of biota, or the increase in carbon dioxide in the atmosphere and oceans. Climate change, in contrast, is one kind of GEC, referring to long-term changes in average weather worldwide, such as those projected to follow from the average Earth temperature increasing by 2°C (3.6°F) by 2050 (Vitousek 1992) (**Q52, Q53**). Such changes interact with other GECs to amplify or attenuate both. For example, forest losses or gains (**Q17**) increase or decrease atmospheric emissions of CO_2 (**Q44**). The climatic consequences, in turn, may enhance or diminish the conditions supporting forest growth (Anderson-Teixeira *et al.* 2013; IPPC 2014). Not all areas or regions of the Earth may experience similar consequences from GEC or climate change, but both create fundamental changes in the structure and function of the Earth system, with implications for the biosphere and human welfare (IPCC 2014).

Text box 8.1 States/stocks and flows/cycles: cumulative and systemic global environmental change

Cumulative and systemic GEC refers to the basic process by which the states and flows of the Earth system change owing to human actions (Turner *et al.* 1990). *Cumulative* refers to changes in specific states/stocks that reach a sufficient magnitude to endanger their role in the functioning of the Earth system. For example, worldwide tropical deforestation degrades global biodiversity, leading to mass extinctions **(Q53)**, and impacts land–atmosphere dynamics, resulting in climate change. Typically, the magnitude of this kind of change is reached by different human activities repeated worldwide across tropical forests. *Systemic*, in contrast, refers to changes in flows/cycles that reach sufficient magnitude that the Earth system at large changes. For instance, large-scale greenhouse gas emissions flow through the spheres, adding heat to atmosphere and oceans, altering the average temperature of the planet. Hypothetically, only a few huge sources of these emissions could create this change, owing to the movement of the gases through the spheres of the Earth system.

References

Anderson-Teixeira, K. *et al.* (2013). "Altered dynamics of forest recovery under a changing climate". *Global Change Biology* 19(7): 2001–21. doi.org/10.1111/gcb.12194.

IPCC (2014). *Climate Change 2014: Impacts, Adaptation, and Vulnerability. Summary for Policymakers.* Contribution of Working Group II to the Intergovernmental Panel on Climate Change, C. Field *et al.* (eds). Cambridge: Cambridge University Press.

Pachauri, R. *et al.* (2014). *Climate Change 2014: Synthesis Report.* Contribution of Working Groups I, II and III to the Fifth Assessment Report of the Intergovernmental Panel on Climate Change. Geneva: Intergovernmental Panel on Climate Change. epic.awi.de/id/eprint/37530.

Turner, B. II *et al.* (1990). "Two types of global environmental change: definitional and spatial-scale issues in their human dimensions". *Global Environmental Change* 1(1): 14–22. doi.org/10.1016/0959-3780(90)90004-S.

Vitousek, P. (1992). "Global environmental change: an introduction". *Annual Reviews in Ecology and Systematics* 23: 1–14. doi.org./10.1146/annurev.es.23.110192.000245.

ARE THE TOTALITY OF HUMAN IMPACTS ON THE EARTH SYSTEM NOVEL?

Short answer: most probably, owing to the variety, magnitude and pace of human-induced changes in the Earth system, although global-scale changes are evidenced in the distant past.

Humankind has long altered environments to sustain and improve well-being (Section II). Deforestation, the burning of grasslands, wetland manipulation and the construction of irrigation systems, for example, changed local to regional environments and enhanced provisioning services for prolonged periods. Over the long haul, however, such efforts become increasingly costly in upkeep as environmental services are degraded and, in some cases, local to regional climate changes have followed (Pielke 2005; Turner & Sabloff 2012). These consequences have long been recognized (Koelsch 2012; Tufano 2017), although their possible global-level implications have been challenged. For example, were early peoples partly responsible for the extinction of the Holocene megafauna (**Q10**) and for warming the otherwise projected natural cooling of the Earth system about 8,000 to 5,000 years ago owing to deforestation and wetland expansion for agriculture (**Q11**)? These claims and challenges are explored in Section II.

Notwithstanding the past consequences of human actions or debates about the appropriate beginning of the Anthropocene, the accumulation of human impacts on the Earth system in terms of global climate change (**Q46, Q47**), ocean acidification (**Q41**) and mass extinction of biota (**Q55**), among many others, are unique to human history. With the advent of the Industrial Revolution, especially from the mid-twentieth century onwards – the period of "Great Acceleration" – the variety, magnitude and pace of human impacts on both the states and flows of the Earth system (**Q4**) have been unprecedented, including documented global climate change consequences, and increasingly challenge the functioning of the planetary environment (Steffen *et al.* 2011a, 2015). In these ways, human-induced global environmental change is novel (Steffen *et al.* 2011b), and that portion involving climate change ignites the recognition of the Anthropocene for many in the expert community (**Q1**).

References

Koelsch, W. (2012). "The legendary 'rediscovery' of George Perkins Marsh". *Geographical Review* 102(4): 510–24. doi.org/10.1111/j.1931-0846.2012.00172.x.

Pielke, R. (2005). "Land use and climate change". *Science* 310(5754): 1625–6. doi:10.1126/science.1120529.

Steffen, W. *et al.* (2011a). "The Anthropocene: conceptual and historical perspectives". *Philosophical Transactions of the Royal Society A: Mathematical, Physical and Engineering Sciences* 369(1938): 842–67. doi.org/10.1098/rsta.2010.0327.

Steffen, W. *et al.* (2011b). "The Anthropocene: from global change to planetary stewardship". ***Ambio*** **40(7): 739–61. doi.org/10.1007/s13280-011-0185-x.**

Steffen, W. *et al.* (2015). "The trajectory of the Anthropocene: the great acceleration". ***Anthropocene Review*** **2(1): 81–98. doi.org/10.1177/2053019614564785.**

Tufano, A. (2017). "Man's role in *Changing the Face of the Earth* (1955): from the rediscovery of Marsh to the emergence of urban metabolism". *Global Environment* 10(2): 335–62. doi.org/10.3197/ge.2017.100204.

Turner II, B. & J. Sabloff (2012). "Classic period collapse of the Central Maya Lowlands: insights about human–environment relationships for sustainability". *Proceedings of the National Academy of Sciences* 109(35): 13908–14. doi.org/10.1073/pnas.1210106109.

SECTION II

THE EMERGENCE OF THE ANTHROPOCENE

Homo sapiens have always explored ways to secure food, fuel and shelter in the face of the vagaries of nature's provisions. This exploration has led to technological innovations from the control of fire and tools, through the domestication of plants and animals, to the use of fossil fuels and nuclear energy (Steffen *et al.* 2011) and beyond. Synthetics and genomics, or, respectively, the laboratory creation of compounds, including those not native to the Earth system, and the direct manipulation of the genetic structure of biota (**Q22**), portend a new technological era (Fig. II.1). Each revolution has taken place in shorter time spans, increased our capacity to support rising levels of population and material consumption, and enlarged our capacities to change environments and the Earth system at large, taking us to the Anthropocene.

Understanding the emergence of the Anthropocene is important for several reasons. It reminds us that consequential human impacts on the biosphere are not the product alone of the "Great Acceleration" of the mid-twentieth century (Steffen *et al.* 2015), and raises questions about the antiquity of human-induced global environmental change, and hence, the starting point of the Anthropocene (**Q2**). Such global-scale impacts from human activities can be traced back to the latter portions of the Pleistocene glacial period, with subsequent escalating impacts associated with each advance in technology.

Increases in the global population and in changes in social organization and political economy accompanied the technological eras in question. Closely intertwined, the

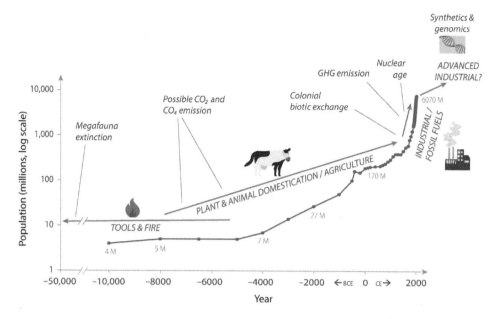

Figure II.1 Technological and population changes linked to Earth system impacts

Note the log-linear scale emphasizing slow growth in the early periods but escalating in pace with new technologies. The arrow for the tools and fire era indicates origins earlier than 50,000 BCE. The arrow for the advanced industrial era indicates that we may be engaging in a new era and that will continue into the future. The Earth system impacts noted are detailed in this section.

Source: adapted from Waldir (2010).

identification of the primacy driver of the eras involves various interpretations (see Sections VII and VIII). Technological innovation, however, offers a coherent tracking of human impacts on our planet (Görg *et al.* 2020).

References

Görg, C. *et al.* (2020). "Scrutinizing the Great Acceleration: the Anthropocene and its analytic challenges for social-ecological transformations". *Anthropocene Review* 7(1): 42–61. doi.org/10.1177/2053019619895034.

Steffen, W. *et al.* (2011). The Anthropocene: conceptual and historical perspectives. *Philosophical Transactions of the Royal Society A: Mathematical, Physical and Engineering Sciences, 369* (1938): 842–67. doi.org/10.1098/rsta.2010.0327.

Steffen, W. *et al.* (2015). "The trajectory of the Anthropocene: the great acceleration". *Anthropocene Review* 2(1): 81–98. doi:10.1177/2053019614564785.

Waldir (2010). "World population growth (lin-log scale)". commons.wikimedia.org/wiki/File:World_population_growth_(lin-log_scale).png.

DID STONE AGE PEOPLE CHANGE
THE EARTH SYSTEM?

Short answer: possibly. They appear to have assisted in eradicating the late Pleistocene and early Holocene megafauna, perhaps in conjunction with climate change.

As long ago as 400,000 years, if not before, our ancestors learned to control fire (Marlon *et al.* 2013; Pinter, Fiedel & Keely 2011; Roebroeks & Villa 2011) subsequently leading to significant anthropogenic burning in Africa and Eurasia (MacDonald *et al.* 2021). Landscape-level changes from anthropogenic fire have been detected 85,000 years ago in south-central Africa (Thompson *et al.* 2021), if not earlier, with firm evidence of ecological landscape alterations by 12,000 years ago (Ellis 2021). Stone Age hunters learned to use fire for hunting (Saltré *et al.* 2016; Sandom *et al.* 2014), driving large animals into "killing zones" and altering the landscapes that supported multiple megafauna (see text, Fig. 10.1) worldwide. The consequences were dramatic; various studies demonstrate a strong likelihood that Stone Age activities contributed to the late Pleistocene and early Holocene megafauna extinctions.

Much of this extinction, however, occurred as the last glacial period gave way to the interglacial period (~14,700 years ago) in which much smaller species also became extinct (Melzer 2020; Wroe *et al.* 2013). Changes in climate altered the capacity of ecosystems to support large animals (Seersholm *et al.* 2020; Stewart, Carleton & Groucutt 2021; Wang *et al.* 2021), including a cooler interlude (~12,900 years ago) that disrupted the longer-term warming. Indeed, recent evidence in eastern Beringia (eastern Alaska and western Yukon, Canada) indicates that vegetation changes from climate change preceded the extinctions of the mammoth and horse (Monteath *et al.* 2021). Such evidence need not eliminate the role of Stone Age hunting and landscape burning in the extinction event, however. For example, the dates associated with landscapes significantly modified by human burning keep getting older, extending as early as 125,000 years ago in deciduous forests across northern Germany and Poland (Roebroeks *et al.* 2021). Landscape burning associated with the arrival of Aboriginal hunters to Australia about 46,000 years ago (Fig. 10.1) has been supported in multiple studies examining the extinction of giant emu, kangaroo and many other species (David *et al.* 2021; Hocknull *et al.* 2020). The arrival of hunter-gatherers in the Americas about 13,000–15,000 years ago is linked in various studies to the loss of

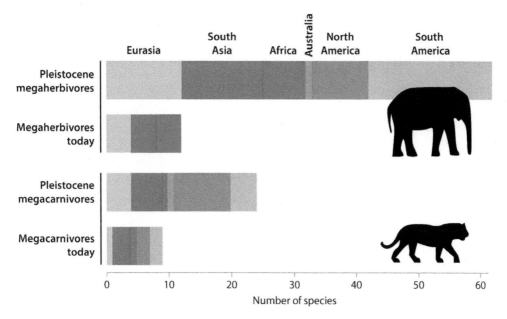

Figure 10.1 Megafauna extinction by continent

The size (i.e. weight) constituting a megafauna varies by herbivore and carnivore and their categorization as either "mega" or large. Megaherbivores = ≥1,000 kg; large herbivores = 45–999 kg; megacarnivores = ≥100 kg; large carnivores = 21.5–99 kg. North and South America lost all megaherbivores by these measures, although the bison comes close to the size of a megaherbivore.

Source: Mantecón and TNF (2020).

mammoth, mastodon, sabre-toothed tiger and horse, among other species. The use of fire to burn landscapes in the hunt altered vegetation, perhaps decreasing forage for grazing animals (Pinter, Fiedel & Keeley 2011). Reduced numbers of large herbivories had cascading impacts on large carnivores.

The number of studies supporting the role of early hunters in the megafauna extinction, however, is challenged by other studies from Australia and the Americas that find no anthropogenic signal or a reduced signal in the data and analysis, or a stronger signal of vagaries in climate conditions (Hocknull *et al.* 2020; Mottl *et al.* 2021). Evidence may exist that an impact event (meteorite) may have played a role in the extinction (Powell 2022), and a recent study even proposes that a shift in the Earth's magnetic fields ~42 ka created drier conditions in Australia, contributing to the megafauna extinction there (Cooper *et al.* 2021).

It is important to recognize that the megafauna extinction is not considered a mass extinction **(Q55)** because the proportion of biota lost was relatively small, not in the 60–95 per cent range associated with events designated as mass extinctions. Nevertheless, a substantial number of large land animals went extinct across the continents (Fig. 10.1), especially in the mid- and northern latitudes. From a global perspective, assessments of fossil pollen indicate that the rates of global vegetation changes accelerated at the beginning of the Late Holocene, probably owing to anthropogenic burning, which has continued to this day

(Mottl *et al.* 2021), and perhaps assisted in the megafauna extinction event. Challenges exist, however, as noted above. Other studies indicate some combination of human and climate change causes were at play (Broughton & Weitzel 2018; Fordham *et al.* 2021; Saltré *et al.* 2016). Finally, the loss of grazing megafauna may have increased fuel loads in grasslands and savannas, increasing wildfires within them (Karp 2021), and perhaps assisted in the shift of more open landscapes to those dominated by woody species (Dantas & Pausas 2022).

The causes of megafauna extinction hold significance regarding the beginning of the Anthropocene. If the Anthropocene is determined by human-induced changes in the Earth system, and if early hunters were involved in the megafauna extinction, then the global state of biota, both the large animals and landscape-level vegetation, were altered, making a case that the Anthropocene began more or less in concert with the current geological epoch of the Holocene, 12,000 years ago or earlier.

References

Broughton, J. & E. Weitzel (2018). "Population reconstructions for humans and megafauna suggest mixed causes for North American Pleistocene extinctions". *Nature Communications* 9(1): 1–12. / doi.org/10.1038/s41467-018-07897-1.

Cooper, A. *et al.* (2021). "A global environmental crisis 42,000 years ago". *Science* 371(6531): 811–18. doi:10.1126/science.abb8677.

Dantas, V. & J. Pausas (2022). "The legacy of the extinct Neotropical megafauna on plants and biomes". *Nature Communications* 13: 129. doi.org/10.1038/s41467-021-27749-9.

David, B. *et al.* (2021). "Late survival of megafauna refuted for Cloggs Cave, SE Australia: implications for the Australian Late Pleistocene megafauna extinction debate". *Quaternary Science Reviews* 253: 106781. doi.org/10.1016/j.quascirev.2020.106781.

Ellis, E. (2021). "Land use and ecological change: a 12,000-year history". *Annual Review of Environment and Resources* **46: 1–33. doi.org/10.1146/annurev-environ-012220-010822.**

Fordham, A. *et al.* (2021). "Process-explicit models reveal pathways to extinction for woolly mammoth using patter-oriented validation." *Ecological Letters* 25(1): 1–13. doi:10.1111/ele.13911.

Hocknull, S. *et al.* (2020). "Extinction of eastern Sahul megafauna coincides with sustained environmental deterioration". *Nature Communications* 11(1): 1–14. doi.org/10.1038/s41467-020-15785-w.

Karp, A. *et al.* (2021). "Global responses of fire activity to late Quaternary grazer extinctions". *Science* 374(6571): 1145–8. doi:10.1126/science.abj1580.

MacDonald, K. *et al.* **(2021). "Middle Pleistocene fire use: the first signal of widespread cultural diffusion in human evolution".** *Proceedings of the National Academy of Sciences* **118(31): e2101108118. doi.org/10.1073/pnas.2101108118.**

Mantecón, L. & TNF (Trun Nature Foundation) (2020). "What is megafauna?" truenaturefoundation.org/what-is-megafauna/.

Marlon, J. *et al.* (2013). "Global biomass burning: a synthesis and review of Holocene paleofire records and their controls". *Quaternary Science Reviews* 65: 5–25. doi.org/10.1016/j.quascirev.2012.11.029.

Melzer, D. (2020). "Overkill, glacial history, and the extinction of North America's Ice Age megafauna". *Proceedings of the National Academy of Sciences* 117: 28555–63. doi.org/10.1073/pnas.2015032117.

Monteath, A. *et al.* (2021). "Late Pleistocene shrub expansion preceded megafauna turnover and extinctions in eastern Beringia". *Proceedings of the National Academy of Sciences* 118(52): e2107977118. doi.org/10.1073/pnas.2107977118.

Mottl, O. *et al.* (2021). "Global acceleration in rates of vegetation change over the past 18,000 years". *Science* 372: 860–64. doi:10.1126/science.abg1685.

Pinter, N., S. Fiedel & J. Keeley (2011). "Fire and vegetation shifts in the Americas at the vanguard of Paleoindian migration". *Quaternary Science Reviews* 30(3/4): 269–72. doi.org/10.1016/j.quascirev.2010.12.010.

Powell, J. (2022). "Premature rejection in science: the case of the Younger Dryas Impact Hypothesis." *Science Progress* 105(1): 1–43. doi:10.117/00368504211064272.

Roebroeks, W. & O. Villa (2011). "On the earliest evidence for habitual use of fire in Europe". *Proceedings of the National Academy of Sciences* 108(13): 5209–14. doi.org/10.1073/pnas.1018116108.

Roebroeks, W. *et al.* (2021). "Landscape modification by Last Interglacial Neanderthals". *Science Advances* 7(51): eabj5567. doi:10.1126/sciadv.abj5567.

Saltré, F. *et al.* (2016). "Climate change not to blame for late Quaternary megafauna extinctions in Australia". *Nature Communications* 7: 10511. doi:10.1038/ncomms10511.

Sandom, C. *et al.* (2014). "Global late Quaternary megafauna extinctions linked to humans, not climate change". *Proceedings of the Royal Society B: Biological Sciences* 281(1787): 20133254. doi.org/10.1098/rspb.2013.3254.

Seersholm, F. *et al.* (2020). "Rapid range shifts and megafaunal extinctions associated with late Pleistocene climate change". *Nature Communications* 11(1): 1–10. doi.org/10.1038/s41467-020-16502-3.

Stewart, M., W. Carleton & H. Groucutt (2021). "Climate change, not human population growth, correlates with Late Quaternary megafauna declines in North America". *Nature Communications* 12(965). doi.org/10.1038/s41467-021-21201-8.

Thompson, J. *et al.* (2021). "Early human impacts and ecosystem reorganization in southern-central Africa". *Science Advances* 7: eabf9776. doi:10.1126/sciadv.abf9776.

Wang, Y. *et al.* (2021). "Late Quaternary dynamics of Arctic biota from ancient environmental genomics". *Nature* 600: 86–92. doi.org/10.1038/s41586-021-04016-x.

Wroe, S. *et al.* (2013). "Climate change frames debate over the extinction of megafauna in Sahul (Pleistocene Australia-New Guinea)". *Proceedings of the National Academy of Sciences* 110(22): 8777–81. doi.org/10.1073/pnas.1302698110.

Q11

DID EARLY AGRICULTURALISTS CHANGE THE EARTH SYSTEM AS PROPOSED BY THE RUDDIMAN HYPOTHESIS?

Short answer: perhaps. By reducing forests and creating wetlands for agriculture, emissions of greenhouse gases may have countered an otherwise cooling trend of the atmosphere.

Animal domestication began as early as 12,000 years ago, followed by plant domestication about 8,000 BP. Societal shifts to cultivation and animal rearing, especially in conjunction with burning landscapes to improve forage (Bowman 2011), led to significant changes in land cover that not only escalated through time but may have had Earth system consequences earlier than previously thought (Ellis *et al.* 2021). The expansion of deforestation and flooding fields for cultivation by early agriculturalists was initially proposed to have been sufficient by 8,000 to 5,000 years ago to constitute cumulative state changes (**Q8**) in the vegetative cover of the global terrestrial surface that generated systemic changes through biogeochemical cycling.

The Ruddiman hypothesis (Ruddiman 2007) proposes that early land changes for cultivation triggered increased greenhouse gas emissions, effectively stalling and redirecting the natural temperature trend of the Earth. The original hypothesis claims that the Milankovitch cycle (Section V) should have been leading to a global climate cooling beginning about 10,000 years ago. Instead, carbon dioxide (CO_2) began to increase in the atmosphere about 8,000 years ago and methane (CH_4) about 5,000 years ago (Fig. 11.1 A, B). The increases in these two greenhouse gases in the atmosphere countered the cycle's natural cooling processes (the difference between red and blue lines, Fig. 11.1 C), warming the Earth (Fig. 11.1 C), a prelude to current systemic change (**Q8**) and global warming.

This hypothesis has been challenged on a number of grounds, including that the amount of CO_2 required for the proposed warming is too large for the amount of deforestation that may have taken place 8,000 years ago (Broecker & Stocker 2006; Marlon *et al.* 2013). Reassessment of the hypothesis (Ruddiman *et al.* 2014) adjusts the rising CO_2 emissions to ~7,000 years ago but retains the original CH_4 date (Fig. 11.2). Subsequent modelling efforts suggest that early land clearance contributed more CO_2 to the atmosphere than was once thought (Kaplan *et al.* 2011), and recent fossil pollen assessments indicate an acceleration

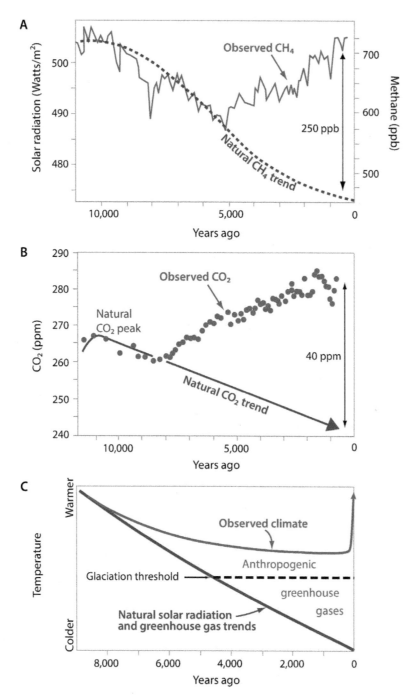

Figure 11.1 Evidence for the original Ruddiman hypothesis

A–B: amount of observed (red line and dots) atmospheric methane (CH_4) and carbon dioxide (CO_2). The natural trend in CH_4 (blue line in A) is projected from decreasing solar radiation (left axis) beginning 11,000 years ago. The natural trend in CO_2 (blue line in B) is projected from past interglacial periods similar to the Holocene. C: Temperature projected from natural forces (blue), such as those in the Milankovitch cycle (Section V), versus observed climate (red) created by greenhouse gases from human activities. Glacial threshold is the temperature at or below which glacial period conditions arise.

Source: Ruddiman (2007).

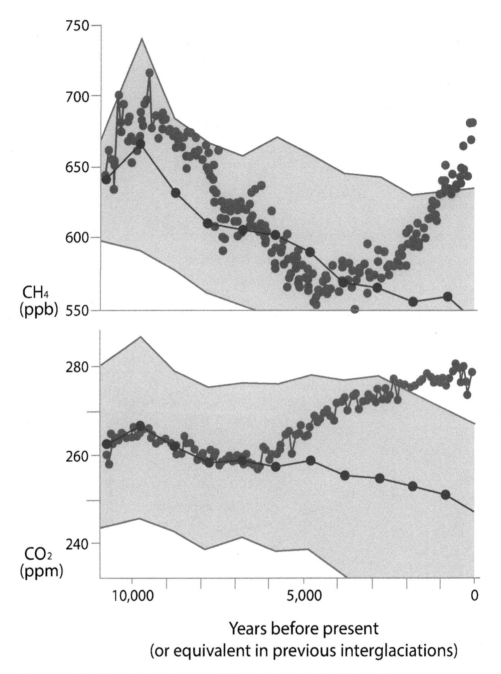

Figure 11.2 Ruddiman's revised methane (CH₄) and carbon dioxide (CO₂) emissions
Red dots are actual measures of the two gases. Blue dots are the average of gases calculated from other interglacial periods; light blue = standard deviation of calculations.
Source: Ruddiman *et al.* (2014).

in global vegetation changes beginning in the Late Holocene (~4,500 to 27,000 years ago), probably owing to anthropogenic burning of landscapes (Mottl *et al.* 2021). Other research supports Ruddiman's CH_4 trends from flooded field cultivation (Mitchell *et al.* 2013). The revised Ruddiman hypothesis proposes that the land-based, human activities of 7,000 to 5,000 years ago led to an anthropogenic warming impact of 0.7°C to 1.2°C **(Q5)** from projected deforestation (Ruddiman *et al.* 2014). This range of temperature increase accounts for the cooling effect from changes in land surface albedo **(Q5)** owing to the new land covers that followed for agriculture.

A major proposition holds that the marker for the beginning of the Anthropocene **(Q1)** is human-induced changes in biogeochemical cycling (systemic change, **Q8**), such as that leading to climate change. Proponents of this view tend to point to the first Industrial Revolution and the subsequent large-scale burning of fossil fuels as the beginning of the Anthropocene. If the Ruddiman hypothesis is correct, however, cycling changes took place in human antiquity. Consistent with the case for global state changes in biota via the world-wide spread of early hunters **(Q10)**, the systemic changes of the hypothesis could mark the Anthropocene as almost as old as the Holocene **(Q2)**.

References

Bowman, D. (2011). "The human dimension of fire regimes on Earth". *Journal of Biogeography* 38(12): 2223–36. doi:10.1111/j.1365-2699.2011.02595.x.

Broecker, W. & T. Stocker (2006). "The Holocene CO_2 rise: anthropogenic or natural?" *Eos* 3: 27–8. doi.org/10.1029/2006EO030002.

Ellis, E. *et al.* (2021). "People have shaped most of terrestrial nature for at least 12,000 years". *Proceedings of the National Academy of Sciences* 118(17): e2023483118. doi.org/10.1073/pnas.2023483118.

Kaplan, J. *et al.* (2011). "Holocene carbon emissions as a result of anthropogenic land cover change". *The Holocene* 21: 775–91. doi:10.1177/0959683610386983.

Marlon, J. *et al.* (2013). "Global biomass burning: a synthesis and review of Holocene paleofire records and their controls". *Quaternary Science Reviews* 65: 5–25. doi.org/10.1016/j.quascirev.2012.11.029.

Mitchell, L. *et al.* (2013). "Constraints on the Late Holocene anthropogenic contribution to the atmospheric methane budget". *Science* 342: 964–6. doi:10.1126/science.1238920.

Mottl, O. *et al.* (2021). "Global acceleration in rates of vegetation change over the past 18,000 years". *Science* 372: 860–64. doi:10.1126/science.abg1685.

Ruddiman, W. (2007). "The early anthropogenic hypothesis: challenges and responses". *Reviews of Geophysics* 45(4): RG 4001. doi:10.1029/2006RG000207.

Ruddiman, W. *et al.* (2014). "Does pre-industrial warming double the anthropogenic total?" *Anthropocene Review* 1: 147–53. doi.org/10.1177/2053019614529263.

DID THE EARLY COLONIAL ERA CHANGE THE EARTH SYSTEM?

Short answer: yes. The redistribution of domesticated biota, pests, weeds and diseases changed ecosystem and landscape states globally.

Humankind has long made innumerable cumulative changes **(Q8)** in local to regional environments, some reaching continental dimensions (Denevan 1992; Turner & Butzer 1992) and beyond (Ellis *et al.* 2021) **(Q10)**. The biotic exchanges and resulting land changes that took place during the colonial era, especially from the sixteenth through the eighteenth centuries, were worldwide in scope. Plants, animals, cropping systems, pests and diseases were transported around the world, intentionally and not (Crosby 2015) (Fig. 12.1), creating state changes globally, with especially large impacts in the Western hemisphere, Australia and Africa.

With the exception of alpaca and llama – which were geographically restricted to the high Andes of western South America – the Americas, Australia and New Zealand had no domesticated herd or draft animals to support livestock herding and transport (although dogs pulled sleds and *travois* in North America) or to pull ploughs before the arrival of Europeans. Herd animals introduced to the American continents often ran rampant across landscapes, especially feral horses and cattle, which were introduced during the early Spanish occupation of the New World (Crosby 2015). Feral horses were taken by Native Americans, transforming their livelihoods into more nomadic modes, especially across the Great Plains (Hämäläinen 2003). The introduction of the steel plough and associated draft animals (e.g. oxen, mules) to the Americas allowed the cultivation of certain ecosystems, such as grasslands, which were difficult to cultivate without this technology. Eurasian cultivars, such as wheat, barley and oats, followed the plough in the Western hemisphere and Australia, whereas non-Eurasian food crops, such as potato and maize (corn), commonly transformed diets and cropping practices in the Eastern hemisphere.

In addition, the Americas had been largely sheltered from Eurasian weeds, pests and diseases before the colonial period. Of these, none was more dramatic than human disease, such as smallpox, which severely reduced the number of indigenous peoples, perhaps by 56 million (Koch *et al.* 2019), especially among those in densely settled portions

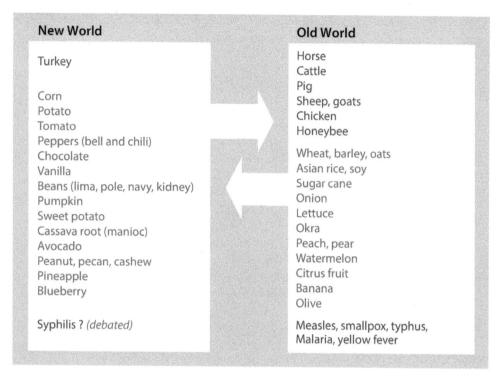

Figure 12.1 Domesticates and diseases exchanged between the Old and New Worlds since the sixteenth century
The transfer of biota and diseases between the Old and New Worlds had major ecosystem and human impacts worldwide. .
Source: author.

of the Americas (Crosby 2015). This loss, assisted by forced labour, led to abandoned or altered land uses, in many cases rendering reforestation and other landscape changes that reduced global carbon dioxide emissions to the atmosphere (Lewis & Maslin 2015; Nevle *et al.* 2011). One estimate holds that 55.8 MH were reforested, absorbing 7.4 Pg of CO_2 from the atmosphere (Koch *et al.* 2019). In addition, the loss of Native Americans led to the movement of enslaved Africans to throughout various parts of the Americas, estimated to have approached 20 million Africans taken from their homes. Such losses surely reduced human pressures on African landscapes and may have contributed to trapping much of sub-Saharan Africa in a low production equilibrium that is exhibited in its economic conditions to this day (Nunn 2008).

The sheer scale of these biotic exchanges had dramatic impacts directly on ecosystems and habitats globally, and indirectly through cropping systems, especially the distribution of the steel plough, and through the loss of indigenous landscapes to colonial ones (Nevle *et al.* 2011). Taken as a whole, the changes in the states of the terrestrial surface accumulated worldwide, with sufficient alterations of ecosystems to be considered global environmental change (**Q8**), not accounting for the impacts on biogeochemical cycles. As such, the timing of the beginning of the Anthropocene is again called into question.

References

Crosby, A. (2015). *Ecological Imperialism*. Cambridge: Cambridge University Press.

Denevan, W. (1992). "The pristine myth: the landscape of the Americas in 1492". *Annals of the Association of American Geographers* 82(3): 426–43. doi.org/10.1111/j.1467-8306.1992.tb01965.x.

Ellis, E. *et al.* (2021). "People have shaped most of terrestrial nature for at least 12,000 years". *Proceedings of the National Academy of Sciences* 118(17): 2023483118. doi.org/10.1073/pnas.2023483118.

Hämäläinen, P. (2003). "The rise and fall of Plains Indian horse cultures". *Journal of American History* 90(3): 833–62. doi.org/10.2307/3660878.

Koch, A. *et al.* (2019). "Earth system impacts of the European arrival and Great Dying in the Americas after 1492". *Quaternary Science Reviews* 207: 13–36. doi.org/10.1016/j.quascirev.2018.12.004.

Lewis, S. & M. Maslin (2015). "Defining the Anthropocene". *Nature* 519(7542): 171–80. doi.org/10.1038/nature14258.

Nevle, R. *et al.* (2011). "Neotropical human–landscape interactions, fire, and atmospheric CO$_2$ during European conquest". *The Holocene* 21(5): 853–64. doi.org/10.1177/0959683611404578.

Nunn, N. (2008). "The long-term effects of Africa's slave trades". *Quarterly Journal of Economics* 123(1): 139–76. doi.org/10.1162/qjec.2008.123.1.139.

Turner II, B. & K. Butzer (1992). "The Columbian encounter and land-use change". *Environment* 43(8): 16–20, 37–44. doi.org/10.1080/00139157.1992.9931469.

HAS THE INDUSTRIAL ERA AFFECTED THE EARTH SYSTEM?

Short answer: yes. The technologies of this era, along with major increases in population and affluence, have transformed virtually all dimensions of the states and flows of the Earth system, especially biogeochemical cycles.

Industrial technology, coupled with the magnitude of global population supported at unprecedented levels of per capita consumption, has directly and indirectly transformed the Earth system. This transformation includes changes in virtually all of the states of the environment on the terrestrial land surface and oceans, and many biogeochemical cycles, save, perhaps, those operating in the crust and below (i.e. the lithosphere) (Ellis 2011). Indeed, the extent of human impacts on the terrestrial land surface is such that about two-thirds of this surface can be considered anthropogenic biomes or anthromes (Ellis 2013) (Fig. 13.1). These biomes/anthromes have emerged from the worldwide changes to forests and grasslands (Foley *et al.* 2005) and arid-land degradation (desertification; Reynolds *et al.* 2007) to rainfed and irrigated agriculture, expansion of wetland cultivation (wet rice) and pastures, water withdrawals and depletion of aquifers (Wada *et al.* 2010) (**Q17, Q20, Q24**).

The collapse of major ecosystems has followed from some of these changes, such as the desiccation of one of the larger water sources in the world, the Aral Sea (**Q33**) (Micklin 2007). Moreover, the soil nutrients of prime agricultural lands (**Q21**) have been drawn down substantially (Tilman *et al.* 2002), as have global fishing stocks (**Q58**) (Worm *et al.* 2009). In addition, large portions of the terrestrial surface are being taken up by large-scale urbanization, with various environmental impacts (**Q31**). Water, air and ocean pollution are increasing in extent. Details on these changes follow in Sections IV–VII.

These changes notwithstanding, the increased human capacity to intercede directly in the biogeochemical cycling of many elements (**Q4**) marks the industrial era as distinctive. Compared with the earlier impacts on such flows, which are contested (**Q11**), those of the industrial era are not. For many experts, the systemic change (**Q8**) associated with fossil fuel burning constitutes the emergence of the Anthropocene (**Q1**). This change began to escalate from the middle of the eighteenth century as fossil fuels became an energy source in western Europe and North America (Fig. 20.2). In order, coal, oil and natural gas – all carbon-based

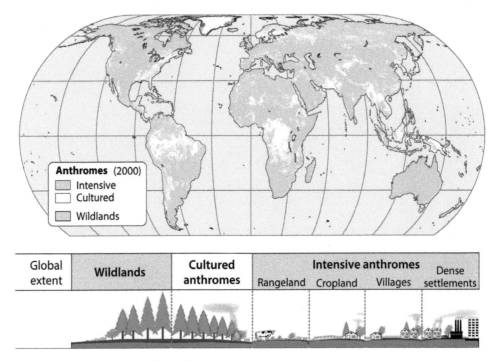

Figure 13.1 Anthropogenic biomes (anthromes)
Source: Ellis (2013); Ellis *et al.* (2021).

compounds stored in deposits in the lithosphere – became mainstays of modern society, the use of which releases greenhouse gas emissions into the atmosphere and oceans (**Q44**). By 2014, fossil fuel use, including that for cement production, produced 91 per cent of CO_2 emissions, totalling 545 Gt emitted since the advent of their use (Wrigley 2013). Additional emissions, about 9 per cent, have come from deforestation and other land-use changes. In sum, CO_2 concentrations in the atmosphere rose from about 280 ppm from the onset of fossil fuel burning to more than ~410 ppm currently (Canadell *et al.* 2007). In addition to these emissions, other greenhouse gas sources escalated from the middle of the twentieth century, especially methane (CH_4) from irrigated agriculture and nitrogen compounds from fertilizer. Today, agricultural fertilizers are now the largest source of nitrous oxide (N_2O) in the Earth system (Tilman *et al.* 2002), surpassing that emitted by nature.

Systemic global environmental change – that of biogeochemical flows (**Q4**) – emphatically increased with fossil fuel usage, especially from the middle of the twentieth century, altering biogeochemical cycling and the Earth system. The consequences include climate warming, the reduction of continental glaciers and snow and ice cover, and ocean acidification among others, the consequences of which are detailed in subsequent parts of this text. The significance of these changes has been proposed as the indicator of the Anthropocene.

References

Canadell, J. *et al.* (2007). "Contributions to accelerating atmospheric CO_2 growth from economic activity, carbon intensity, and efficiency of natural sinks". *Proceedings of the National Academy of Sciences* 104(47): 18866–70. doi.org/10.1073/pnas.0702737104.

Ellis, E. (2011). "Anthropogenic transformation of the terrestrial biosphere". *Philosophical Transactions of the Royal Society A: Mathematical, Physical and Engineering Sciences* 369(1938): 1010–35. doi.org/10.1098/rsta.2010.0331.

Ellis, E. (2013). "Sustaining biodiversity and people in the world's anthropogenic biomes". *Current Opinion in Environmental Sustainability* 5(304): 368–72. doi:10.1016/J.COSUST.2013.07.002.

Ellis, E. *et al.* (2021). "People have shaped most of terrestrial nature for at least 12,000 years". *Proceedings of the National Academy of Sciences* 118: e2023483118. doi.org/10.1073/pnas.2023483118.

Foley, J. *et al.* (2005). "Global consequences of land use". *Science* I (5734): 570–74. doi:10.1126/science.1111772.

Micklin, P. (2007). "The Aral Sea disaster". *Annual Review of Earth and Planetary Science* 35: 47–72. doi.org/10.1146/annurev.earth.35.031306.140120.

Reynolds, J. *et al.* (2007). "Global desertification: building a science for dryland development". *Science* 316(5826): 847–51. doi:10.1126/science.1131634.

Tilman, D. *et al.* (2002). "Agricultural sustainability and intensive production practices". *Nature* 418(6898): 671–7. doi.org/10.1038/nature01014.

Wada, Y. *et al.* (2010). "Global depletion of groundwater resources". *Geophysical Research Letters* 37(20). doi.org/10.1029/2010GL044571.

Worm, B. *et al.* (2009). "Rebuilding global fisheries". *Science* 325(5940): 578–85. doi:10.1126/science.1173146.

Wrigley, E. (2013). "Energy and the English Industrial Revolution". *Philosophical Transactions of the Royal Society A: Mathematical, Physical and Engineering Sciences* 371(1986): 20110568. doi.org/10.1098/rsta.2011.0568.

Q14

ARE WE ENTERING A NEW TECHNOLOGICAL ERA BEYOND THE INDUSTRIAL ONE?

Short answer: perhaps. Synthetic structures and genomics may signal a new technological era.

Identifying transformations of human–environment relationships is difficult in the midst of the process. The technologies of synthetics and genomics, however, promise substantial changes in those relationships, as those advances emerge in tandem with the digital technological (or advanced-industrial) era and service-based economies. The term synthetics refers to industrially produced chemical compounds that differ from those in natural biogeochemical cycles in one of two ways: (1) adding human-made compounds to those existing in nature, altering the magnitude of their biogeochemical flows; and (2) creating compounds that do not exist in nature and adding them to the cycling process.

Synthetic ammonia (NH_3), created in the Haber–Bosch process and used throughout industry, especially for agricultural fertilizers (Brightling 2018), is a compound added to the cycling of nitrogen in the Earth system. Fertilizers stimulate plant growth but also release nitrous oxide (N_2O) to the atmosphere, which is the third-most important greenhouse gas (**Q44**). Synthetic ammonia, the use of which continues to increase (Fig. 14.1), is the largest source of this gas in the atmosphere (Park *et al.* 2012; Shcherbak, Millar & Robertson 2014).

Chlorofluorocarbons or CFCs, in contrast, do not exist in nature. Their innovation, used in many industrial processes but especially as a refrigerant, were originally thought to be environmentally benign (**Q53, Q98**). CFCs, however, not only turned out to serve as a greenhouse gas, but, more importantly, to destroy stratospheric ozone that protects the biosphere from ultraviolet radiation (**Q53**). Plastics are synthetics as well, produced at more than 300 Mt per year (IUCN 2018). Their waste – about 60–80 per cent of their production – becomes global litter, leaches potential toxic chemicals into landfills and breaks down in oceans, making their way to plants and animals, including humans (Thompson *et al.* 2009) (**Q42**). About 5.25 trillion plastic particles are estimated to float in oceans of the world (Eriksen *et al.* 2014), holding various implications for marine life. By 2050, landfill sites are projected to hold 12,000 Mt of plastic (Geyer, Jambeck & Law 2017).

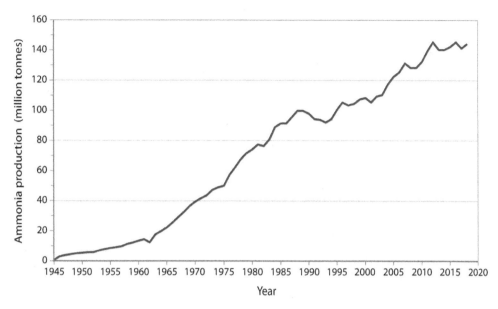

Figure 14.1 Increasing global production of synthetic ammonia
The large increase in ammonia (NH_3) since 1945 made through the Haber–Bosch process, most of which has been used for industrial fertilizer production.
Source: USGS (2018).

The newest synthetics resulting from synbio (synthetic biology) constitute novel life and life systems by way of genetic engineering. Examples include artificial DNA, mimicking natural DNA used to reconfigure the genetic qualities of crop plants (Sonah *et al.* 2011), for instance, to improve their disease and drought resistance (Shcherbak, Millar & Robertson 2014). The longer-term consequences, both positive and negative, of these new varieties of plants and the technology underpinning them are not yet known. The escalating increase in synbio companies and the billions of dollars directed to them, signifies, perhaps, a new era of the Anthropocene, with unknown Earth system impacts.

References

Brightling, J. (2018). "Ammonia and the fertiliser industry: the development of ammonia at Billingham". *Johnson Matthey Technology Review* **62(1): 32–47. doi.org/10.1595/205651318X696341.**

Eriksen, M. *et al.* (2014). "Plastic pollution in the world's oceans: more than 5 trillion plastic pieces weighing over 250,000 tons afloat at sea". *PLoS One* 9(12): e111913. doi.org/10.1371/journal.pone.0111913.

Geyer, R., J. Jambeck & K. Law (2017). "Production, use, and fate of all plastics ever made". *Science Advances* **3(7): e1700782. doi: 10.1126/sciadv.1700782.**

IUCN (International Union for Conservation of Nature) (2018). "Marine plastics". IUCN Issue Briefs. lncn.org/resources/issues-briefs.

Park, S. *et al.* (2012). "Trends and seasonal cycles in the isotopic composition of nitrous oxide since 1940". *Nature Geoscience* 5(4): 261–5. doi.org/10.1038/ngeo142.

Shcherbak, I., N. Millar & G. Robertson (2014). "Global metaanalysis of the nonlinear response of soil nitrous oxide (N2O) emissions to fertilizer nitrogen". *Proceedings of the National Academy of Sciences* 111(25): 9199–204. doi.org/10.1073/pnas.1322434111.

Sonah, H. *et al.* (2011). "Genomic resources in horticultural crops: status, utility and challenges". *Biotechnology Advances* 29(2): 199–209. doi.org/10.1016/j.biotechadv.2010.11.00.

Thompson, R. *et al.* (2009). "Plastics, the environment and human health: current consensus and future trends". *Philosophical Transactions of the Royal Society of London B, Biological Sciences* 364(1526): 2153–66. doi.org/10.1098/rstb.2009.0053.

USGS (US Geological Survey) (2018). "Nitrogen statistics [through 2018l last modified 30 March 2020]" in *Historical Statistics for Mineral and Material Commodities in the United States* (2015 version), T. Kelly & G. Matos, comps. US Geological Survey Data Series 140. www.usgs.gov/centers/national-minerals-information-center/historical-statistics-mineral-and-material-commodities.

IS THE ANTHROPOCENE CONCEPT APPLICABLE
IN THE DISTANT PAST?

Short answer: the beginning of the Anthropocene depends on the criteria used to define the Anthropocene concept, despite the possibility that the label may become a formal geological time unit with a prescribed starting date.

As noted in **Question 1**, the Anthropocene concept involves human impacts on the Earth system that begin to match nature's forcings on the planet's condition and functioning, generating conditions that differ from the Holocene (Waters *et al.* 2016). When did such impacts begin? As this section has demonstrated, various examples of human-induced, global environment changes have taken place in the distant past, providing different interpretations of the Anthropocene's beginning based on attention given to changes in the states or flows of the Earth system (Table 15.1) (Ellis *et al.* 2021; Lewis & Maslin 2015; Stephens *et al.* 2019). Cumulative changes **(Q8)** in land-cover states are linked to various times up to and through the colonial era, with possible links to systemic changes in biogeochemical cycling (Stephens *et al.* 2019). Cases can made for the start of an informal (conceptual use) Anthropocene at any of the dates noted in Table 15.1.

That any of the early changes in Table 15.1 might be considered as constituting the beginning of a formal geological unit of time, however, preferably requires a stratigraphic marker, Global Boundary Stratotype Section and Point (GSSP) and a criterion for defining the boundaries in the stratigraphy (Ellis *et al.* 2021). In this regard, several of the early cases prove problematic regarding a GSSP. The Pleistocene–Holocene megafauna extinction and its subsequent landscape impacts constitute one possible case **(Q10)**, although much larger extinctions in the distant past are designated as events, not geological time units. In addition, this extinction took place over millennia as noted in **Question 10**. An even more significant case can be made for the redistribution of domesticated plants and animals during the colonial era, some 400–500 years ago, with its dramatic global-scale landscape and ecosystem impacts **(Q12)**. The impacts led to a massive loss of indigenous populations of the Americas **(Q12)**, which led to reforestation and other landscape recoveries that appear to have decreased, if momentarily, atmospheric carbon dioxide emissions (Lewis & Maslin 2015; Stephens *et al.* 2019). Evidence of major systemic changes – various impacts on biogeochemical cycles **(Q4)** – also provide several possibilities for consideration. Perhaps the

Table 15.1 Potential beginning of the Anthropocene

Event	Date	Geographical extent	Primary stratigraphic marker
Megafauna extinction	50,000–10,000 yr BP	Near global	Megafauna fossils
Origins of farming	~11,000 yr BP	Southwest Asia, becoming global	Fossil pollen or phytoliths
Extensive farming	~8,000 yr BP to present	Eurasian event, becoming global	CO_2 inflection in ice cores
Rice farming	~6,500 yr BP to present	Southeast Asia with global impact	CH_4 inflection in ice cores
Anthropogenic soils	~3,000–500 yr BP	Local event and impact with global impact	Dark high organic matter in soils
New–Old World collision	1492–1800	Eurasian–Americas event with global impact	Low point of CO_2 in glacial ice
Industrial Revolution	1760 to present	Northwest Europe becoming global	Fly ash from coal burning
Nuclear weapon detonation	1945 to present	Local events with global impacts	Radionuclides (^{14}C) in tree rings
Persistent industrial chemicals	~1950 to present	Local events with global impacts	Examples include sulphur hexafluoride (SF_6) peak in glacial ice

Source: after Lewis & Maslin (2015). Reprinted by permission from Springer Nature. © (2019).

most striking is the increase of CH_4 about 5,000 years ago, consistent with the Ruddiman hypothesis **(Q11)**. The Great Acceleration in industry and science following the Second World War is associated with nuclear materials, radiocarbon (^{14}C) or plutonium, providing a possible GSSP (Lewis & Maslin 2015).

Setting aside the formal time unit considerations, the beginning of the Anthropocene conceptually, that in which human impacts on the Earth system began to match natural drivers, depends on the criteria employed, is debatable, and involves the role changes in states, flows, or both, that are used in the assessment. As the criteria differ, so might the beginning of the Anthropocene. Regardless of the designation or not of the Anthropocene as a geological time unit **(Q2)** (Fig. 2.2), the significance of the conceptual use of the Anthropocene is hard to dismiss.

References

Ellis, E. *et al.* (2021). "People have shaped most of terrestrial nature for at least 12,000 years". *Proceedings of the National Academy of Sciences* 118(17): e2023483118. doi.org/10.1073/pnas.2023483118.

Lewis, S. & M. Maslin (2015). "Defining the Anthropocene". *Nature* 519(7542): 171–80. doi.org/10.1038/nature14258.

Stephens, L. *et al.* (2019). "Archaeological assessment reveals Earth's early transformation through land use". *Science* 365(6456): 897–902. doi:10.1126/science.aax1192.

Waters, C. *et al.* (2016). "The Anthropocene is functionally and stratigraphically distinct from the Holocene". *Science* 315(6269): aad2622-8. doi:10.1126/science.aad2622.

SECTION III

HUMAN CHANGES TO THE LAND SURFACE/LITHOSPHERE

The lithosphere constitutes the crust and upper mantle of the Earth, including the abiotic surface materials such as rock, sand and soil. Crust is destroyed and created, but apparently has grown in area episodically in 500 to 700 million-year cycles (Garçon 2021). The ductile upper portion – the asthenosphere – moves the oceanic and continental crust (i.e. tectonic plates), shaping the terrestrial surface of the Earth, shifting the distributions of oceans and continents over millions of years at rates of ~10 cm yr^{-1} (Fig. III.1). This extremely

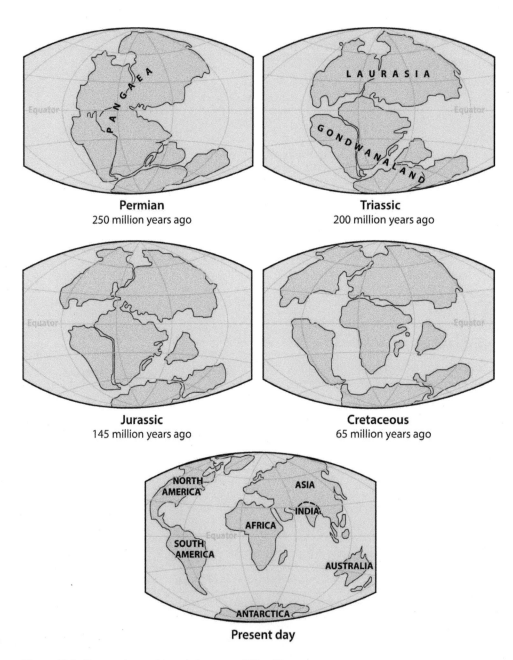

Figure III.1 Changes in continental plates over 250 million years
Source: USGS (2012).

slow-paced movement, created by the immense heat from the Earth's core and perhaps by gravitational interactions of Sun, Moon and Earth (Hofmeister, Criss & Criss 2022), changes ocean and atmospheric dynamics and outcomes profoundly.

In the distant past (280–230 Mya), the supercontinent of Pangaea existed (Fig. III.1) (Frizon de Lamotte et al. 2015; Nance, Murphy & Santosh 2014; Stern 2018), subsequently breaking up into different configurations of continents and oceans that have moved into the distributions that are present today. The various distributions culminating from this movement altered the amount of solar energy received on land and oceans, the albedo of Earth (e.g. snow and ice cover), and circulations in the oceans and atmosphere. For example, changes in ocean circulation owing to the loss of land bridges linking Antarctica to South America and Australia 34 Mya generated the conditions giving rise to the Antarctica ice sheet (Sauermilch et al. 2021). In general, the spatial concentration of the continental plates, as in the Pangaea case, is thought to be associated with lower sea levels and cooler and more arid climates, whereas the opposite conditions occur with dispersed continental plates witnessed now (Hasterok et al. 2022).

In addition, major volcanism – prolonged periods with significant volcanic activity associated with plate tectonics – has changed global climate and, in some cases, the biota in the distant past. Prolonged, multiple eruptions can place large amounts of aerosols and particulate matter in the upper stratosphere, capable of creating a volcanic winter (Rampino & Self 1993; Robock 2000) and associated with decadal periods of cooling in the Northern hemisphere over the past 2,500 years (Sigl et al. 2015). Large tropical volcanoes appear to create consistent decadal-level climate anomalies: dry conditions over tropical Africa, Central Asia and the Middle East and wet conditions in much of the monsoon portions of Oceania and South America (Tejedor et al. 2021). In some cases, eruptions can indirectly strengthen the polar vortex (**Q52**), affecting climate in the northern latitudes (Robock 2000). Over the long term, however, continental volcanic arcs may have moderated atmospheric CO_2 through its drawdown, owing to chemical weathering associated with the volcanism (Gernon et al. 2021).

Volcanic aerosols and sulphur dioxide (SO_2) can also break down ozone, with impacts on the stratospheric ozone layer (**Q53**) (McComick et al. 1995; Robock 2013). The supervolcano Toba (Sumatra) eruption ~74 kya may have reduced this layer by 50 per cent, possibly contributing to the population reduction of *Homo sapiens* migrating from Africa about 50 kya (Osipov et al. 2021). Recent research hints that major volcanic eruptions, global in scale, may take place about every 25.5 My (Rampino, Caldeira & Zhu 2021).

More recently, isolated volcanic eruptions have created short-term cooling. For example, the huge eruption of Mt Tambora, Indonesia, in 1815 emitted 41 km³ of materials, dropping global average temperatures by 0.4°–0.7°C (Marshall et al. 2018). The Mt Pinatubo eruption in the Philippines in 1991 cooled the Earth for two years by about 0.6°C and assisted in reducing ozone in the stratosphere (Gernon et al. 2021; Minnis et al. 1993). Indeed, various eruptions in the first decade of the twenty-first century created a radiative forcing of −0.1° W/m², countering the warming from greenhouse gases (Robock 2013).

Our species cannot begin to match the consequences of plate tectonics on the Earth system, although human-induced emissions now have much longer time consequences on the atmosphere than do sporadic volcanic eruptions. Beyond these emissions, our impacts have a long history of abiotic and biotic changes in the land surface of the lithosphere, our evolutionary home, and now appear to be the major driver of sediment changes globally (Syvitski *et al.* 2022). These changes involve soils (i.e. the pedosphere), water (Section IV), mining and, most profoundly, surface vegetation. Vegetation, of course, is also part of the biosphere, but its predominant location on, and key element of the "cover" of the surface of the Earth, and the criticality of our sweeping changes in the vegetative cover in the functioning of the Earth system, make it appropriate to be treated in this section.

Human impacts on surface hydrology, including snow-ice albedo, are treated in the hydrosphere section (V), while the biosphere section focuses on changes in fauna. That addressing fossil fuels taken from the lithosphere are treated as greenhouses gasses and reside in the atmosphere section (VI). Finally, the role of the deep interior of the Earth (the geosphere) and its impacts on, for example, precession (gyration of the axis of a spinning body, such as Earth) and solar radiation are not treated per se, although precession is treated in the atmosphere introduction (VI).

References

Frizon de Lamotte, D. *et al.* (2015). "Style of rifting and the stages of Pangea breakup". *Tectonics* 34(5): 1009–29. doi:10.1002/2014TC003760.

Garçon, M. (2021). "Episodic growth of felsic continents in the past 3.7 Ga". *Science Advances* 7(39): eabj1807. doi:10.1126/sciadv.abj1807.

Gernon, T. *et al.* (2021). "Global chemical weathering dominated by continental arcs since the mid-Palaeozoic". *Nature Geoscience* 14: 690–96. doi.org/10.1038/s41561-021-00806-0.

Hasterok, D. *et al.* (2022). "New maps of global geological provinces and tectonic plates." *Earth-Science Reviews,* 321: 104069. doi.10.1016/j.earscirev.2022.104069

Hofmeister, A., R. Criss & E. Criss (2022). "Links of planetary energetics to moon size, orbit, and planet spin: a new mechanism for plate tectonics". In G. Folger *et al.* (eds), *In the Footsteps of Warren B. Hamilton: New Ideas in Earth Science.* The Geological Society of America Special Papers 553. doi.org/10.1130/2021.2553(18).

Marshall, L. *et al.* (2018). "Multi-model comparison of the volcanic sulfate deposition from the 1815 eruption of Mt. Tambora". *Atmospheric Chemistry and Physics* 18(3): 2307–28. doi.org/10.5194/acp-18-2307-2018.

McComick, M. *et al.* (1995). "Atmospheric effects of the Mt Pinatubo eruption". *Nature* 373(6513): 399–404. doi.org/10.1038/373399a0.

Minnis P. *et al.* (1993). "Radiative climate forcing by the Mount Pinatubo eruption". *Science* 259(5100): 1411–15. doi:10.1126/science.259.5100.1411.

Nance, R., J. Murphy & M. Santosh (2014). "The supercontinent cycle: a retrospective essay". *Gondwana Research* 25(1): 4–29. doi.org/10.1016/j.gr.2012.12.026.

Osipov, S. *et al.* (2021). "The Toba supervolcano eruption caused severe tropical stratospheric ozone depletion". *Communications Earth & Environment* 2(1): 1–7. doi.org/10.1038/s43247-021-00141-7.

Rampino, M., K. Caldeira & Y. Zhu (2021). "A pulse of the Earth: a 25.5 Myr underlying cycle in coordinated geological events over the last 260 Myr". *Geoscience Frontiers* 12(6): 101245. doi.org/10.1016/j.gsf.2021.101245.

Rampino, M. & S. Self (1993). "Climate-volcanism feedback and the Toba eruption of 74,000 years ago". *Quaternary Research* 40(3): 269–80. doi.org/10.1006/qres.1993.1081.

Robock, A. (2000). "Volcanic eruptions and climate". *Reviews of Geophysics* 38(2): 191-219. doi.org/10.1029/1998RG000054.

Robock, A. (2013). "The latest on volcanic eruptions and climate". *Eos, Transactions American Geophysical Union* 94(35): 305–6. doi.org/10.1002/2013EO350001.

Sauermilch, I. *et al.* (2021). "Gateway-drive weakening of ocean gyres leads to Southern Ocean cooling". *Nature Communications* 12: 64–5. doi.org/10.1038/s41467-021-26658-1.

Sigl, M. *et al.* (2015). "Timing and climate forcing of volcanic eruptions for the past 2,500 years". *Nature* 523(7562): 543–9. doi.org/10.1038/nature14565.

Stern, R. (2018). "The evolution of plate tectonics". *Philosophical Transactions of the Royal Society A: Mathematical, Physical and Engineering Sciences* 376(2132): 20170406. dx.doi.org/10.1098/rsta.2017.0406.

Syvitski, J. *et al.* (2022). "Earth's sediment cycle during the Anthropocene". *Nature Reviews Earth & Environment* 3: 179–96. doi.org/10.1038/s43017-021-00253-w.

Tejedor, E. *et al.* (2021). "Global hydroclimatic response to tropical volcanic eruptions over the last millennium". *Proceedings of the National Academy of Sciences* 118(12): e2019145118. doi.org/10.1073/pnas.2019145118.

USGS (United States Geological Service) (2012). "Historical perspective". pubs.usgs.gov/gip/dynamic/historical.html.

Q16

HAS HUMAN ACTIVITY CHANGED
THE LAND SURFACE OF THE EARTH?

Short answer: yes. About one-half of the Earth's ice-free land surface has been changed directly and indirectly from its natural condition at the end of the last glaciation and another 25 per cent affected in various ways, such as changing albedo, environmental functioning and global climate.

Human activities range in the severity and permanency of their impacts on the land surface of the Earth. Subtle or modest change in which many, if not most, of the natural (pre-human) conditions and functions remain constitutes modification. For instance, abandoned swidden (i.e. slash-and-burn or shifting) cultivation generates landscapes of secondary forests, altering tree species composition but retaining much of the functioning of the original forest biome. Various forests formally designated as native are now recognized as relics of past indigenous practices propagating "useful" tree species for the managers (Armstrong *et al.* 2021; Ter Steege *et al.* 2013). Radical change, such as creating irrigated agricultural land in deserts or draining coastal wetlands for urban expansion or aquaculture, represents transformation (Murray 2022) (Fig. 16.1). In such cases, a return to the original land cover may be impeded for lengthy periods or permanently.

Various estimations of human alteration of the ice-free surface of the Earth, either directly or indirectly, range as high as 75–83 per cent (Fig. 16.2) (Ellis & Ramankutty 2008; Foley *et al.* 2005; Sanderson *et al.* 2002). A comprehensive assessment for the year 2007 calculated that direct human impacts covered 53.5 ± 5.1 per cent of ice-free land, from slightly modified meadow-pastures to transformed metropolitan areas (Hooke, Martín-Duque & Pedraza Gilsanz 2012).

Agricultural lands – those given to cultivation and pasture or modified grazing lands – comprise the overwhelming majority of this human-affected area. They continue to expand at a high rate; about 1 Mkm^2 of cropland (e.g. rice and wheat) were added to the global total in the first two decades of this century (Potapov *et al.* 2021). These land uses are a major cause of deforestation (**Q17, Q18**), arid land degradation (**Q26**), soil erosion and nutrient losses (**Q21**), increases in surface runoff, and decreases in water infiltration. They also appropriate between 20 and 40 per cent of potential primary productivity

Figure 16.1 Land covers and their change: an example

Cultivated lands significantly alter former land covers and their ecosystem functions. Abandonment of that land use may lead to tree recovery (the forest in this view), although as modified in species and function from the original forest cover. Suburbia, in contrast, especially its impervious surfaces, radically changes ecosystem functioning and may impede vegetative recovery for a century or more once abandoned.

Source: imagery © Google 2021.

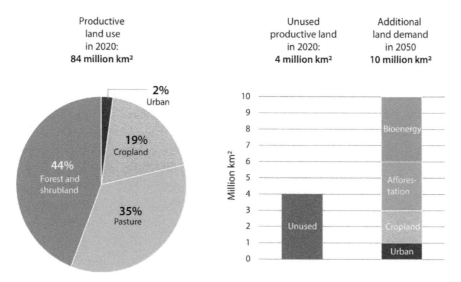

Figure 16.2 Global land use and cover projection for 2050

Source: adapted from Creutzig (2017).

(biomass generated) in the Earth system (Haberl *et al.* 2007), largely through impacts on the photosynthesis process. Metrics of the intensity of all types of land uses have been proposed but have yet to gain wide acceptance in the research community. Only ~3 per cent of the global land area has functionally intact ecosystems undisturbed in some way by human activity (Plumptre *et al.* 2021).

The scale of changes in the land surface has elevated human activity to the principal driver of changes in global sediments: their mobilization, transport and sequestration. As a recent review estimates, between 1950 and 2010, human activities increased global soil erosion by ~467 per cent and sequestered or consumed sediments by ~2,550 per cent, among other impacts (Syvitski *et al.* 2022). Indeed, the net global sediment load is in excess of 300 Gt yr^{-1} of which less than 6 per cent is a result of natural processes (Syvitski *et al.* 2022).

Not all land changes are unidirectionally and increasingly moving from natural to modified to transformed states. Land uses can be abandoned, and given sufficient time, land cover may recover to conditions somewhat similar to their native state. For example, large areas of tropical forests in Central America–Yucatán and Amazonia originally thought to have been pristine are now known to have recovered from major deforestation, land-use changes and abandonment by previous occupants (Demetrio *et al.* 2021), although the scale of past deforestation in Amazonia is contested (Piperno *et al.* 2021). Even today (1982–2016), considerable reforestation (restoration) and afforestation (converting non-forest lands to forests), especially in temperate biomes, have actually added tree cover globally (Table 16.1) (Song *et al.* 2018). Such additions, however, do not imply that ecosystems and their functions have fully recovered. For example, added tree cover includes monocultures for commercial uses. For the most part, land changes degrade ecosystems of their prehuman structure and function, which may prove difficult to recover (Williams *et al.* 2020).

Direct land-cover changes involve land uses such as forest cleared for cultivation, pastures planted for livestock grazing or creation of urban areas (Creutzig 2017). One assessment indicates that 1.9 Mkm^2 of intact ecosystems globally, an area approximately the size of Mexico, was highly modified between 2000 and 2013 (Houghton 2010). Another assessment of land-cover conversion globally from 1990 to 2015 estimated 1.6 Mkm^2, about 178 km^2 daily or over 12 ha min^{-1} (Theobold *et al.* 2020). Such conversions are projected for the future as well. By 2050 about 10 Mkm^2 of additional land, an area about the size of China, may be required to provision society with food and fibre (Fig. 16.2) (Creutzig 2017).

Indirect land change follows from human-induced climate change – to which land change itself contributes (Houghton 2010) – acid rain (**Q43**) or other unintended consequences of human uses. Between 28–40 per cent of anthropogenic carbon dioxide in the atmosphere comes from land changes since the dawning of the industrial era (Houghton 2010). Land conversion, such as biomass burning, livestock production, wet rice cultivation and landfills, produces a large measure of methane (CH_4) emissions (**Q44**). Fertilized agricultural soils and livestock manures are the largest human source of nitrous oxide (N_2O) in the atmosphere (**Q44**) (Reay *et al.* 2012). Land changes also impact the albedo (**Q5**) of the terrestrial surface, affecting ecosystem functioning, habitats, biotic diversity, and local to

Table 16.1 Global land-cover changes, 1982–2016

Land-cover category	Area change	% change	Commentary (natural and human-induced)
Tall vegetation (tree canopy – forest)	+2.24 Mkm²	+7.1	Significant loss in tropics to agriculture; but major gains in non-tropical reforestation/afforestation
Short vegetation (cultivated, grassland, pasture)	−0.88 Mkm²	−1.4	Tropical net gain but non-tropical net loss, largely agricultural gain and loss, respectively
Bare land (desert, land degradation)	−1.16 Mkm²	+3.1	Subtropical net gain offsets net losses elsewhere, especially to grassland degradation

Source: adapted from Song *et al.* (2018).

regional climate change through feedbacks to the atmosphere from altered solar and terrestrial radiation (Alkama & Cescatti 2016; Yan, Liu & Wang 2017). Climate change, in turn, has consequences for land cover in terms of phenology (cyclical or seasonal characteristics of biota), temperature and precipitation. In addition to global climate impacts, human uses of the land affect local to regional surface and near-surface temperatures and precipitation, and fossil fuel burning can degrade downwind tree and other vegetation growth through acid rain **(Q43)**. The array of direct impacts is detailed in the remainder of this section.

References

Alkama, R. & A. Cescatti (2016). "Biophysical climate impacts of recent changes in global forest cover". *Science* 351(6273): 600–04. doi:10.1126/science.aac8083.

Armstrong, C. *et al.* (2021). "Historical indigenous land-use explains plant functional trait diversity". *Ecology and Society* 26(2): 6. doi.org/10.5751/ES-12322-260206.

Creutzig, F. (2017). "Govern land as a global commons". *Nature News* 546: 28–9. doi:10.1038/546028a.

Demetrio, W. *et al.* (2021). "A 'dirty' footprint: macroinvertebrate diversity in Amazonia anthropic soils". *Global Change Biology* 27: 4564–74. doi:10.1111/gcb.15735.

Ellis, E. & N. Ramankutty (2008). "Putting people in the map: anthropogenic biomes of the world". *Frontiers in Ecology and Environment* 6(8): 439–47. doi.org/10.1890/070062.

Foley, J. *et al.* (2005). "Global consequences of land use". *Science* 309(5734): 570–74. doi:10.1126/science.111177.

Haberl, H. *et al.* (2007). "Quantifying and mapping the human appropriation of net primary production in earth's terrestrial ecosystems". *Proceedings of the National Academy of Sciences* 104(31): 12942–7. doi10.1073pnas.0704243104.

Hooke, R., J. Martín-Duque & J. Pedraza Gilsanz (2012). "Land transformation by humans: a review". *GSA Today* 22(12): 4–10. doi:10.1130/GSAT151A.1.

Houghton, R. (2010). "How well do we know the flux of CO_2 from land-use changes?" *Tellus B: Chemical and Physical Meteorology* 62(5): 337–51. doi.org/10.1111/j.1600-0889.2010.00473.x.

Murray, N. J. (2022). "High-resolution mapping of losses and gains of Earth's tidal wetlands". *Science* 376(6594): 744–9. doi.10.1126/science.abm9583.

Piperno, D. *et al.* (2021). "A 5,000-year vegetation and fire history for *tierra firme* forests in Medio Putumayo-Algodón watersheds, northeastern Peru". *Proceedings of the National Academy of Sciences* 118(40): e2022213118. doi.org/10.1073/pnas.2022213118.

Plumptre, A. *et al.* (2021). "Where might we find ecologically intact communities?" *Frontiers in Forests and Global Change* 4: 26. doi.org/10.3389/ffgc.2021.626635.

Potapov, P. *et al.* (2021). "Global maps of cropland extent and change show accelerated cropland expansion in the twenty-first century". *Nature Food* 3: 19–28. doi.org/10.1038/s43016-021-00429-z.

Reay, D. *et al.* (2012). "Global agriculture and nitrous oxide emissions". *Nature Climate Change* 2(6): 410–16. doi.org/10.1038/nclimate1458.

Sanderson, E. *et al.* (2002). "The human footprint and the last of the wild". *BioScience* 52(10): 891–904. doi.org/10.1641/0006-3568(2002)052[0891:THFATL]2.0.CO;2.

Song, X. *et al.* (2018). "Global land change from 1982 to 2016". *Nature* 560(7720): 639–43. doi:10.1038/s41586-018-0411-9.

Syvitski, J. *et al.* (2022). "Earth's sediment cycle during the Anthropocene". *Nature Reviews Earth & Environment* 3: 179–96. doi.org/10.1038/s43017-021-00253-w.

Ter Steege, H. *et al.* (2013). "Hyperdominance in the Amazonian tree flora". *Science* 342(6156): 1243092. doi:10.1126/science.1243092.

Theobald, D. *et al.* (2020). "Earth transformed: detailed mapping of global human modification from 1990 to 2017". *Earth System Science Data* 12(3): 1953–72. doi.org/10.5194/essd-12-1953-2020.

Williams, B. *et al.* (2020). "Change in terrestrial human footprint drives continued loss of intact ecosystems". *OneEart*h 3: 371–82. doi.org/10.1016/j.oneear.2020.08.009.

Yan, M., J. Liu & Z. Wang (2017). "Global climate response to land use and land cover changes over the past two millennia". *Atmosphere* 8(4): 64. doi:10.3390/atmos8040064.

HAS HUMAN ACTIVITY CHANGED FORESTS?

Short answer: yes. Almost one-half of all trees on Earth and up to 11 Mkm² of land lost to forests over the three last centuries have increased greenhouse gases emitted to the atmosphere and degraded ecosystem function and biodiversity; recent tree cover gains are under way, however, except in tropical biomes.

Forests were estimated to cover ~40 Mkm² in 2015 or 31 per cent of the terrestrial surface of the Earth (Grainger 2008; Keenan *et al.* 2015). Over the past 300 years, 7–11 Mkm² of forest cover has been lost to human activity (Foley *et al.* 2005). The current rate of this loss is estimated to be ~192,000 km² yr^{-1} (Hansen *et al.* 2013), perhaps approaching 15.3 billion trees per year (Crowther *et al.* 2015). The current rates include selective logging of certain species or tree sizes, which leads to a decline in the quantity of favoured species to be logged and of large, old-growth trees (Lindenmayer, Laurence & Franklin 2012). The broader historic levels of forest loss and degradation have been offset over the past 40 years by gains in forests for all biomes except those within the tropics, increasing global tree cover by 2.24 Mkm² (Hansen *et al.* 2013; Song *et al.* 2018) (Fig. 17.1). Importantly, much of this gain involves secondary growth and reforestation which does not immediately create, and may never achieve, the functions the forest lost.

The largest amount of deforestation historically has taken place in the mid-latitudes, including China, eastern Russia and much of Europe, and the eastern and mid-western parts of North America. In the distant past, significant areas of tropical forests were once cut, such as the greater Yucatán peninsular region, parts of Amazonia and Southeast Asia, and possibly Central Africa, but forests returned, although altered in species abundance, with abandonment of the land long before the modern era **(Q16)** (Evans *et al.* 2007; Garcin *et al.* 2018; McMichael *et al.* 2017; Turner & Sabloff 2012). Given this history, an estimated three trillion trees remain on Earth (Crowther *et al.* 2015). Global-level deforestation, however, continues, with decadal loss of 7.8 per cent, 5.2 per cent and 4.7 per cent respectively for 1990–2000, 2000–10 and 2010–20 (FAO 2020), much of it involving tropical forests as noted. During the 1990s alone, estimates of tropical forest loss and degradation took place on approximately 81,000 km² (Achard *et al.* 2002). In addition, an estimated 35 per cent of

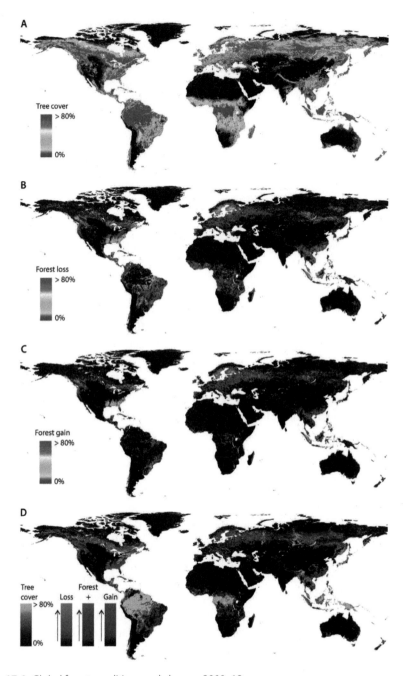

Figure 17.1 Global forest conditions and change, 2000–12

(A) 2012 tree cover. (B) Forest loss and (C) forest gain from 2000–12. (D) Colour composite of tree cover (green), forest loss (red), forest gain (blue), forest loss and gain co-located (magenta). Mangrove forest not included.
Source: Hansen *et al.* (2013).

the world's mangrove forests have been lost, largely to mariculture (e.g. shrimp production), salt production and other near-shore activities (Valela, Bowen & York 2001) **(Q28)**.

Despite logging and burning from natural or unintended fires, boreal forests cover more land area than any other forest biome and have gained in land area (Keenan *et al.* 2015; Pearce 2022; Song *et al.* 2018), as have temperate and subtropical forests (Fig. 17.2). This gain in the temperate forest, if associated with the forest transition **(Q19)** process, reduces forest lost to cultivation and may add timber or tree crop plantations with their impacts on biodiversity (Gurevitch 2022). The boreal or Arctic gains, in contrast, warm the atmosphere by reducing the spatio-temporal dimensions of the albedo **(Q5)** of snow and ice cover (Pearce 2022).

Currently, forest loss and degradation (i.e. thinning of the canopy cover or loss of species) contributes about 12 per cent of human-activated carbon emissions to the atmosphere annually (van der Werf *et al.* 2009), although between 2001–19 forests were a net carbon sink of $-7.6 \pm 49\,\mathrm{GtCO_2e\,yr^{-1}}$ ($-15.6 \pm 49\,\mathrm{GtCO_2e\,yr^{-1}}$ carbon removed and $8.1 \pm 2.5\,\mathrm{GtCO_2e}$ $\mathrm{yr^{-1}}$ carbon emitted) (Harris *et al.* 2021). Boreal forests maintain major areas of peatlands indirectly influenced by human activity through climate warming. Rising temperatures increase evapotranspiration, lowering the water table and releasing CH_4 to the atmosphere, especially from peatlands and permafrost interspersed among boreal forests **(Q29, Q37)**. On the other hand, biomass burning in general, but largely forest burning, also releases aerosols (black and organic carbon) to the atmosphere **(Q27)**, contributing to a decrease in global surface temperatures by 0.13°C and to slight declines in precipitation (Tosca, Randerson & Zender 2003).

Beyond atmospheric impacts, forest loss leads to various ecological consequences, including the loss of "foundational" tree species, with impacts on dependent species and the structure and functioning of the forest ecosystem integrity (Ellison *et al.* 2005; Reich *et al.* 2012). A recent assessment documents that only about 40.5 per cent of remaining forests have high levels of ecosystem integrity, owing to human interventions within them. These

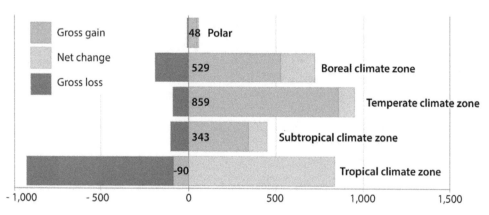

Figure 17.2 Forest gains/losses by climatic zone
Source: Butler (2018) based on Song *et al.* (2018).

forests cover an area of 17.4 Mkm², largely located in the boreal lands of Canada and Russia and in the more remote tropics of Amazonia, Central Africa and New Guinea (Grantham *et al.* 2020). In tropical areas, such forest changes decrease soil moisture, increase fire frequency and lower primary productivity, the outcome of which is the loss of various ecosystem services, with impacts on water balance and flows and regional climate. These same forests host a high percentage of global biodiversity, much of which is lost with deforestation, although the scale of such impacts may be related to areas or hotspots of diversity, rather than tropical forests in general. Loss of forest in general amplifies habitat loss, changing the configuration of landscapes, typically with losses in biodiversity. In addition, evidence also mounts that human-altered environments, which include the habitats of forests, influences evolutionary change (Smith & Bernatchez 2007).

A final note, trees also grow in "non-forest" habitats. Recent high-resolution satellite data and machine-learning technology (software learns to make identifications), for example, identified 1.8 billion individual trees scattered across an area of 1.3 Mkm² (500,000 mi²) West African arid lands (Brandt *et al.* 2020). Many estimates of forest cover area and tree abundance have not accounted for these highly dispersed, individual trees.

References

Achard F. *et al.* (2002). "Determination of deforestation rates of the world's humid tropical forests". *Science* 297: 999–1002. doi:10.1126/science.1070656.

Brandt, M. *et al.* (2020). "An unexpectedly large count of trees in the West African Sahara and Sahel". *Nature* 587(7832): 78–82. doi.org/10.1038/s41586-020-2824-5.

Butler, R. (2018). "The Planet now has more trees than it did 35 years ago". psmag.com/environment/the-planet-now-has-more-trees-than-it-did-35-years-ago.

Crowther, T. *et al.* (2015). "Mapping tree density at a global scale". *Nature* 525(7568): 201–05. doi.org/10.1038/nature14967.

Ellison, A. *et al.* (2005). "Loss of foundation species: consequences for the structure and dynamics of forested ecosystems". *Frontiers in Ecology and the Environment* 3(9): 479–86. doi.org/10.1890/1540-9295(2005)003[0479:LOFSCF]2.0.CO;2.

Evans, D. *et al.* (2007). "A comprehensive archaeological map of the world's largest preindustrial settlement complex at Angkor, Cambodia". *Proceedings of the National Academy of Sciences* 104(36): 14277–82. doi10.1073pnas.0702525104.

FAO (Food and Agricultural Organization) 2020. *Global Forest Resource Assessment 2020: Main Report*. Rome.

Foley, J. *et al.* (2005). "Global consequences of land use". *Science* 309(5734): 570–74. doi:10.1126/science.1111772.

Garcin, Y. *et al.* (2018). "Early anthropogenic impact on Western Central African rainforests 2,600 y ago". *Proceedings of the National Academy of Sciences* 115(13): 3261–66. doi.org/10.1073/pnas.1715336115.

Grainger, A. (2008). "Difficulties in tracking the long-term global trend in tropical forest area". *Proceedings of the National Academy of Sciences* 105(2): 818–23. doi.org/10.1073/pnas.0703015105.

Grantham, H. *et al.* (2020). "Anthropogenic modification of forests means only 40 per cent of remaining forests have high ecosystem integrity". *Nature Communications* 11(1): 1–10. doi.org/10.1038/s41467-020-19493-3.

Gurevitch, J. (2022). "Managing forests for competing goals". *Science* 376(6595): 792–3. doi:10.1126/science.abp8463.

Hansen, M. *et al.* (2013). "High-resolution global maps of 21st-century forest cover change". *Science* 342(6160): 850–53. doi:10.1126/science.1244693.

Harris, N. *et al.* (2021). "Global maps of twenty-first century forest carbon fluxes". *Nature Climate Change* 11: 234–40. doi.org/10.1038/s41558-020-00976-6.

Keenan, R. *et al.* (2015). "Dynamics of global forest area: results from the FAO Global Forest Resources Assessment (2015)". *Forest Ecology and Management* 352: 9–20. doi.org/10.1016/j.foreco.2015.06.014.

Lindenmayer, D., H. Laurence & J. Franklin (2012). "Global decline in large old trees". *Science* 338(6112): 1305–06. doi:10.1126/science.1231070.

McMichael, C. *et al.* (2017). "Ancient human disturbances may be skewing our understanding of Amazonian forests". *Proceedings of the National Academy of Sciences* 114(3): 522–7. doi.org/10.1073/pnas.1614577114.

Pearce, F. (2022). "The forest forecast". *Science* 316(6595): 789–91. doi:10.1126/science.adc9867.

Reich, P. *et al.* (2012). "Impacts of biodiversity loss escalate through time as redundancy fades". *Science* 336(6081): 589–92. doi:10.1126/science.1217909.

Smith, T. & L. Bernatchez (2007). "Evolutionary change in human-altered environments". *Molecular Ecology* 17(1): 1–8. doi:10.1111/j.1365-294X.2007.03607.x.

Song, X. *et al.* (2018). "Global land change from 1982 to 2016". *Nature* 560(7720): 639–43. doi:10.1038/s41586-018-0411-9.

Tosca, M., J. Randerson & C. Zender (2003). "Global impact of smoke aerosols from landscape fires on climate and the Hadley circulation". *Atmospheric Chemistry and Physics* 13(10): 5227–41. doi.org/10.5194/acp-13-5227-2013.

Turner II, B. & J. Sabloff (2012). "Classic Period collapse of the Central Maya Lowlands: insights about human–environment relationships for sustainability". *Proceedings of the National Academy of Sciences* 109(35): 13908–14. doi.org/10.1073/pnas.1210106109.

Valiela, I., J. Bowen & J. York (2001). "Mangrove forests: one of the world's threatened major tropical environments". *BioScience* 51(10): 807–15. doi.org/10.1641/0006-3568.

van der Werf, G. *et al.* (2009). "CO_2 emissions from forest loss". *Nature Geoscience* 2(11): 737–8. doi.org/10.1038/ngeo671.

Q18

WHY DOES TROPICAL DEFORESTATION ATTRACT SO MUCH ATTENTION?

Short answer: with the highest level of biotic diversity and stored carbon per unit area on the terrestrial surface of Earth, the rates of tropical forest loss have large Earth system impacts.

Tropical forests, ranging from wet (rainforests) to seasonally dry conditions, cover ~17.7 Mkm^2 of the land surface of the Earth, an area about the size of South America (Fig. 18.1, dark green). These forests experience the highest rate of loss of all forest biomes, estimated to have ranged from 55,000 km^2 to 95,000 km^2 per annum during the years 1990 to 2015 (Keenan *et al.* 2015; Sloan & Sayer 2015; Tyukavina *et al.* 2015). These estimates do not include tropical forest degradation and the impacts from selective logging (Asner *et al.* 2005). Protected areas of tropical forests have grown in total number and area, but are affected by the land-uses/covers adjacent to them (DeFries *et al.* 2005). Forests located within the tropics, especially those within less-developed economies, are a major source of forestland products. These products are exported worldwide, including to some countries that are conserving forests or reforesting (**Q23**).

The loss of tropical forests is of concern for several important reasons, ranging from biodiversity to global climate (Lewis 2006). It is well known that tropical forests have the highest biotic diversity of land-based biomes. A systematic assessment of 6 Mkm^2 of Amazonia estimated the presence of 16,000 tree species and a median tree density of 565/ha (Ter Steege *et al.* 2013). Among the tropical forest types, primary forests are the most sensitive to the loss of this diversity through disturbances (Gibson *et al.* 2011). Such disturbances are estimated to have reduced species by 41 per cent globally, constituting a central part of the mass extinction under way in the Anthropocene (Alroy 2017) (**Q55**). This loss in biota degrades the structure and function of ecosystems, and creates land–climate interactions that potentially change forest cover (Brodie, Post & Laurence 2012). Beyond biodiversity, tropical forests are highly significant to the global carbon cycle, storing more carbon than any other forest biome (Fig. 18.1), especially accounting for below-ground carbon (i.e. the root system) in the forest floor (Malhi & Grace 2000). Notably, 2.6 GtCO$_2$ yr^{-1} emanates from tropical deforestation, a significant portion (~29–39 per cent) of which is associated with land used for international trade (Pendrill *et al.* 2019). These forests may cease to serve as a

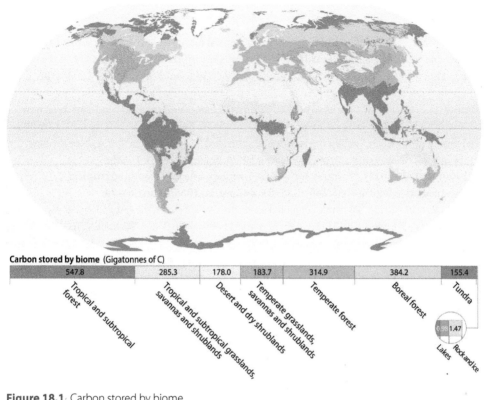

Figure 18.1 Carbon stored by biome
Source: Pravettoni (2014).

major carbon sink, however, owing to changing climate, biodiversity loss and land changes (Lewis 2006). Indeed, recent research suggests that from 2000 to 2012 the annual tropical forest above-ground carbon loss rose from 58 to 62 per cent or 588–631 TgC yr^{-1} (Asner *et al.* 2005). The largest expanse of tropical forest in the world, that of Amazonia, and one of the largest carbon sinks of above-ground biomass, emitted more carbon than it consumed between 2010–19, owing to forest degradation and deforestation (Qin *et al.* 2021).

Tropical forests also generate evapotranspiration and organic molecules that react with compounds in the atmosphere to create clouds, generating up to 50–80 per cent of the rainfall these forests receive (Pöschl *et al.* 2010). As such, loss of tropical forests can change regional precipitation patterns. For example, one study indicates that large-scale deforestation in Southeast Asia from the eighteenth to mid-nineteenth centuries changed atmospheric dynamics sufficiently to have weakened the Asian monsoon for a period of 150 years (Takata, Saito & Yasunari 2009). In another example, various models indicate that 25–40 per cent of deforestation in Amazonia generates land–atmosphere feedbacks that reduce precipitation and increase drought throughout its expansive region, affecting its environmental functioning and the Earth system at large (Lovejoy & Nobre 2018). Amazonia has already lost some 1 Mkm^2 or ~20 per cent of its tree cover.

References

Alroy, J. (2017). "Effects of habitat disturbance on tropical forest biodiversity". *Proceedings of the National Academy of Sciences* 114(23): 6056–61. doi.org/10.1073/pnas.1611855114.

Asner, G. *et al.* (2005). "Selective logging in the Brazilian Amazon". *Science* 310(5747): 480–82. doi:10.1126/science.1118051.

Brodie, J., E. Post & W. Laurance (2012). "Climate change and tropical biodiversity: a new focus". *Trends in Ecology & Evolution* 27(3): 145–50. doi.org/10.1016/j.tree.2011.09.008.

DeFries, R. *et al.* (2005). "Increasing isolation of protected areas in tropical forests over the past twenty years". *Ecological Applications* 15(1): 19–26. doi.org/10.1890/03-5258.

Gibson, L. *et al.* (2011). "Primary forests are irreplaceable for sustaining tropical biodiversity". *Nature* 478(7369): 378–81. doi.org/10.1038/nature10425.

Keenan, R. *et al.* (2015). "Dynamics of global forest area: results from the FAO Global Forest Resources Assessment 2015". *Forest Ecology and Management* 352: 9–20. doi.org/10.1016/j.foreco.2015.06.014.

Lewis, S. (2006). "Tropical forests and the changing earth system". *Philosophical Transactions of the Royal Society B: Biological Sciences* 361(1465): 195–210. doi.org/10.1098/rstb.2005.1711.

Lovejoy, T. & C. Nobre (2018). "Amazon tipping point". *Science Advances* 4(2): eaat2340. doi:10.1126/sciadv.aat2340.

Malhi, Y. & J. Grace (2000). "Tropical forests and atmospheric carbon dioxide". *Trends in Ecology & Evolution* 15(8): 332–7. doi.org/10.1016/S0169-5347(00)01906-6.

Pendrill, F. *et al.* (2019). "Agricultural and forestry trade drives large share of tropical deforestation emissions". *Global Environmental Change* 56: 1–10. doi.org/10.1016/j.gloenvcha.2019.03.002.

Pöschl, U. *et al.* (2010). "Rainforest aerosols as biogenic nuclei of clouds and precipitation in the Amazon". *Science* 329(5998): 1513–16. doi:10.1126/science.1191056.

Pravettoni, R. (2014). "Carbon stored by biome". www.grida.no/resources/7554.

Qin, Y. *et al.* (2021). "Carbon loss from forest degradation exceeds that from deforestation in the Brazilian Amazon". *Nature Climate Change* 11(5): 442–48. doi.org/10.1038/s41558-021-01026-5.

Sloan, S. & J. Sayer (2015). "Forest Resources Assessment of 2015 shows positive global trends but forest loss and degradation persist in poor tropical countries". *Forest Ecology and Management* 352: 134–45. doi.org/10.1016/j.foreco.2015.06.013.

Takata, K., K. Saito & T. Yasunari (2009). "Changes in the Asian monsoon climate during 1700–1850 induced by preindustrial cultivation". *Proceedings of the National Academy of Sciences* 106(24): 9586–9. doi.org/10.1073/pnas.0807346106.

Ter Steege, H. *et al.* (2013). "Hyperdominance in the Amazonian tree flora". *Science* 342(6156): 1243092. doi:10.1126/science.1243092.

Tyukavina, A. *et al.* (2015). "Above ground carbon loss in natural and managed tropical forests from 2000 to 2012". *Environmental Research Letters* 10(7): 074002. doi.org/10.1088/1748-9326/10/7/074002.

WHAT IS THE FOREST TRANSITION THESIS AND ITS RELEVANCE TO GLOBAL FORESTS?

Short answer: country-level forestation is explained by high levels of economic development, but it is unclear what percentage of reforestation can be gained globally.

Much of the once deforested land in mid-latitudes, especially in North America, Europe, Russia and China, has gained forest cover owing to forestation processes, forest recovery on abandoned agricultural lands and reforestation projects (Song *et al.* 2018). Global- and regional-scale assessments reveal that forestation occurs with economic development that generates non-farm employment (industrial and service economies) and advances in crop yields, providing the opportunity for forest regrowth and recovering some of their environmental services (Redo *et al.* 2012). In China, for example, flood protection and climate amelioration have been gained with reforestation. The forest transition theory asserts that country-level economic development leads to country-level forest recovery (Fig. 19.1) (Michinaka 2018). The empirical support for this relationship is strong (Rudel *et al.* 2005), as noted by regional forest cover worldwide (Fig. 19.2). Various political–economic changes, however, may also lead to forest gains similar to the later stages of the transition model. A case in point is the return of forests with the abandonment of farm lands in Eastern Europe and European Russia with the collapse of government-subsidized communal farms and land privatization (Alix-Garcia *et al.* 2016; Kuemmerle *et al.* 2009).

It is a mistake, however, to project that global economic development will lead to a global forest transition. Such development actually increases demand for forest products and for resources obtained by clearing forests for agriculture or converting native forests to commercial tree cover, such as oil palm or cacao plantations. In some cases, high-return plantation forestry may be expanded in relatively well-developed countries (Michinaka 2018). More commonly, however, foreign companies and governments from economies focused on industry and services purchase large tracts of land in agrarian-based countries to service the demand in question (Meyfroidt *et al.* 2013; Meyfroidt & Lambin 2009), a process sometimes labelled "land-grabbing" (**Q23**). In addition, decisions in some developing economies to conserve and regrow forests drives deforestation in other countries. For example, Vietnam's forest regrowth programme between 1987 and 2006 displaced

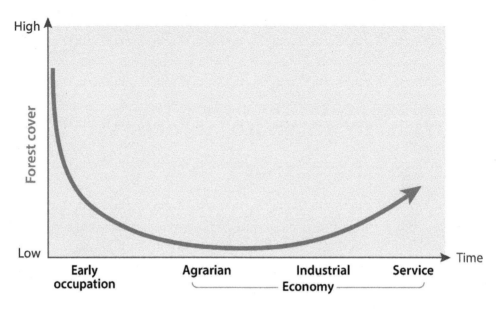

Figure 19.1 The forest transition model

As an area is initially occupied, the forest cover begins to fall and continues to do so as an agrarian economy dominates. The movement into an industrial economy tends to reduce agrarian demands of land uses, the produce of which can be obtained from afar, permitting forest to return. This return continues to escalate as the area advances in industrial and service economies.

Source: author.

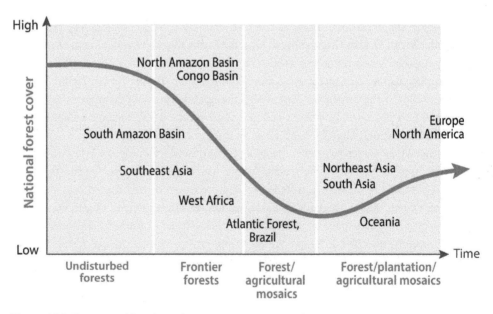

Figure 19.2 Forest transitions by region

Source: Gregerson *et al.* (2011).

about 49 million m³ of wood products which were obtained by importation (Meyfroidt & Lambin 2009). Overall, whether advances in agricultural technology associated with economic advances hold the potential to reduce the total amount of forestland to be opened or remain open remains unclear **(Q22)**.

References

Alix-Garcia, J. *et al.* (2016). "Drivers of forest cover change in Eastern Europe and European Russia, 1985–2012". *Land Use Policy* 59: 284–97. doi.org/10.1016/j.landusepol.2016.08.014.

Gregerson *et al.* (2011). "The greener side of REDD+". Washington, DC: Rights and Resource Commission.

Kuemmerle, T. *et al.* (2009). "Land use change in southern Romania after the collapse of socialism". *Regional Environmental Change* 9(1): 1–12. doi:10.1007/s10113-008-0050-z.

Meyfroidt, P. & E. Lambin (2009). "Forest transition in Vietnam and displacement of deforestation abroad". *Proceedings of the National Academy of Sciences* **(38): 16139–44. doi10.1073pnas.0904942106.**

Meyfroidt, P. *et al.* (2013). "Globalization of land use: distant drivers of land change and geographic displacement of land use". *Current Opinion in Environmental Sustainability* 5(5): 438–44. doi.org/10.1016/j.cosust.2013.04.003.

Michinaka, T. (2018). "Approximating forest resource dynamics in Peninsular Malaysia using parametric and nonparametric models, and its implications for establishing forest reference (emission) levels under REDD+". *Land* 7(2): 70. doi.org/10.3390/land7020070.

Redo, D. *et al.* (2012). "Asymmetric forest transition driven by the interaction of socioeconomic development and environmental heterogeneity in Central America". *Proceedings of the National Academy of Sciences* 109(23): 8839–44. doi.org/10.1073/pnas.1201664109.

Rudel, T. *et al.* (2005). "Forest transitions: towards a global understanding of land use change". *Global Environmental Change* **15(1): 23–31. doi.org/10.1016/j.gloenvcha.2004.11.001.**

Song, X. *et al.* (2018). "Global land change from 1982 to 2016". *Nature* 560(7720): 639–43. doi:10.1038/s41586-018-0411-9.

Q20

HOW MUCH LAND IS CULTIVATED AND WHAT ARE ITS ENVIRONMENTAL CONSEQUENCES?

Short answer: about 14 per cent of the ice-free surface of the Earth has been cultivated, with consequences for water use and quality, land degradation, albedo and greenhouse gas emissions.

Estimates based on satellite (remote sensing) data target 18 Mkm² (an area larger than Russia) or 13.8 per cent of ice-free land surface as given to crop-pasture production (Ramankutty & Foley 1998; Ramankutty *et al.* 2008a) (Fig. 20.1). This figure includes land cultivated for plant food and fibre as well as fallow lands employed for grazing. It does not include native grasslands or those managed by improved grasses for grazing. As much as 30 per cent of cultivated land is devoted to the production of animal feed (Bouwman *et al.* 2013). Of the

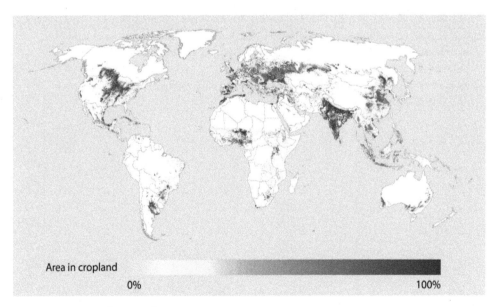

Figure 20.1 Global fraction of cropland, 2000
Source: Ramankutty *et al.* (2008b).

total cultivated area, about 11.4 Mkm^2 (4.4 million mi^2) have been taken from forests and 6.7 Mkm^2 (2.6 million mi^2) from grasslands (Ramankutty & Foley 1999).

The figures on cropland change, of course, over time and as the quality of remote sensing data increases. The most recent global assessment registers 1,244 MH for croplands in 2019 with a net primary production (NPP) of 5.5 PgC yr^{-1} (Potapov *et al.* 2021). This area increased global cropland by 1 MH or 9 per cent between 2003 and 2019, largely by way of expansion in the tropical world. Interestingly, the global per capita cropland decreased by 10 per cent but per capita annual cropland NPP increased owing to increased intensity of cultivation (Potapov *et al.* 2021).

About 301 MH (over 3 Mkm^2) or 16–18 per cent of all cultivated lands are irrigated (Boutraa, 2010), dominated in order of area irrigated by India, China and the United States (Siebert *et al.* 2010; Salmon *et al.* 2015). The estimated global area of rice is estimated to be 167 MH, most of which is paddy (or *padi*, flooded fields) rice (Siebert *et al.* 2010) (Fig. 20.2).

Remote sensing data indicates that shifting cultivation consumes about 280 MH, mostly in the tropical Americas and Africa (Heinimann *et al.* 2017). This form of agriculture commonly involves long fallow periods in which secondary forest growth appears. About 43 per cent of cultivated lands maintain at least 10 per cent tree cover – agroforestry, tree monocropping or trees attendant to other cultivated land uses – that may hold up to 75 per cent of 45.3 PgC stored on agricultural land (Zomer *et al.* 2016).

Significant variation in the amount, gains and losses of cultivated land by time period and geographical location exist, as do the data sources employed. In addition, sustained climate change will alter the amount and location of cultivable lands. Models project that the tropical world (i.e. much of the Global South) will be the most sensitive to degrading cropping conditions, whereas the northern high latitudes may gain 5.6 Mkm^2 of additional cropland, especially in Canada, China and Russia (Zabel, Putzenlechner & Mauser 2014).

Croplands provide most of the world's food intake for the global population, notwithstanding the inequalities in food access and distribution. This success, however, has come with significant environmental impacts. Soil degradation (e.g. erosion, nutrient loss, compaction, salinization) is a well-known local to regional outcome of agriculture (**Q21**). Given that much cropland is taken from lands once in forest cover, and that in some parts of the world annual burning is part of cropping, a significant amount of carbon emissions and aerosols are derived from cultivation (**Q44, Q45**). Carbon emissions follow from the vegetational changes that ensue and from the soil in the cultivation process (**Q21**). Other uses, such as surface mining peatlands for fertilizer, have played major emission roles as well. Overall, as much as 17,150 ± 1,750 TgCO$_2$eq yr^{-1} of emissions may follow from agricultural production, about 30 per cent of all human-induced greenhouse gas emissions (Xu *et al.* 2021), of which the majority is accounted for by animal-based production.

As a percentage of global emissions, agriculture contributes a significant proportion of nitrogen-based greenhouse gases. Around 50 per cent of all such gases are derived from nitrogen fertilizers (Reay *et al.* 2012) and 40 per cent of methane emissions are from irrigated

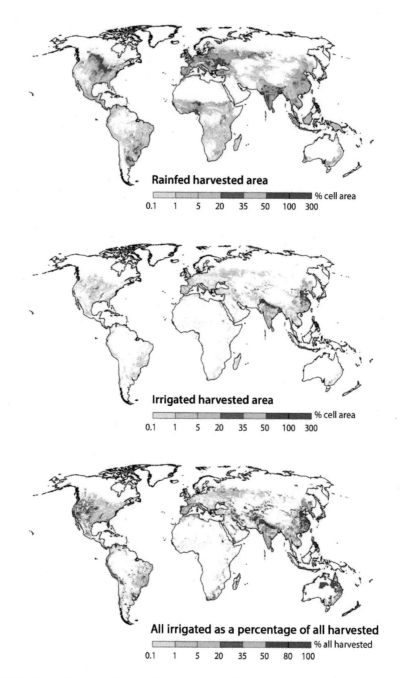

Figure 20.2 Global rainfed and irrigated harvested area, 1998–2002
Irrigated includes wet rice cultivation. Cell = 5 arc minute grid cell, about 8.3 km at equator.
Source: Portman, Siebert & Döll (2010).

Table 20.1 Global food system: Mt yr^{-1} carbon dioxide equivalent (CO$_2$e) emissions, 2018

Activity	CO$_2$e (Mt yr^{-1})
Land use	852
Land-use change	3,238
Net forest conversion	3,034
Crop and livestock	5,218
On-farm energy use	946
Food transport	511
Waste disposal	996

Source: Tubiello *et al.* (2021).

cultivation, foremost wet rice. Emissions also produce aerosols, with cooling climate effects and negative stratospheric ozone impacts (Galloway *et al.* 2008).

Significantly, irrigated crop production withdraws about 3,900 km^3 (ca. 936 mi^3) of water annually, of which 1,800–2,300 km^3 is consumed and not returned to the immediate hydrological cycle (Fig. 20.2). Of this consumption, about 43 per cent annually or 545 km^3 yr^{-1} is derived from groundwater (Siebert *et al.* 2010), commonly fossil water collected in aquifers over millennia. Croplands account for about 70 per cent of global water withdrawals and 85 per cent of global consumptive use (Siebert *et al.* 2010) **(Q33, Q34)**. Irrigation waters returned to the hydrological cycle are often degraded in quality owing to mixing with pesticides and fertilizers. In terms of CO$_2$ emission equivalents for 2018, food systems produced about one-third of all anthropogenic emissions globally (Tubiello *et al.* 2021) (Table. 20.1).

The historic increases in land-based cultivated foods notwithstanding, concerns exist about the next generation of food technology to feed 9–10 billion people sustainably **(Q22)**. This concern is amplified by the inability of global society to reduce food waste and to problems of over- and under-consumption of foods owing to global vagaries in food access.

References

Boutraa, T. (2010). "Improvement of water use efficiency in irrigated agriculture: a review". *Journal of Agronomy* 9(1): 1–8. doi:10.3923/ja.2010.1.8.

Bouwman, L. *et al.* (2013). "Exploring global changes in nitrogen and phosphorus cycles in agriculture induced by livestock production over the 1900–2050 period". *Proceedings of the National Academy of Sciences* 110(52): 20882–7. doi.org/10.1073/pnas.1012878108.

Galloway, J. *et al.* (2008). "Transformation of the nitrogen cycle: recent trends, questions, and potential solutions". *Science* 320(5878): 889–92. doi:10.1126/science.1136674.

Heinimann, A. *et al.* (2017). "A global view of shifting cultivation: recent, current, and future extent". *PLoS One* 12(9): e184479. doi.org/10.1371/journal.pone.0184479.

Portman, F., S. Siebert & P. Döll (2010). "MIRCA2000-Gloal monthly irrigated and rainfed crop area around the year 2000: a new high-resolution data set for agricultural and hydrological modeling". *Global Biogeochemical Cycles* 24(1). GB10011. doi.10.1029/2008GB003435.

Potapov, P. *et al.* (2021). "Global maps of cropland extent and change show accelerated crop-land expansion in the twenty-first century". *Nature Food* 3: 19–28. doi.org/10.1038/s43016-021-00429-z.

Ramankutty, N. & J. Foley (1998). "Characterizing pattern of global land use: an analysis of global croplands data". *Global Biogeochemical Cycles* 12(4): 667–85. doi.org/10.1029/98GB02512.

Ramankutty, N. & J. Foley (1999). "Estimating historical changes in global land cover: croplands from 1700 to 1992". *Global Biogeochemical Cycles* 13(4): 997–1027. doi.org/10.1029/1999GB900046.

Ramankutty, N. *et al.* (2008a). "Cropland and pasture area in 2000. Farming the planet: 1. Geographic distribution of global agricultural lands in the year 2000". *Global Biogeochemical Cycles* 22: GB1003. doi:10.1029/2007GB002952.

Ramankutty, N. *et al.* (2008b). "Cropland and pasture area in 2000". www.earthstat.org/cropland-pasture-area-2000/ (based on Ramankutty 2008a).

Reay, D. *et al.* (2012). "Global agriculture and nitrous oxide emissions". *Nature Climate Change* 2(6): 410–16. doi.org/10.1038/nclimate1458.

Salmon, J. *et al.* (2015). "Global rain-fed, irrigated, and paddy croplands: a new high resolution map derived from remote sensing, crop inventories and climate data". *International Journal of Applied Earth Observation and Geoinformation* 38: 321–34. doi.org/10.1016/j.jag.2015.01.014.

Siebert, S. *et al.* (2010). "Groundwater use for irrigation – a global inventory". *Hydrology and Earth System Sciences* 14(10): 1863–80. doi.org/10.5194/hess-14-1863-2010.

Tubiello, F. *et al.* (2021). "Greenhouse gas emissions from food systems: building the evidence base". *Environmental Research Letters* 16(6): 065007. doi.org/10.1088/1748-9326/ac018e.

Xu, X. *et al.* (2021). "Global greenhouse gas emissions from animal-based foods are twice those of plant-based foods". *Nature Food* 2: 724–32. doi.org/10.1038/s43016-021-00358-x.

Zabel, F., B. Putzenlechner & W. Mauser (2014). "Global agricultural land resources – a high resolution suitability evaluation and its perspectives until 2100 under climate change conditions". *PLoS One* 9(9): e107522. doi.org/10.1371/journal.pone.0107522.

Zomer, R. *et al.* (2016). "Global tree cover and biomass carbon on agricultural land: the contribution of agroforestry to global and national carbon budgets". *Scientific Reports* 6(1): 1–12. doi.org/10.1038/srep29987.

HAS HUMAN ACTIVITY ERODED AND DEGRADED SOILS GLOBALLY?

Short answer: yes. Soil erosion and degradation, primarily from agricultural uses, are global in scope, threatening food production and linked to numerous environmental problems, including emissions of greenhouse gases.

Humankind has long cultivated prime soils for agriculture and in many cases improved soil conditions for this use, such as the transformation of wetlands or the expansion of irrigated rice infrastructure that have sustained cropping for hundreds of years. Such improvements, however, have been surpassed by the worldwide amount of wind and water soil erosion and degradation from human activity. This degradation involves soil nutrient losses, compaction, salinization and contamination. These processes commonly follow from inappropriate land-use management (e.g. inadequate cultivation or overgrazing practices) and land-cover transformation, such as the impervious cover of paved-over cities that compacts soils (**Q31**), and industrial and commercial uses that may render soils toxic.

Soil degradation through cultivation and grazing has a history extending back to antiquity: soil salinization in ancient Mesopotamia, erosion in ancient Greece, central Mexico and China (González-Arqueros, Mendoza & Vázquez-Selem 2017; McNeill & Winiwarter 2004). Soil erosion is registered in lake deposition globally, dating back ~35 kya (Jenny *et al.* 2019). Degraded soils and their landscape conditions, in many cases interacting with prolonged drought, have had serious consequences for societies worldwide (Dotterweich 2013; Turner & Sabloff 2012).

Despite the antiquity of soil degradation and the understanding that it is currently a worldwide problem that is likely to exist into the future, systematic documentation of the process is difficult. No sensor can register soil degradation from satellites, for example. As a result, estimates rely on various expert assessments or on proxies, such as NDVI (normalized difference vegetation index) which registers the health of vegetation through the lens of visible and near-infrared light. The results constitute estimates that infer some dimension of soil degradation, ranging from 1–6 billion ha (Gibbs & Salmon 2015), depending on methods used. As a result, *The World Atlas of Desertification* (Cherlet *et al.* 2018) uses a "convergence of evidence" to map land and soil degradation. Various estimates conclude that at least 33 per cent of global soils show some sign of degradation and 60 per cent have lost some environmental services (**Q7**) between 1950 and 2010 (Lal 2015).

Perhaps 40 per cent of cultivated land has incurred soil erosion and other degrading processes (Foley *et al.* 2005), whereas all land uses are estimated to lose 24 billion tonnes yr^{-1} through erosion (UNCCD 2014). Such losses lead to yield gaps – the difference between optimal and actual yields in cultivation – which tend to be large for rainfed cultivation, especially in developing countries. In sub-Saharan Africa, for example, this gap ranges up to 30 per cent (Bindraban *et al.* 2012). These gaps may be reduced through appropriate management, although they are likely to require a substantial increase in irrigation (Rosa *et al.* 2018). This increase, in turn, may lead to soil salinization if irrigation practices are inappropriate. Currently, ~1.5 MH yr^{-1} of cultivable land are salinized (Lal 2015). Of all soil degradation, however, perhaps none has received more attention than arid land degradation or desertification (**Q26**), estimated to affect about 25 per cent of arid lands worldwide, typically from overgrazing livestock (Foley *et al.* 2005; Reynolds *et al.* 2007). In contrast, an estimated 10 per cent of soils across all biomes has improved (Lal 2015), owing to restoration or sustained maintenance.

The impacts of soil degradation are profound and have a new ingredient, microplastics (≤5 mm) from waste, and for which research on soil, ecosystem processes and the Earth system has begun in earnest only recently (Rillig & Lehmann 2020; Stubbins *et al.* 2021). Soil compaction increases surface water runoff, contributing to flooding, especially in built-up areas, such as cities, where soil contamination is common (Alaoui *et al.* 2018). The lower the capacity of soils to maintain vegetation, and the degradation of vegetation by its use, amplifies wind and water erosion of soils. The degraded land surface, in turn, may amplify climatic conditions, for instance via feedbacks that increase climatic drought intensity and wind erosion, as took place in the Dust Bowl across the Great Plains of the United States in the 1930s (Baveye *et al.* 2011; Cook, Miller & Seager 2009) (Fig. 21.1), owing to inappropriate land management practices. "Dust bowls" continue throughout the world (Ma *et al.* 2021), such as that experienced in northern China in 2001 in which Asian dust crossed the Pacific Ocean to North America.

Soils also store huge amounts of carbon (Fig. 21.2) because ~24 per cent of plant biomass is located in them (e.g. roots). Recent calculations indicate this biomass holds as much as 113 Gt of carbon, equivalent to about 10 years of current emissions of global CO_2 (Ma *et al.* 2012). This amount is a significantly larger amount than previous estimates (Scharlemann *et al.* 2014). All assessments, however, indicate that the amount of soil carbon is larger than that in the atmosphere and above-ground biota (vegetation) (Lal 2007), indicating that how soils are treated amplifies or attenuates carbon emissions. Soil also holds large amounts of nitrogen. In general, the intensity of soil use impacts nitrogen oxide emissions as well as CO_2 emissions to the atmosphere.

Soils are also indirectly affected by human activity through climate change. Significant attention has been given to the release of CO_2 and CH_4 from tundra as the thawing of the subsurface soil storing these greenhouse gases releases them to the atmosphere (**Q29**). Global warming, however, will also release carbon from all soils, including that from

Figure 21.1 The Dust Bowl
Inappropriate cultivation practices coupled with drought created huge losses of wind-blown soil in the 1930s in the Southern Great Plains of the United States.
Source: Album & NOAA (1935).

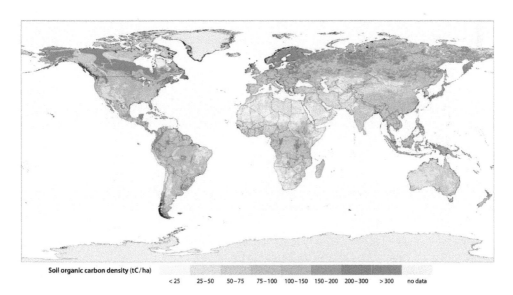

Soil organic carbon density (tC/ha)

< 25   25 – 50   50 – 75   75 – 100   100 – 150   150 – 200 200 – 300   > 300   no data

Figure 21.2 Distribution of global soil organic carbon (SOC)
Source: Scharlemann *et al.* (2014).

fractured and weathered rock between bedrock and soil (Tune *et al.* 2020). A recent estimate indicates that a 2°C increase in the global average temperature will release 232 ± 52 Pg of soil carbon from the top one metre of soils worldwide (Varney *et al.* 2020). Notably, land-based carbon flows not only to the atmosphere, but via rivers and groundwater to the oceans. Recent estimates indicate that rivers deposit $43^{+9.2/-7}$ Pg yr^{-1} of carbon to the oceans (Borrelli *et al.* 2020), whereas another 0.8±0.5 GtC yr^{-1} flows from groundwater and coastal vegetation (Kwon *at al.* 2012). Beyond these assessments, new baseline estimates of potential global soil losses from agriculture are about 43 Pg yr^{-1}, but this rate could be reduced by about 5 per cent following conservation agriculture (Varney *et al.* 2020). Large losses exist even in technologically advanced agricultural systems, such as the Corn Belt in the middle of the United States, where nearly 40.5 MH (100 million acres) of cultivated land has lost its topsoil, reducing yields by 6 per cent (Kwon *et al.* 2021; Thaler, Larsen & Yu 2021). Potential soil erosion increases of 30–66 per cent are forecast globally owing to increases in rainfall duration and intensity associated with climate warming (Varney *et al.* 2020).

Overall, soils – a critical medium for the provisioning of humankind – have been significantly altered from their natural state worldwide. Soil losses and degradation have innumerable environmental consequences that accumulate to affect the Earth system at large and increase the amount and costs of inputs to maintain the environmental services and livelihoods they support.

References

Alaoui, A. *et al.* (2018). "Does soil compaction increase floods? A review". *Journal of Hydrology* 557: 631–42. doi.org/10.1016/j.jhydrol.2017.12.052.

Album, G. & NOAA (National Oceanic and Atmospheric Administration) (1935). "Dust storm approaching Stratford, Texas, 1935". commons.wikimedia.org/wiki/File:Dust-storm-Texas-1935.png.

Baveye, P. *et al.* (2011). "From dust bowl to dust bowl: soils are still very much a frontier of science". *Soil Science Society of America Journal* 75(6): 2037–48. doi.org/10.2136/ sssaj2011.0145.

Bindraban, P. *et al.* (2012). "Assessing the impact of soil degradation on food production". *Current Opinion in Environmental Sustainability* 4(5): 478–88. doi.org/10.1016/j.cosust.2012.09.015.

Borrelli, P. *et al.* (2020). "Land use and climate change impacts on global soil erosion by water (2015–2070)". *Proceedings of the National Academy of Sciences* 117(36): 219994–22001. doi.org/ 10.1073/pnas.2001403117.

Cherlet, M. *et al.* (eds) (2018). *World Atlas of Desertification*. Third edition. Luxembourg: Publication Office of the European Union.

Cook, B., R. Miller & R. Seager (2009). "Amplification of the North American 'Dust Bowl' drought through human-induced land degradation". *Proceedings of the National Academy of Sciences* 106(13): 4997–5001. doi.org/10.1073/pnas.0810200106.

Dotterweich, M. (2013). "The history of human-induced soil erosion: geomorphic legacies, early descriptions and research, and the development of soil conservation – A global synopsis". *Geomorphology* 201: 1–34. doi.org/10.1016/j.geomorph.2013.07.021.

Foley, J. *et al.* (2005). "Global consequences of land use". *Science* 309(5734): 570–74. doi:10.1126/science.1111772.

Gibbs, H. & J. Salmon (2015). "Mapping the world's degraded lands". *Applied Geography* 57: 12–21.

González-Arqueros, M., M. Mendoza & L. Vázquez-Selem (2017). "Human impact on natural systems modeled through soil erosion in GeoWEPP: a comparison between pre-Hispanic periods and modern times in the Teotihuacan Valley (Central Mexico)". *Catena* 149: 505–13. doi.org/10.1016/j.catena.2016.07.028.

Jenny, J. *et al.* (2019). "Human and climate global-scale imprint on sediment transfer during the Holocene". *Proceedings of the National Academy of Sciences* 116(46): 22972–6. doi/10.1073/pnas.1908179116.

Kwon, E. *et al.* (2012). "Stable carbon isotopes suggest large terrestrial carbon inputs to the global ocean". *Global Biogeochemical Cycles* 35(4): e2020GB006684. doi.org/10.1029/2020GB006684.

Lal, R. (2007). "Carbon management in agricultural soils". *Mitigation and Adaptation Strategies for Global Change* 12(2): 303–22. doi.org/10.1007/s11027-006-9036-7.

Lal, R. (2015). "Restoring soil quality to mitigate soil degradation". *Sustainability* 7(5): 5875–95. doi.org/10.3390/su7055875.

Ma, H. *et al.* (2021). "The global distribution and environmental drivers of aboveground versus belowground plant biomass". *Nature Ecology & Evolution* 5(8): 1–13. doi.org/10.1038/s41559-021-01485-1.

McNeill, J. & V. Winiwarter (2004). "Breaking the sod: humankind, history, and soil". *Science* 304(5677): 1627–9. doi:10.1126/science.1099893.

Reynolds, J. *et al.* (2007). "Global desertification: building a science for dryland development". *Science* 316(5826): 847–51. doi:10.1126/science.1131634.

Rillig, M. & A. Lehmann (2020). "Microplastics in terrestrial ecosystems". *Science* 368(6498): 1430–31. doi:10.1126/science.abb5979.

Rosa, L. *et al.* (2018). "Closing the yield gap while ensuring water sustainability". *Environmental Research Letters* 13(10): p.104002. doi.org/10.1088/1748-9326/aadeef.

Scharlemann, J. *et al.* (2014). "Global soil carbon: understanding and managing the largest terrestrial carbon pool". *Carbon Management* 5(1): 81–91. doi.org/10.4155/cmt.13.77.

Stubbins, A. *et al.* (2021). "Plastics in the Earth system". *Science* 373(6550): 51–5. doi:10.1126/science.abb0354.

Thaler, E., I. Larsen & Q. Yu (2021). "The extent of soil loss across the US Corn Belt". *Proceedings of the National Academy of Sciences* 118(8): e1922375118. doi.org/10.1073/pnas.1922375118.

Tune, A. *et al.* (2020). "Carbon dioxide production in bedrock beneath soils substantially contributes to forest carbon cycling". *Journal of Geophysical Research: Biogeosciences* 125(12): e2020JG005795. doi.org/10.1029/2020JG005795.

Turner II, B. & J. Sabloff (2012). "Classic Period collapse of the Central Maya Lowlands: insights about human–environment relationships for sustainability". *Proceedings of the National Academy of Sciences* 109(35): 13908–14. doi.org/10.1073/pnas.1210106109.

UNCCD (United Nations Convention to Combat Desertification) (2014). *The Land in Numbers: Livelihoods at a Tipping Point.* Bonn: UNCCD.

Varney, R. *et al.* (2020). "A spatial emergent constraint on the sensitivity of soil carbon turnover to global warming". *Nature Communications* 11(1): 1–8. doi.org/10.1038/s41467-020-19208-8.

WILL AGRICULTURE REQUIRE LESS LAND IN THE FUTURE AS PROPOSED BY THE BORLAUG HYPOTHESIS?

Short answer: the proposition that agricultural intensification has spared land for food and fibre production and will continue to do so in the future is contested, although it underlies future projections.

The long-term future of global cultivation has been projected to consist of high-tech capital and water-intensive practices capable of increasing 2010 production by 50 per cent in order to provision 9–10 billion people on less land than is used today (Searchinger *et al.* 2019). Producing more on less land is consistent with the "Borlaug hypothesis", named after Norman Borlaug, who led the innovation of hybrid crops (cross-pollinated varieties of food crops) and the subsequent "green revolution" that followed from the improved yields. Hybrid crops, combined with fertilizers and irrigation, had the capacity – in principle – to increase food stocks on less land. Modelling estimates suggest that between 1965 and 2004 the green revolution saved anywhere from ~18 to ~27 MH of land from agriculture, of which about 2 MH would have been taken for cultivation from forest lands (Stevenson *et al.* 2013). Remote sensing assessments, in contrast, indicate ~240 MH of forest have been gained globally, largely from the abandonment of agricultural lands in the mid-latitudes, while global crop production has increased worldwide (Pendrill *et al.* 2019; Potapov *et al.* 2021). Evidence that agricultural intensification is linked to land abandonment in particular places, however, is either weakly related or not in various assessments, depending on the study and variations in country conditions (Ewers *et al.* 2009; Rudel *et al.* 2009).

Attention to increasing food, fibre, forage and fodder without excessive expansion of agricultural lands is under way. Advances in smart and vertical farming (i.e. sensors to identify field areas needing inputs, and vertical, stacked layers of planting surfaces, respectively) and crop genomics (or synbio, the genetic manipulation of plants to improve yield and resistances to disease and drought) are expected to raise yields and make them more secure to the vagaries of nature in the future (Byerlee, Stevenson & Villoria 2014; Abberton *et al.* 2016). Assuming these advances are undertaken worldwide, the provisioning of humankind in 2050 is estimated to require ~200–500 MH of added cropland, whereas under current

Net agricultural land expansion (2010–50)

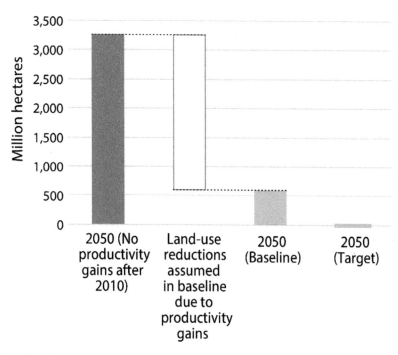

Figure 22.1 Possible net gains in agricultural lands, 2010–50, due to projected productivity gains
Source: Searchinger *et al.* (2019).

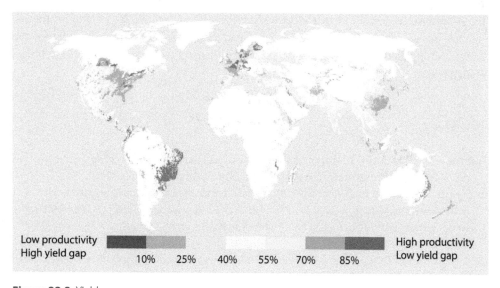

Figure 22.2 Yield gap
The yield gap is the difference between actual and potential yields for major food crops.
Source: data from FAO (2022); basemap: Esri.

techno-managerial strategies ~1–3.25 billion ha would be needed (Searchinger *et al.* 2019; Walter *et al.* 2017). This business-as-usual range in area is ~three to ten times larger than the mid-range (350 MH) area required for the estimated advanced agriculture projection.

To achieve the lower land-added goal (i.e. 200–500 MH) also requires attention to the global "yield gap" – the difference between actual and potential yields – based on current technologies and management practices that underperform in terms of total potential production (Fig. 22.1). Decreasing the yield gap, however, requires the widespread and intensive use of fertilizer, pesticide and irrigation. This use, in turn, has significant consequences for soil acidification (Tilman *et al.* 2011), greenhouse emissions (Guo *et al.* 2010; Reay *et al.* 2012), biodiversity and water pollution. Stress on water withdrawals, accounting for about 90 per cent of global water consumption (Döll & Siebert 2002), raises concerns about water availability for the intensification process. At least one study, however, projects that sustainable supplies of water could be increased by 48 per cent (408 km^3 water yr^{-1}), increasing cropland irrigation by 26 per cent (Rosa *et al.* 2018). In addition, advances in crop breeding, alternative or organic cultivation methods, and smart or precision farming targeting fertilizer use, are required to improve yields and lower nitrogen emissions, while using water more efficiently, perhaps reducing water withdrawals (Döll & Siebert 2002).

These concerns raise the issue of total factor productivity (TFP) and agriculture. TFP is the ratio of all agricultural output to all inputs, including environmental services (**Q7**). The positives of the growth in TFP occurs when the output growth exceeds that of inputs. This growth has taken place in the developing world since about 2003, accounting for about 70 per cent of the total output growth in agriculture (Coomes *et al.* 2019). Presumably, worldwide improvement in TFP should lower the per capita area for agricultural land, but will require attention to the environmental impacts of doing so. The global capacity to employ input advances and sustain environmental services will determine the scale of future sustainable agriculture (Scherr, Shames & Friedman 2012), as will reducing food losses and waste, shifting diets, lowering fertility rates, and perhaps phasing out biofuels (Searchinger *et al.* 2019).

References

Abberton, M. *et al.* (2016). "Global agricultural intensification during climate change: a role for genomics". *Plant Biotechnology Journal* 14(4): 1095–8. doi.org/10.1111/pbi.12467.

Byerlee, D., J. Stevenson & N. Villoria (2014). "Does intensification slow crop land expansion or encourage deforestation?" *Global Food Security* 3(2): 92–8. doi.org/10.1016/j.gfs.2014.04.001.

Coomes, O. *et al.* (2019). "Leveraging total factor productivity growth for sustainable and resilient farming". *Nature Sustainability* 2: 22–8. doi.org/10.1038/s41893-018-0200-3.

Döll, P. & S. Siebert (2002). "Global modeling of irrigation water requirements". *Water Resources Research* 38(4): 1–10. doi.org/10.1029/2001WR000355.

Ewers, R. *et al.* (2009). "Do increases in agricultural yield spare land for nature?" *Global Change Biology* 15(7): 1716–26. doi.org/10.1111/j.1365-2486.2009.01849.x.

FAO (Food and Agricultural Organization) (2022). "Yield and production gaps". www.fao.org/nr/gaez/about-data-portal/yield-and-production-gaps/en/.

Guo, J. *et al.* (2010). "Significant acidification in major Chinese croplands". *Science* 327(5968): 1008–10. doi:10.1126/science.1182570.

Pendrill, F. *et al.* (2019). "Deforestation displaced: trade in forest-risk commodities and the prospects for a global forest transition". *Environmental Research Letters* 14(5): 055003. doi.org/10.1088/1748-9326/ab0d41.

Potapov, P. *et al.* (2021). "Global maps of cropland extent and change show accelerated cropland expansion in the twenty-first century". *Nature Food* 3: 19–28. doi.org/10.1038/s43016-021-00429-z.

Reay, D. *et al.* (2012). "Global agriculture and nitrous oxide emissions". *Nature Climate Change* 2(6): 410–16. doi.org/10.1038/nclimate1458.

Rosa, L. *et al.* (2018). "Closing the yield gap while ensuring water sustainability". *Environmental Research Letters* 13(10): p.104002. doi.org/10.1088/1748-9326/aadeef.

Rudel, T. *et al.* (2009). "Agricultural intensification and changes in cultivated areas". *Proceedings of the National Academy of Sciences* 106(40): 20675–80. doi.org/10.1073/pnas.0812540106.

Scherr, S., S. Shames & R. Friedman (2012). "From climate-smart agriculture to climate-smart landscapes". *Agriculture & Food Security* 1(1): 12. doi.org/10.1186/2048-7010-1-12.

Searchinger, T. *et al.* (2019). *Creating a Sustainable Food Future: A Menu of Solutions to Sustainably Feed More Than 10 Billion People by 2050. Final Report.* Washington, DC: World Resource Institute.

Stevenson, J. *et al.* (2013). "Green revolution research saved an estimated 18 to 27 million hectares from being brought into agricultural production". *Proceedings of the National Academy of Sciences* 110(21): 8363–8. doi.org/10.1073/pnas.1208065110.

Tilman, D. *et al.* (2011). "Global food demand and the sustainable intensification of agriculture". *Proceedings of the National Academy of Sciences* 108(50): 20260–4. doi.org/10.1073/pnas.1116437108

Walter A. *et al.* (2017). "Smart farming is key to developing sustainable agriculture". *Proceedings of the National Academy of Sciences* 14(24): 6148–50. doi.org/10.1073/pnas.1707462114.

DOES LAND TAKEN OUT OF CULTIVATION DECREASE THE GLOBAL AREA CULTIVATED OR LEAD TO DISPLACEMENT AND LAND-GRABBING?

Short answer: displacement and land-grabbing tend to offset a significant portion of land taken out of cultivation globally.

Historically, agricultural lands have been degraded and abandoned, in some cases "rewilded" for multiple centuries (Neves & Heckenberger 2019) and in other cases rehabilitated. Today, agricultural lands are lost to the global growth in urbanization (Seto & Ramankutty 2016) and to various attempts by countries across the world to reforest former agricultural lands in order to enhance environmental services, such as flood protection, reduce greenhouse gas emissions, or engage in carbon offsetting programmes. This loss of lands once used to provide food, fibre or fuel by countries or regions has not necessarily led to a global decrease in cultivated lands. Rather, the provisioning from the land taken out of production in the home country is supplied through the processes of displacement and land-grabbing in other countries. Displacement refers to the produce generated on lands in other countries that is used to make up for the lost provisioning in the home country (Meyfroidt, Rudel & Lambin 2010). This phenomenon, largely associated with developed economies, even takes place in countries with developing economies. For example, the lost production to reforestation projects in Vietnam from 1987 to 2006 led to major imports of forest products (Meyfroidt & Lambin 2009).

Land-grabbing, in contrast, refers to large-scale land acquisitions by foreign agents, commonly in hectarages of six figures, for products consumed in the agents' homelands or abroad (Borras & Franco 2012) (Fig. 23.1A and B). Seizing foreign terrain for land resources is an ancient tactic that can be traced back to early empires, especially prominent in the colonial era, and followed in the nineteenth and twentieth centuries by corporate access of foreign lands. In these cases, the focus was on obtaining products requiring climates absent in the consumers' home countries. Current land-grabbing by countries or corporations differs from past grabs because it may involve country-to-country agreements, although the processes involved are not always transparent and can involve products reduced in the home country and supplemented from another country, such as Chinese acquisition of sisal

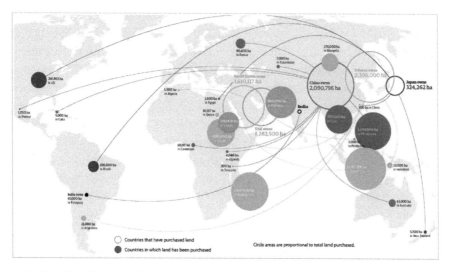

Figure 23.1A Global land-grabbing
Land purchased by governments and private companies from each country where "areas" are known. Data derived from GRAIN 2008 website that no longer exists. Note differences with Fig. 23.1B
Source: Baveye *et al.* (2011).

(hemp) estates in Eastern Africa to supply the product for Chinese production and consumption (Hofman & Ho 2012). Land-grabbing also constitutes "water-grabbing" (surface or groundwater) when the land and water rights are taken (Rulli, Saviori & D'Odorico 2013).

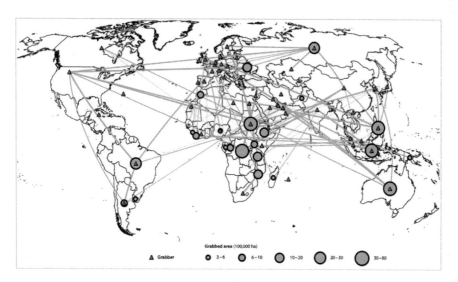

Figure 23.1B Global land-grabbing
Only the 24 major land-grabbed countries considered with linkages to "grabber" country when the area exceeded 100,000 ha. Data from 2012. Notice the differences in grabber and land-grabbed countries with Fig. 23.1A.
Source: Rulli, Saviori & D'Odorico (2013).

Estimates of all "lands grabbed" globally range from 20 to 45 MH as of 2010 (Rulli, Saviori & D'Odorico 2013; Von Braun & Meinzen-Dick 2009), depending on the definition and data employed (illustrated in Fig. 23.1). Uncertainty marks these estimates owing to the robustness of the data on the number of transactions and area involved (Liao *et al.* 2016), among other issues, such as the geopolitical perspective applied to the studies. Indeed, challenges to the entire concept of land grabs exist (Bräutigam & Zhang 2013). Land-grabbing maintains a negative connotation, following from cases in which lands claimed by local inhabitants are accessed by foreign agents. In some cases, however, the foreign investment proves positive for the country and locale in question. Such distinctions notwithstanding, foreign investment in large-scale agricultural systems in the Global South partly explains much of the sustained expansion of cultivated lands there. As a result, land currently taken out of agriculture commonly improves environmental conditions in the accessing country, especially if the abandoned land is permitted to regain some semblance of its former land cover and, hence, its ecological functions. At the global level, however, land-grabbing and displacement lead to land-cover changes, foremost to tropical forests, with various environmental consequences noted in **Question 18.**

References

Baveye, P. *et al.* (2011). "From dust bowl to dust bowl: soils are still very much a frontier of science". *Soil Science Society of America Journal* 75(6): 2037–48. doi.org/10.2136/sssaj2011.0145.

Borras Jr., S. & J. Franco (2012). "Global land grabbing and trajectories of agrarian change: a preliminary analysis". *Journal of Agrarian Change* 12(1): 34–59. doi.org/10.1111/j.1471-0366.2011.00339.x.

Bräutigam, D. & H. Zhang (2013). "Green dreams: myth and reality in China's agricultural investment in Africa". *Third World Quarterly* 34(9): 1676–96. doi.org/10.1080/01436597.2013.843846.

Hofman, I. & P. Ho (2012). "China's 'developmental outsourcing': a critical examination of Chinese global 'land grabs' discourse". *Journal of Peasant Studies* 39(1): 1–48. doi.org/10.1080/03066150.2011.653109.

Liao, C. *et al.* (2016). "Insufficient research on land grabbing". *Science* 353(6295): 131. doi:10.1126/science.aaf6565.

Meyfroidt, P. & E. Lambin (2009). "Forest transition in Vietnam and displacement of deforestation abroad". *Proceedings of the National Academy of Sciences* 106(38): 16139–44. doi.org/10.1073/pnas.0904942106.

Meyfroidt, P., T. Rudel & E. Lambin (2010). "Forest transitions, trade, and the global displacement of land use". *Proceedings of the National Academy of Sciences* 107(49): 20917–22. doi.org/10.1073/pnas.1014773107.

Neves, E. & M. Heckenberger (2019). "The call of the wild: rethinking food production in Ancient Amazonia". *Annual Review of Anthropology* 48: 371–88. doi/full/10.1146/annurev-as-48-themes.

Rulli, M., A. Saviori & P. D'Odorico (2013). "Global land and water grabbing". *Proceedings of the National Academy of Sciences* 110(3): 892–7. doi.org/10.1073/pnas.1213163110.

Seto, K. & N. Ramankutty (2016). "Hidden linkages between urbanization and food systems". *Science* 352(6288): 943–5. doi:10.1126/science.aaf7439.

Von Braun, J. & R. Meinzen-Dick (2009). "Land grabbing by foreign investors in developing countries: risks and opportunities". IFPI Policy Brief 13. Washington, DC: International Food Policy Institute.

Q24

HOW MUCH GRASSLANDS AND PASTURES HAVE BEEN ALTERED OR DEGRADED GLOBALLY?

Short answer: about a third of grassland and pastures globally have been altered by changing land uses, generating significant emissions of greenhouse gases and aerosols, and landscape degradation.

Grasslands, either tropical savannas (aka savannahs) or temperate grasslands (e.g. prairie or steppe), are landscapes dominated by grasses with no more than 10 per cent tree cover. By this definition, they cover about 52.5 Mkm² or about a third of the terrestrial surface (40.5 per cent, excluding Antarctica and Greenland) (Gang *et al.* 2014; Suttie, Reynolds & Batello 2005) (Fig. 24.1). Significant areas of grasslands are cultivated, as on the Great Plains of North America, where many grains, including maize (corn), dominate. Such cultivation commonly leads to the mining (loss) of water from aquifers (**Q34, Q35**). Rangeland (grass-dominated) and sown pasture land, some of which is former forest land converted to pasture (Asner *et al.* 2004), cover about 35 about 35 Mkm², providing about 50 per cent of the biomass for about 1.5 billion domesticated animal units in 1990 (Gao *et al.* 2016; Herrero *et al.* 2013). These units range from intensive livestock grazing on improved pasture to extensive pastoral nomadic activities.

Different environmental impacts follow from the different combinations of land use, biome and economy in question. In terms of global net primary productivity (or fixed energy, largely in biomass generated by photosynthesis), about 50 per cent of grasslands have been degraded, 32.5 per cent by human activity and the rest by climate change (Gang *et al.* 2014), leading to a reduction in productivity of about 15–71 per cent, depending on location (Gao *et al.* 2016). Added to this, overgrazing – too many livestock per unit of grassland – reduces biomass and can trigger desertification (Fig. 24.2) (**Q26**), whereas reduced grassland-burning, anthropogenic or natural, leads to the encroachment of woody species. In the United States this encroachment has taken place at rates of 0.5–2.0 per cent yr^{-1} (Anadón *et al.* 2014). Anthropogenic burning is common in the tropics, a practice that produces new, more nutritious grasses for livestock (**Q27**), but in the process, greenhouse gases and aerosols are emitted to the atmosphere (Abdalla *et al.* 2016). Of late, the grassland area burned globally, from natural and human ignitions, has decreased by ~23 per cent, largely due to grassland conversion to cultivation.

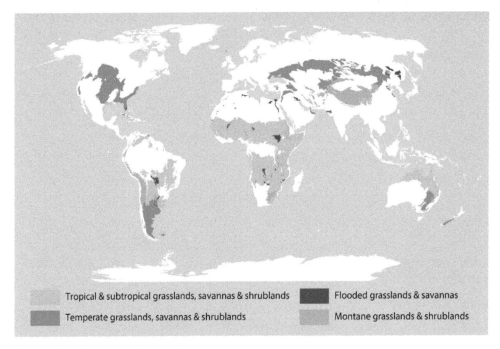

Figure 24.1 Grasslands of the world
What constitutes grasslands are variously defined. Some assessments include tundra **(Q29)** which are not included here.
Source: UNEP-WCMC (2021) based on Dinerstein *et al.* (2017); basemap: Esri.

Importantly, rangeland livestock emit large amounts of nitrous oxide (N_2O) and methane (CH_4) to the atmosphere. The amount of N_2O emitted from animal waste (dung and urine) exceeds that in fertilizers used for plant crops (Bouwman *et al.* 2013). CH_4 emitted by domesticated ruminants (e.g. cattle, sheep, goats) approaches 80 Tg yr^{-1} (Lassey 2007). These emissions could be reduced through improved management practices, such as

Figure 24.2 Overgrazing a grassland
Pictures taken 180° directionally from the same location in the Karoo, South Africa, demonstrating the consequences of overgrazing grasslands.
Source: author.

improved pastures and intensifying ruminant diets (Herrero & Thorton 2013; Thorton & Herrero 2010). In contrast, initial modelling work indicates that in some regions of the world, grasslands will prove to be more efficient in the uptake and maintenance of carbon than forests under the stress of climate warming (Das *et al.* 2018). In addition, livestock consume large quantities of water in grassland environments that typically are not water rich. The severe degradation of grasslands, as one type of arid land, is discussed in **Question 26**.

References

Abdalla, K. *et al.* (2016). "Long-term annual burning of grassland increases CO_2 emissions from soils". *Geoderma* 282: 80–86. doi.org/10.1016/j.geoderma.2016.07.009.

Anadón, J. *et al.* (2014). "Effect of woody-plant encroachment on livestock production in North and South America". *Proceedings of the National Academy of Sciences* 111(35): 12948–53. doi.org/10.1073/pnas.1320585111.

Asner, G. *et al.* (2004). "Grazing systems, ecosystem responses, and global change". *Annual Review of Environment and Resources* 29: 261–99. doi.org/10.1146/anurev.enegy.29.062403.102142.

Bouwman, L. *et al.* (2013). "Exploring global changes in nitrogen and phosphorus cycles in agriculture induced by livestock production over the 1900–2050 period". *Proceedings of the National Academy of Sciences* 110(52): 20882–7. doi.org/10.1073/pnas.1012878108.

Dass, P. *et al.* (2018). "Grasslands may be more reliable carbon sinks than forests in California". *Environmental Research Letters* 13(7): 074027. doi.org/10.1088/1748-9326/aacb39.

Dinerstein, E. *et al.* (2017). "An ecoregion-based approach to protecting half the terrestrial realm". BioScience 67(6): 534–54. doi.org/10.1093/biosci/bix014.

Gang, C. *et al.* (2014). "Quantitative assessment of the contributions of climate change and human activities on global grassland degradation". *Environmental Earth Sciences* 72(11): 4273–82. doi.org/10.1007/s12665-014-3322-6.

Gao, Q. *et al.* (2016). "Climatic change controls productivity variation in global grasslands". *Scientific Reports* 6: 26958. doi.org/10.1038/srep26958.

Herrero, M. *et al.* (2013). "Biomass use, production, feed efficiencies, and greenhouse gas emissions from global livestock systems". *Proceedings of the National Academy of Sciences* 110(52): 20888–93. doi.org/10.1073/pnas.1308149110.

Herrero, M. & P. Thornton (2013). "Livestock and global change: emerging issues for sustainable food systems". *Proceedings of the National Academy of Sciences* 110(52): 20878–81. doi.org/10.1073/pnas.1321844111.

Lassey, K. (2007). "Livestock methane emission: from the individual grazing animal through national inventories to the global methane cycle". *Agriculture and Forest Meteorology* 142(2–4): 12–132. doi.org/10.1016/j.agrformet.2006.03.028.

Suttie, J., S. Reynolds & C. Batello (eds). 2005. *Grasslands of the World*. Rome: FAO.

Thornton, P. & M. Herrero (2010). "Potential for reduced methane and carbon dioxide emissions from livestock and pasture management in the tropics". *Proceedings of the National Academy of Sciences* 107(46): 19667–72. doi.org/10.1016/j.agrformet.2006.03.028.

UNEP-WCMC (UN Environmental Programme-World Monitoring Conservation Centre) (2021). The RESOLVE Ecoregions dataset, updated in 2017, offers a depiction of the 846 terrestrial ecoregions that represent our living planet. data-gis.unep-wcmc.org/portal/home/item.html?id=601f6615a2dc492d9b74c5ab38782549.

WHAT ARE CO-ADAPTED LANDSCAPES?

Short answer: they are naturally appearing landscapes partly created and maintained by human activity with effects on ecosystem composition and functioning.

Two or more species adapting as a group constitutes co-adaptation. As applied here, a co-adapted landscape refers to biota adapted to human activity. Foremost, these landscapes are flora and fauna complexes that are partly the product of long-term human activity that has subtly altered natural conditions, helping to create heterogeneous landscapes (Ellis 2021). Unlike iconic transformed landscapes, such as those observed in the Mediterranean hillsides of vineyards and villages (Geri, Amici & Rocchini 2010) or covered bridges and dairy fields in Vermont (Morse *et al.* 2014), co-adaptive landscapes appear to be native or natural but originate from regular human management, often begun in antiquity and practised to this day. One such landscape is the fire climax savanna or grasslands intermixed with open woodlands present in eastern and western Africa (Salvatori *et al.* 2001). Evidence exists that the frequency of burning – natural burns amplified by human-set fires (Bird & Cali 1998), the latter to create conditions beneficial for hunting, herding, and cultivation and village fire protection – has expanded and maintains these grasslands (aka secondary grasslands in ecology) (Archibald, Staver & Levin 2012). Pastoralists in parts of Africa, for example, have regularly burned open woodland-savannas from at least 4,000 BP (Archibald, Staver & Levin 2012) in order to propagate fresh growth in grasses to improve forage for livestock (Fig. 25.1).

Agriculturalists also burn to generate a landscape mosaic for protection against biomass build-up conducive to catastrophic fires (Laris *et al.* 2015; Laris & Wechsler 2018). In eastern Africa, this burning reduces the expansion of woody growth, maintaining the very ecosystem that supports the great wild herd animals that share the lands with the pastoralists' livestock, perhaps by affecting the plant diversity supporting the grazing and browsing animals (Kartzinel *et al.* 2015).

In another example, some tropical forests maintain an abundance of economically useful tree species that appear to be incongruent with native vegetation. These forests, such as those existing in parts of Central America and Yucatán, are thought to be the relics of the distant past and, in some cases, present forest management (Alcorn 1984; Ford & Nigh

Figure 25.1 Livestock and wildlife grazing in an African savanna landscape burned by herders
Source: © Robert M. Pringle, Princeton University.

2009). "Hyperdominant" species in Amazonian forests may have associations with past anthropogenic forest uses (Ter Steege *et al.* 2013). Pine tree-savannas in parts of tropical forest biomes, especially in Central America (Denevan 1961), and the open or removed undergrowth of pine forests encountered by Anglo settlers in the southern United States, are products of long-term human burning of those landscapes (Carter & Foster 2004).

Co-adaptive landscapes are not global in reach but are significant because they demonstrate the antiquity of the modification of landscapes by humans. The Earth system consequences of the global scale of landscape burning are noted in **Question 27**.

References

Alcorn, J. (1984). *Huastec Mayan Ethnobotany.* Austin, TX: University of Texas Press.

Archibald, S., A. Staver & S. Levin (2012). "Evolution of human-driven fire regimes". *Proceedings of the National Academy of Sciences* 109(3): 847–52. doi.org/10.1073/pnas.1118648109.

Bird, M. & J. Cali (1998). "A million-year record of fire in sub-Saharan Africa". *Nature* 394: 767–9. doi.org/10.1038/29507.

Carter, M. & C. Foster (2004). "Prescribed burning and productivity in southern pine forests: a review". *Forest Ecology and Management* 191(1–3): 93–109. doi.org/10.1016/j.foreco.2003.11.006.

Denevan, W. (1961). *The Upland Pine Forest of Nicaragua: A Study in Cultural Plant Geography*. Berkeley, CA: University of California Press.

Ellis, E. (2021). "Land use and ecological change: a 12,000-year history". *Annual Review of Environment and Resources* **46: 1–33. doi.org/10.1146/annurev-environ-012220-010822.**

Ford, A. & R. Nigh (2009). "Origins of the Maya forest garden: Maya resource management". *Journal of Ethnobiology* 29(2): 213–36. doi.org/10.2993/0278-0771-29.2.213.

Geri, F., V. Amici & D. Rocchini (2010). "Human activity impact on the heterogeneity of a Mediterranean landscape". *Applied Geography* 30(3): 370–79. doi.org/10.1016/j.apgeog.2009.10.006.

Kartzinel, T. *et al.* (2015). "DNA metabarcoding illuminates dietary niche partitioning by African large herbivores". *Proceedings of the National Academy of Sciences* 112(26): 8019–24. doi.org/10.1073/pnas.1503283112.

Laris, P. *et al.* (2015). "The human ecology and geography of burning in an unstable savanna environment". *Journal of Ethnobiology* 35(1): 111–39. doi.org/10.2993/0278-0771-35.1.111.

Laris, P., A. Jo & S. Wechsler (2018). "Effects of landscape pattern and vegetation type on the fire regime of a mesic savanna in Mali". *Journal of Environmental Management* 227: 134–45. doi.org/10.1016/j.jenvman.2018.08.091.

Morse, C. *et al.* (2014). "Performing a New England landscape: viewing, engaging, and belonging". *Journal of Rural Studies* 36: 226–36. doi.org/10.1016/j.jrurstud.2014.09.002.

Salvatori, V. *et al.* (2001). "The effects of fire and grazing pressure on vegetation cover and small mammal populations in the Maasai Mara National Reserve". *African Journal of Ecology* 39(2): 200–204. doi.org/10.1046/j.1365-2028.2001.00295.x.

Ter Steege, H. *et al.* (2013). "Hyperdominance in the Amazonian tree flora". *Science* 342(6156): 12433092. doi:10.1126/science.1243092.

WHAT IS DESERTIFICATION AND HOW MUCH HAS OCCURRED?

Short answer: human activities have degraded the biomass, soils and ecosystem functioning of 10–20 per cent of arid lands globally, a process known as desertification.

Arid lands extend across ~41–46 per cent of Earth's surface, depending on the assessment criteria used (IPCC 2019; Reynolds *et al.* 2007). A common criterion is an aridity index – precipitation divided by potential evapotranspiration – of <0.65. Deserts are the most arid of drylands, receiving less than 250 mm of precipitation annually and usually maintaining an aridity index <0.5, dominated by sand, rock, ice, or extremely sparse vegetation. Using the 250 mm measure, about 33 per cent of the terrestrial surface of the Earth is climatically classified as desert. This figure includes the polar deserts of Antarctica and the Arctic, which comprise about 39 per cent of all desert areas or more than 29 Mkm². The ice covering of climatically polar deserts has accumulated over millennia owing to the minimal loss of snow that falls on them.

Arid lands, including many portions otherwise labelled deserts, maintain scrubby vegetation and grasses which historically have supported livestock rearing, such as that by pastoralists in Africa, Asia and the Middle East, and ranching elsewhere. The margins or wetter parts of arid lands may even be used for rainfed cultivation, and access to water permits irrigated cultivation. Overgrazing, overcultivation or abandoned irrigated cultivation, in some cases combined with extended drought, commonly reduces native vegetation and impacts soils, creating a more xeric (arid) land-cover condition known as arid land degradation or desertification (Middleton 2018). Such conditions commonly follow the loss of groundwater for irrigation as in the Badain Jaran Desert of China or the Sonoran Desert of Arizona (US) (Jiao, Zhang & Wang 2015). Alternatively, significant impacts on surface hydrology may ensue where water is accessed from lakes, reservoirs and rivers. For example, large-scale desert irrigation has led to the desiccation of the Aral Sea (**Q33**), between Kazakhstan and Uzbekistan, and to the significantly low flow of the Colorado River as it reaches the Sea of Cortez in Mexico. Another important degrading process associated with irrigation is salinization, a process in which water-soluble salts accumulate in the soil, ultimately rendering it non-productive for cultivation (Thomas & Middleton 1993). Salinization

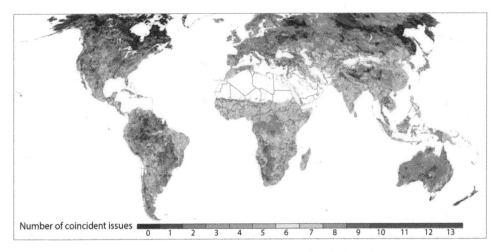

Number of coincident issues

0 1 2 3 4 5 6 7 8 9 10 11 12 13

Figure 26.1 Convergence of evidence of land degradation to identify desertification
Coincident issues refers to the number of global environmental incidents converging in locales. Much of the area marked yellowish-orange to red are dryer lands, especially in the Sahel of Africa, Middle East and Central Asia, that have major signs of aridification. Other areas are more mesic lands that display various signs of degradation, some of which lead to more xeric ecological conditions.
Source: JRC-European Commission (2019).

has been advanced and challenged as contributing to the collapse of past civilizations, as in the case of the Indus Valley and in Mesopotamia (Borsch 2004; Alam, Sahota & Jeffrey 2007; Allen & Heldring 2021).

The aridification of arid lands is a global concern (Cherlet *et al.* 2018; IPCC 2019), including the expansion of shifting or moving sands (aka sandification) (Wu & Ci 2002). It reduces ecosystem functioning and environmental services of arid lands, including their provisioning services (**Q7**) (D'Odorico *et al.* 2013), affecting evapotranspiration with positive feedbacks on drought conditions. Such are the impacts of arid land degradation, especially in the face of climate change (IPCC 2019), that the United Nations Convention to Combat Desertification was established in 1994 to address the problem. Despite the longstanding attention, the amount of desertification globally is not clear for several reasons: the absence of standard measures, past calculations using precipitation limits beyond those used for arid lands or deserts, and arguable claims that the large majority of drylands worldwide are affected or potentially affected by the process (Cherlet *et al.* 2018; Verón *et al.* 2006).

Importantly, desertification tends to focus on soil and vegetation degradation, whether or not precipitation is reduced. Use of the aridity index as a measure of the process presupposes a decline in precipitation, but it fails to capture land surface dynamics. The ecohydrological index, however, captures plant behaviour with atmospheric CO_2, and its use in models yields minimal changes in the global area of arid lands (Berg & McColl 2021; Higginbottom & Symeonakis 2014). For these and other reasons, the *World Atlas of Desertification* (Fig. 26.1) assesses the process of arid land degradation globally by the number of coincident issues indicative of the desertification process. A relatively safe assessment is that 10–20 per cent of arid lands have been "desertified", amounting to 6–12 Mkm² of land (Verón *et al.* 2006), although it is not clear if this area includes extant desert lands per se. A recent study combining climate change and human activity indicates that 9–13 per cent of the areas adjacent to current deserts or barren lands are at moderate to high risk of desertification currently, a percentage that may increase to 23 per cent by the end of this century (Huang *et al.* 2020).

References

Alam, U., P. Sahota & P. Jeffrey (2007). "Irrigation in the Indus basin: a history of unsustainability? *Water Science and Technology: Water Supply* 7(1): 211–18. doi:10.2166/ws.2007.024.

Allen, R. & L. Heldring (2021). "The collapse of civilization in Southern Mesopotamia. *Cliometrica*: 1–36. doi.org/10.1007/s11698-021-00229-2.

Berg, A. & K. McColl (2021). "No projected global drylands expansion under greenhouse warming". *Nature Climate Change* **11: 331–7. doi.org/10.1038/s41558-021-01007-8.**

Borsch, S. (2004). "Environment and population: the collapse of large irrigation systems reconsidered". *Comparative Studies in Society and History* 46(3): 451–68. doi.org/10.1017/S0010417504000234.

Cherlet, M. *et al.* (eds) (2018). *World Atlas of Desertification.* Luxembourg: Publication Office of the European Union.

D'Odorico, P. *et al.* (2013). "Global desertification: drivers and feedbacks". *Advances in Water Resources* 51: 326–44. doi.org/10.1016/j.advwatres.2012.01.013.

Higginbottom, T. & E. Symeonakis (2014). "Assessing land degradation and desertification using vegetation index data: current frameworks and future directions". *Remote Sensing* **6: 9552–75. doi.org/10.3390/rs6109552.**

Huang, J. *et al.* **(2020). "Global desertification vulnerability to climate change and human activities".** *Land Degradation and Development* **31(11): 1380–91. doi.org./10.1002/ldr.3556.**

IPCC (Intergovernmental Panel on Climate Change) (2019). *Special Report on Climate Change, Desertification, Land Degradation, Sustainable Land Management, Food Security, and Greenhouse Gas Fluxes in Terrestrial Ecosystems* (SR2). P. Shukla *et al.* (eds). www.ipcc.ch/site/assets/uploads/2018/07/sr2_background_report_final.pdf.

Jiao, J., X. Zhang & X. Wang (2015). "Satellite-based estimates of groundwater depletion in the Badain Jaran Desert, China". *Scientific Reports* 5: 8960. doi.org/10.1038/srep08960.

JRC (Joint Research Centre)-European Commission (2019). "Convergence of global change issues". wad.jrc.ec.europa.eu/countryreport.

Middleton, N. (2018). "Rangeland management and climate hazards in drylands: dust storms, desertification and the overgrazing debate". *Natural Hazards* 92(1): 57–70. doi.org/10.1007/s11069-016-2592-6.

Reynolds, J. *et al.* (2007). "Global desertification: building a science for dryland development". *Science* 316(5826): 847–51. doi:10.1126/science.1131634.

Thomas, D. & N. Middleton (1993). "Salinization: new perspectives on a major desertification issue". *Journal of Arid Environments* 24(1): 95–105. doi.org/10.1006/jare.1993.1008.

Verón, S. *et al.* (2006). "Assessing desertification". *Journal of Arid Environments* 66: 751–63. doi.org/10.1016/j.jaridenv.2006.01.021.

Wu, B. & L. Ci (2002). "Landscape change and desertification development in the Mu Us Sandland, northern China". *Journal of Arid Environments* 50(3): 429–44. doi.org/10.1006/jare.2001.0847.

DOES HUMAN LANDSCAPE BURNING HAVE EARTH SYSTEM IMPACTS?

Short answer: yes. Landscape burning is a common practice globally, emitting substantial greenhouse gases and aerosols to the atmosphere.

Biomass burning is the consumption of organic material, overwhelmingly vegetation, by fire. It involves the use of wood fuel and dung for heat and cooking, with significant human health consequences from indoor use (Naeher *et al.* 2007), and landscape burning, with significant consequences for the Earth system. Landscapes burn naturally through lightning strikes on dry vegetation and through intentional and unintentional human actions. In the first decade of this century, ~464 MH or ~3.5 per cent of the ice-free land surface burned annually (Hodshire *et al.* 2019). Landscape burning has negative impacts on air quality and various consequences on ecosystem to Earth system functions, including generating co-adapted landscapes **(Q25)**. Modelling indicates that increasingly warmer conditions and drier climates will increase severe forest fires (both natural and unintentional human actions), especially from the mid-latitudes poleward (Barbero *et al.* 2015). In addition, a major part of land-use management worldwide involves burning landscapes, commonly small patches (small fires in Fig. 27.1). Such intentional burning is significant globally, accounting for ~2.2 PgC yr^{-1} emitted between 1997 and 2016 (van der Werf *et al.* 2017).

Humankind has used fire to burn landscapes since the early Holocene **(Q10)** and subsequently to clear lands for cultivation (Pyne 2019), foremost in the temperate and boreal forests. Biomass burning decreased during the Little Ice Age (ca. sixteenth–nineteenth centuries), indicating the role of abiotic climate changes on fire frequency (Marlon *et al.* 2008; Wang *et al.* 2010), with subsequent increases as climates warmed. Entering the twenty-first century, perhaps 2.5 PgC yr^{-1} are emitted from biomass (van der Werf *et al.* 2006), some 90 per cent from human use of fire, especially among tropical savannas **(Q24)** (Mouillot *et al.* 2006). Interestingly, the use of fire to burn landscapes globally has declined over the past 150 years, perhaps owing to changes in land-use management and model projections suggest possible declines in intentional burning linked to increases in population density and urbanization (Wu *et al.* 2022). In contrast, large-scale natural and unintentional fires are forecast in the higher latitude forests, owing to climate warming (Barbero *et al.* 2015; de Groot, Flannigan & Cantin 2013), and attested by 2020 forest fires in Alaska and the

A. Regular burned area (% / yr)

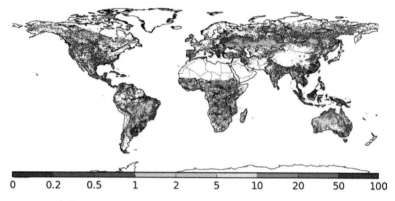

B. Small fire burned area (% / yr)

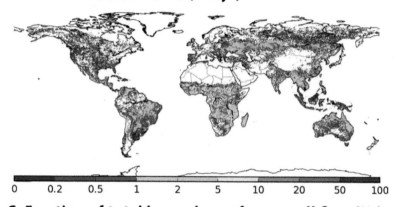

C. Fraction of total burned area from small fires (%)

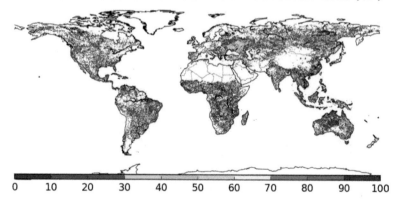

Figure 27.1 Global landscape burning

Small fires tend to indicate human landscape burning; percentages refer to the area.

Source: van der Werf *et al.* (2017).

western continental United States. Importantly, fire in concert with land and climate change is threatening 4,400 species worldwide (Kelly *et al.* 2020).

Biomass burning emits carbon dioxide, carbon monoxide, oxides of nitrogen, among other greenhouse gases (**Q44**), contributing to eight radiative forcing agents, as well as affecting water and energy fluxes (de Groot, Flannigan & Cantin 2013). It also emits aerosols and aerosol precursor vapours (**Q45**). Black carbon aerosols absorb and reradiate heat in the atmosphere, while they and other aerosols stimulate cloud formations that reflect incoming solar radiation as well as stimulate intense rainfall (Hodshire *et al.* 2019). Fire intensity and frequency alter vegetation, such that their interaction determines if the area burned constitutes a source or sink of greenhouse gases (Marlon *et al.* 2008) and the type of aerosols emitted (Hodshire *et al.* 2019). Globally, biomass burning constitutes about 4.4 per cent of all carbon lost annually to the atmosphere (Kaiser *et al.* 2011), estimated to be ~0.4 Pg yr^{-1} (van der Werf *et al.* 2017). If modelling projection of reduced intentional burning practices in the future are correct, however, land-based carbon uptake may increase globally (Wu *et al.* 2022).

Most of the figures presented above were generated from MODIS (remote sensing) data, which have a resolution of 500 m for the area burned, converted to carbon emissions. Recent finer spatial resolution data permits burn detection of sizes as small as 100 m. Applied to sub-Saharan Africa, these data indicate an increase in area burned by 2.02 Mkm^2 over the MODIS data for the year 2016, increasing carbon emissions by 31–101 per cent over the MODIS-based calculations (Ramo *et al.* 2021). As this finer resolution is applied globally, the amount of emissions per year will surely change, most probably increasing the emissions in question.

References

Barbero, R. *et al.* (2015). "Climate change presents increased potential for very large fires in the contiguous United States". *International Journal of Wildland Fire* 24(7): 892–9. dx.doi.org/10.1071/WF15083.

de Groot, W., M. Flannigan & A. Cantin (2013). "Climate change impacts on future boreal fire regimes". *Forest Ecology and Management* 294: 35–44. doi.org/10.1016/j.foreco.2012.09.027.

Hodshire, A. *et al.* (2019). "Aging effects on biomass burning aerosol mass and composition: a critical review of field and laboratory studies". *Environmental Science & Technology* 53(17): 10007–22. doi.org/10.1021/acs.est.9b02588.

Kaiser, J. *et al.* (2011). "Biomass burning emissions estimated with a global fire assimilation system based on observed fire radiative power". *Biogeosciences Discussions* 8(4): 527–54. doi.org/10.5194/bg-9-527-2012.

Kelly, L. *et al.* (2020). "Fire and biodiversity in the Anthropocene". *Science* 370(6519): eabb0355. doi:10.1126/science.abb0355.

Marlon, J. *et al.* (2008). "Climate and human influences on global biomass burning over the past two millennia". *Nature Geoscience* 1(10): 697–702. doi.org/10.1038/ngeo313.

Mouillot, F. *et al.* (2006). "Global carbon emissions from biomass burning in the 20th century. *Geophysical Research Letters* 33(1): L01801. doi.org/10.1029/2005GL024707.

Naeher, L. *et al.* (2007). "Woodsmoke health effects: a review". *Inhalation Toxicology* 19(1): 67–106. doi.org/10.1080/08958370600985875.

Pyne, S. (2019). *Fire: A Brief History*. Seattle, WA: University of Washington Press.

Ramo, R. *et al.* (2021). "African burned area and fire carbon emissions are strongly impacted by small fires undetected by coarse resolution satellite data". *Proceedings of the National Academy of Sciences* 118(9): e2011160118. doi.org/10.1073/pnas.2011160118.

van der Werf, G. *et al.* (2006). "Interannual variability of global biomass burning emissions from 1997 to 2004". *Atmospheric Chemistry and Physics Discussions* 6(2): 3175–226. doi.org/10.5194/acp-6-3423-2006.

van der Werf, G. *et al.* (2017). "Global fire emissions estimates during 1997–2016". *Earth System Science Data* 9(2): 697–720. doi.org/10.5194/essd-9-697-2017.

Wang, Z. *et al.* (2010). "Large variations in Southern hemisphere biomass burning during the last 650 years". *Science* 330(6011): 1663–6. doi:10.1126/science.1197257.

Wu, C. *et al.* (2022). "Reduced global fire activity due ti human demography slows global warming by enhanced land carbon uptake." *Proceedings of the National Academy of Sciences* 119 (20): e2101186119. doi.org/10.1073/pnas.2101186119

Q28

HAS HUMAN ACTIVITY REDUCED MANGROVE FORESTS?

Short answer: yes. Mangroves, one of the most efficient terrestrial carbon sinks, spawning grounds for marine fauna and barriers for inland storm protection, are rapidly disappearing due to human activity.

Mangroves exist as shrubs and small trees occupying saline or brackish waters along coastlines, largely within the tropics (Fig. 28.1). In 2000, mangrove forests covered 127,760 km^2 along the land–marine intersection, a loss of some 35 per cent of their 1980 coverage, generating an attrition rate of ~2.1 per cent per annum (Giri *et al.* 2011; Valiela, Bowen & York 2001). Another assessment of change from 1999–2019 estimates a global loss of 3,700 km^2 of mangrove, accounting for areas lost and gained (Murray *et al.* 2022). This loss is overwhelmingly at the hands of human activities, including agriculture, aquaculture and urban development. These losses have been particularly acute along the coastal Malay Peninsula, Indonesia, Philippines and Central America (Richards & Friess 2016).

Mangroves provide various environmental services. They trap sediments and engage in denitrification, both of which lower nutrient export to the sea, serve as nursery and spawning grounds for fish and shellfish, and protect inland land uses from storm surges. Although they account for only about 1 per cent of all forest carbon sequestration, mangroves are one of most efficient carbon sinks in the Earth system, holding about 937 tC ha^{-1} (Alongi 2012) compared with, for example, 12.8–130 tC ha^{-1} across North America (Hoover & Smith 2021). As such, the restoration of mangrove forests constitutes a means to mitigate a portion of greenhouse gas emissions (Alongi 2014), commonly referred to as "blue" carbon or the removal of carbon dioxide from the atmosphere by coastal environments (Ahmed & Glaser 2016).

Figure 28.1 Mangrove forest and its distribution
Source: Easvur (2014); basemap: Esri, GEBCO, DeLorme, NaturalVue.

References

Ahmed, N. & M. Glaser (2016). "Coastal aquaculture, mangrove deforestation and blue carbon emissions: is REDD+ a solution?" *Marine Policy* 66: 58–66. doi.org/10.1016/j.marpol.2016.01.011.

Alongi, D. (2012). "Carbon sequestration in mangrove forests". *Carbon Management* 3(3): 313–22. doi.org/10.4155/cmt.12.20.

Alongi, D. (2014). "Carbon cycling and storage in mangrove forests". *Annual Review of Marine Science* 6: 195–219. doi.org/10.4155/cmt.12.20.

Easvur, K. (2014). "Pichavaram". commons.wikimedia.org/wiki/File:Pichavaram_1.jpg.

Giri, C. *et al.* (2011). "Status and distribution of mangrove forests of the world using earth observation satellite data (version 1.4, updated by UNEP-WCMC)". *Global Ecology and Biogeography* 20(1): 154–9. doi.org/10.1111/j.1466-8238.2010.00584.x.

Hoover, C. & J. Smith (2021). "Current aboveground live tree carbon stocks and annual net change in forests of conterminous United States". *Carbon Balance and Management* 16(1): 1–12. doi.org/10.1186/s13021-021-00179-2.

Murray, N. *et al.* (2022). "High-resolution mapping of losses and gains of Earth's tidal wetlands". *Science*, 376(6594): 744–9. doi. 10.1126/science.abm9583

Richards, D. & D. Friess (2016). "Rates and drivers of mangrove deforestation in Southeast Asia, 2000–2012". ***Proceedings of the National Academy of Sciences*** **113(2): 344–9. doi. org/10.1073/pnas.1510272113.**

Valiela, I., J. Bowen & J. York (2001). "Mangrove forests: one of the world's threatened major tropical environments". *Bioscience* 51(10): 807–15. doi.org/10.1641/0006-3568(2001)051[0807:MFOOTW]2.0.CO;2.

HAS HUMAN ACTIVITY ON TUNDRA INFLUENCED CLIMATE WARMING?

> *Short answer:* highly likely. Historically a major sink of greenhouse gases, the warming of tundra permafrost promises to emit significant amounts of greenhouse gases.

Tundra refers to a vegetation complex of dwarf shrubs, sedges and grasses with moss and lichen that occurs in environments with permafrost (permanently frozen soils 25–90 cm below the surface) **(Q37)** and surface soils that annually freeze and unfreeze (Walker *et al.* 2016) (Fig. 29.1). While tundra is found at extreme elevations (alpine tundra in high mountains) and on some edges of Antarctica and nearby islands, it is most extensive in the Northern hemisphere, above 60° to 70° N latitude, depending on location, covering about 10 per cent of the land surface of Earth. A short growing season makes tundra a fragile biome. Peat bogs (or peatlands) interspersed in tundra, and the boreal forests to the south of the tundra zone, constitute a large carbon sink, storing as much as one-third of carbon in soils globally and double the amount of carbon than is currently in the atmosphere (Schuur *et al.* 2009). The interspersed peatlands of the northern tundra–boreal zones cover ~3.7 Mkm^2, storing ~415 GtC, about half of which resides in permafrost peatlands, as well as ~10 Pg of nitrogen (Hugelius *et al.* 2020). One calculation estimates that between 1950 and 2010 as much as 72 PgC have been emitted from peatlands, elevating the amount of emissions once stored in the near-surface lithosphere and the amount emitted by cultivation (Qiu *et al.* 2021).

Recently, parts of tundra have been degraded by large-scale oil production and mining operations. The major Earth system concern, however, is the indirect human impacts through global warming of the Arctic, a region that appears to be warming faster than any other region on Earth (Shepherd 2016). Increasing temperatures are not only changing the vegetation complexes affecting biotic diversity, but are leading to the thawing of the permafrost zone and releasing large amounts of CO_2 and CH_4 to the atmosphere (Fig. 29.2) (Frey & Smith 2005; Hugelius *et al.* 2020; Oechel *et al.* 1993; Shepherd 2016; Walker *et al.* 2006). The CO_2 emissions are somewhat offset by increases in tundra uptake of CO_2 (Turetsky *et al.* 2019), owing to vegetation responses to warming, although emissions are larger than uptake (Belshe, Schuur & Bolker 2013). Recent modelling results indicate that the feedbacks from

Peatland carbon storage (kg C m²)

0 · 2 5 8 10 20 40 60 80 100

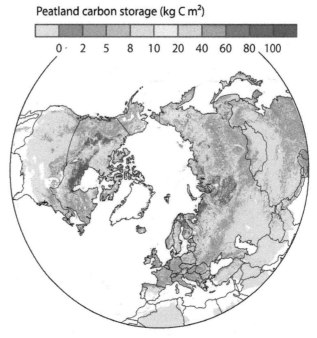

Figure 29.1 Boreal peatland and its distribution in the Northern Hemisphere
Peat exists throughout tundra (pictured) and boreal forest biomes.
Source: map Hugelius *et al.* (2020); photo D. Delso (delso.photo, licence CC-BY-SA 4.0).

the change in albedo **(Q5)** from snow-cover loss, increasing atmospheric water vapour and the accumulation of CO_2 in different parts of the Earth system, have created conditions that will release carbon emissions as "a ticking time bomb", regardless of steps taken to reduce greenhouse gas emissions (Randers & Goluke 2020). Historically resistant to fire, climate warming is expected to increase tundra fire as well, also releasing the long-held carbon sinks of this biome (Hu *et al.* 2015).

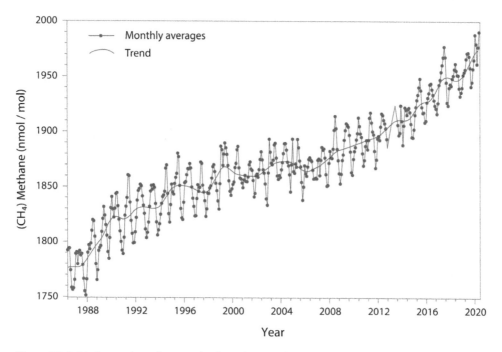

Figure 29.2 Methane release from tundra, Point Barrow, Alaska

Nmol mol⁻¹ = nanomole per mole (see Metrics and measures).

Source: NOAA/Observatoire de Point Barrow (2017).

References

Belshe, E., Schuur, E. & B. Bolker (2013). "Tundra ecosystems observed to be CO_2 sources due to differential amplification of the carbon cycle". *Ecology Letters* 16(10): 1307–15. doi:10.1111/ele.12164.

Frey, K. & L. Smith (2005). "Amplified carbon release from vast West Siberian peatlands by 2100". *Geophysical Research Letters* 32(9). doi.org/10.1029/2004GL022025.

Hu, F. *et al.* (2015). "Arctic tundra fires: natural variability and responses to climate change." *Frontiers in Ecology and the Environment* 13(7): 369–77. doi.org/10.1890/150063.

Hugelius, G. *et al.* (2020). "Large stocks of peatland carbon and nitrogen are vulnerable to permafrost thaw". ***Proceedings of the National Academy of Sciences*** **117(34): 20438–46. doi.org/10.1073/pnas.1916387117.**

NOAA (National Oceanic and Atmospheric Administration)/Observatoire de Point Barrow (2017). Methane concentrations in Barrow, Alaska. commons.wikimedia.org/wiki/File:CH4.BRW. Monthly.png.

Oechel, W. *et al.* (1993). "Recent change of Arctic tundra ecosystems from a net carbon dioxide sink to a source". *Nature* 361(6412): 520–3. www.nature.com/articles/361520a0.

Qiu, C. *et al.* (2021). "Large historical carbon emissions from cultivated northern peatlands". *Science Advances* 7(23): eabf1332. doi:10.1126/sciadv.abf1332.

Randers, J. & U. Goluke (2020). "An earth system model shows self-sustained melting of permafrost even if all man-made GHG emissions stop in 2020". *Scientific Reports* 10(1): 1–9. doi.org/10.1038/s41598-020-75481-z.

Schuur, E. *et al.* (2009). "The effect of permafrost thaw on old carbon release and net carbon exchange from tundra". *Nature* 459(7246): 556–9. doi.org/10.1038/nature08031.

Shepherd, T. (2016). "Effects of a warming Arctic". *Science* 353(6303): 989–90. doi:10.1126/science.aag2349.

Turetsky, M. *et al.* (2019). "Permafrost collapse is accelerating carbon release". *Nature* 569(7754): 32–4. doi:10.1038/d41586-019-01313-4.

Walker, M. *et al.* (2006). "Plant community responses to experimental warming across the tundra biome". *Proceedings of the National Academy of Sciences* 103(5): 1342–6. doi.org/10.1073/pnas.0503198103.

Walker, D. *et al.* (2016). "Circumpolar Arctic vegetation: a hierarchic review and roadmap toward an internationally consistent approach to survey, archive and classify tundra plot data". *Environmental Research Letters* 11(5): 055005. doi:10.1088/1748-9326/11/5/055005.

HAS HUMAN ACTIVITY CREATED LAND SUBSIDENCE WORLDWIDE?

Short answer: yes. Land surface subsidence has increased globally owing to large-scale surface construction and the depletion of aquifers.

The combination of various types of mineral extraction, the sheer weight of the "built" environment on the terrestrial surface, and especially fresh water withdrawals from aquifers, has created land surface subsidence as a global issue. Cities themselves, rapidly increasing in

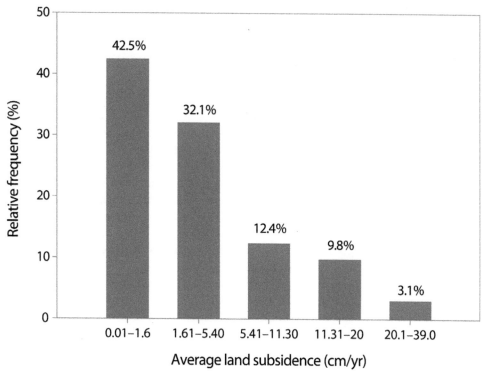

Figure 30.1 Frequency of average rates of land city subsidence for 140 cities worldwide
Source: Bagheri *et al.* (2021).

population and infrastructure weight worldwide, add huge pressures on the land surface. For instance, the estimated building infrastructure of San Francisco alone tops in at 1.6 billion tonnes, compressing the soil by 5 mm to 80 mm over a hundred-year period (Parsons 2021). A study of 290 cities across the world reveals that ~77 per cent exhibited land subsidence in which ~60 per cent also involved groundwater extraction (Bagheri *et al.* 2021). The weight of construction compacting soils, commonly combined with depleting aquifers, reduces the subterranean capacity to support the surface weight. Calculations from 140 cities worldwide provide average land subsidence rates ranging from 0.01–1.60 cm yr^{-1} to 20.1–39 cm yr^{-1} (Bagheri *et al.* 2021) (Fig. 30.1). Mexico City constitutes an iconic case. Much of this metropolis of 22 million people, including its city centre, occupies a series of drained lake beds in which the aquifers below the beds were drawn down by rates as high as 33 cm yr^{-1} in the past. The overall subsidence today leaves ~1.5 million inhabitants occupying subsidence zones with high to very high surface faulting that creates unstable occupational spaces (Cigna & Tapete 2021).

Mining of various minerals and rocks, particularly those close to the surface, induces subsidence as well, especially in areas with urban development on the surface above the mined materials (Bell, Stacey & Genske 2000). Subsidence owing to groundwater depletion, however, affects a much larger area globally. As much as 2.2 Mkm^2, an area slightly larger than Greenland, is estimated to be at high risk of all types of subsidence (Herrera-Garciá *et al.* 2021). A large number of major cities worldwide adjacent to shorelines and endangered by subsidence are also at increasing risk of floods from storm surges associated with sea-level rise and increases in storm intensities (**Q39, Q51**).

References

Bagheri, M. *et al.* (2021). "Land subsidence: a global challenge". *Science of the Total Environment*: **146193. doi.org/10.1016/j.scitotenv.2021.146193.**

Bell, F., T. Stacey & D. Genske (2000). "Mining subsidence and its effect on the environment: some differing examples". *Environmental Geology* 40(1): 135–52. doi.org/10.1007/s002540000140.

Cigna, F. & D. Tapete (2021). "Present-day land subsidence rates, surface faulting hazard and risk in Mexico City with 2014–2020 Sentinel-1 IW InSAR". *Remote Sensing of Environment* 253: 112161. doi.org/10.1016/j.rse.2020.112161.

Herrera-García, G. *et al.* (2021). "Mapping the global threat of land subsidence". *Science* **371(6524): 34–6. doi:10.1126/science.abb8549.**

Parsons, T. (2021). "The weight of cities: urbanization effects on Earth's subsurface". *AGU Advances* 2(1): e2020AV000277. doi.org/10.1029/2020AV000277.

DOES URBANIZATION IMPACT THE EARTH SYSTEM?

> *Short answer:* yes. Urban areas, the smallest of all major land covers globally, create large Earth system impacts through the magnitude of material and energy consumption associated with them.

Settlements, from hamlets to mega-metro areas, represent a comparatively small land use in terms of global area but hold enormous environmental consequences. Disregarding Antarctica and Greenland, the proportion of the urbanized Earth surface is: ~3 per cent based on urban administrative boundaries; ~0.65 per cent for the actual "built up" area, eliminating unused lands within the urban boundaries; and ~0.45 per cent accounting for impervious surfaces (e.g. buildings, houses, streets) only (Liu *et al.* 2014). The latest data and methods suggest low and high estimates at 657,000 km^2 and 1,289,000 km^2, respectively, for the global urban area (Esch *et al.* 2017). As a proportion of the global population, urban population growth is so large that by 2030 the added urban area will significantly increase in size (Fig. 31.1). The specific increase depends on the 2000 base area used in the projection. Estimates range as high as 1.2 Mkm^2 (Seto *et al.* 2012), about the size of the Republic of South Africa. This added area follows from the increasing proportion of the global population residing in urban areas, estimated to be ~70 per cent by 2050 (DESA 2018).

Urban land constitutes a transformation (radical change) of the previous land cover, in many cases consuming prime agricultural lands (Seto *et al.* 2000), leading to a large array of environmental changes (Grimm *et al.* 2008; Seto & Shepherd 2009). Foremost, it replaces vegetation and soil with huge expanses of impervious surfaces (e.g. buildings, houses, roads), changing the surface albedo, surface hydrology and aquifer recharge. Much urban surface is composed of heterogeneous, small-sized land parcels, and the vegetation on these parcels tends to be non-native species suited to land users' preferences, commonly reducing genetic diversity as registered in Belgium and the Netherlands (Broeck *et al.* 2018). In some cases, urbanization actually increases vegetative cover, especially in arid environments where they are added by residents by way of irrigation, as well as increasing phenotypic changes (Alberti *et al.* 2017). In other cases, however, the built landscape of urban areas and settlements reduce tree cover. For example, between 2009 and 2014, the United States lost nearly 70,820 ha yr^{-1} of trees or 36 million trees yr^{-1}, largely to urban expansion in forest biomes (Nowak &

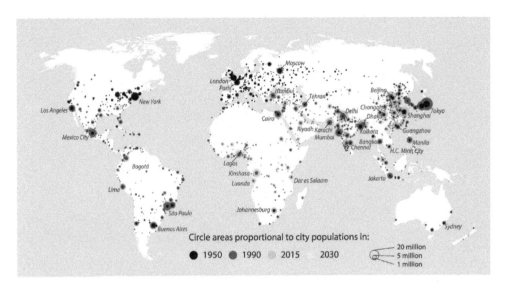

Figure 31.1 Global urban growth, 1950–2030
Minimum city population threshold: 300,000.
Source: UN DESA-PD (2018).

Greenfield 2018). Such losses, and the concentration of industry, transportation and energy use in urban areas, create a major source for carbon emissions (Moran *et al.* 2018). Yet cities have open spaces that could be planted with trees to ameliorate the urban heat island effect **(Q54)** and serve as carbon sinks; globally, this space covers about ~17.6 per cent of city areas or 10.9 ± 2.8 MH, calculated to account for 82.4 ± 25.7 MtCO$_2$e yr^{-1} (Teo *et al.* 2021).

Surprisingly, biogenic CO$_2$ from biofuels, people and vegetation, especially the latter, composes a significant proportion of all urban-based carbon emissions, in part following from the "oasis effect" from irrigated urban vegetation (Vivoni *et al.* 2020). This proportion has been recorded as 33 per cent of the annual mean carbon emissions from fossil fuel combustion in the Los Angeles metro-area, for example (Miller *et al.* 2020). Cities are also the source of major soil contamination (Li *et al.* 2014), especially lead, and tropospheric pollution (atmosphere near the Earth's surface), particularly ozone pollution **(Q53)**. Energy use, in tandem with impervious surfaces and albedo change, create the urban heat island effect **(Q54)**. The effect of urban agglomerations is to amplify temperatures (Georgescu *et al.* 2014). The average of 50 cities in the United States from the 1960s to 2010 indicates increases in the number of heatwaves per year from two to six, lasting longer with increased intensity (EPA 2021). An examination of 242 urban areas in eastern China demonstrates major increases in mean heat stress and its frequency as global warming and urbanization increased between 1971 and 2014 (Luo & Lau 2018). Modelling of 571 cities in Europe indicates an increase in heatwaves across all cities, with more heatwave days in southern Europe but the largest increases in temperature in central Europe (Guerreiro *et al.* 2018).

Cities are also the major source of the consumption of world resources, placing demands on resources globally through their environmental footprint **(Q96)**. Given that the majority of the global population resides in cities, and the percentage doing so will continue to grow, the environmental impacts of urbanization well beyond cities will escalate in the future (Seto & Shepherd 2009), despite the many efforts to make cities more sustainable. Furthermore, as urbanization escalates in the Global South, attention to changes in bundles of environmental services and their impacts on people raises concern (Lapointe, Gurney & Cumming 2021).

References

Alberti, M. *et al.* (2017). "Global urban signatures of phenotypic change in animal and plant populations". *Proceedings of the National Academy of Sciences* 114(34): 8951–6. doi.org/10.1073/pnas.1606034114.

Broeck, A. *et al.* (2018). "Genetic diversity loss and homogenization in urban trees: the case of *Tilia × europaea* in Belgium and the Netherlands". *Biodiversity and Conservation* 27(14): 3777–92. doi.org/10.1007/s10531-018-1628-5.

DESA (Department of Economics and Social Affairs) (2018). *The 2018 Revision of World Urbanization Prospects.* Population Division of the United Nations DESA. New York: United Nations.

EPA (Environmental Protection Agency US) (2021). "Climate change indicators: heat waves". www.epa.gov/climate-indicators/climate-change-indicators-heat-waves.

Esch, T. *et al.* (2017). "Breaking new ground in mapping human settlements from space – the Global Urban Footprint". *ISPRS Journal of Photogrammetry and Remote Sensing* 134: 30–42. doi.org/10.1016/j.isprsjprs.2017.10.012.

Georgescu, M. *et al.* (2014). "Urban adaptation can roll back warming of emerging megapolitan regions". *Proceedings of the National Academy of Sciences* 111(8): 2909–14. doi.org/10.1073/pnas.1322280111.

Grimm, N. *et al.* (2008). "Global change and the ecology of cities". *Science* 319(5864): 756–60. doi:10.1126/science.1150195.

Guerreiro, S. *et al.* (2018). "Future heat-waves, droughts and floods in 571 European cities". *Environmental Research Letters* 13(3): 034009. doi.org/10.1088/1748-9326/aaaad3.

Lapointe, M., G. Gurney & G. Cumming (2021). "Urbanization affects how people perceive and benefit from ecosystem service bundles in coastal communities of the Global South". *Ecosystems and People* 17(1): 57–68. doi.org/10.1080/26395916.2021.1890226.

Li, L. *et al.* (2014). "Release of cadmium, copper and lead from urban soils of Copenhagen." *Environmental Pollution* 187: 90–97. doi.org/10.1016/j.envpol.2013.12.016.

Liu, Z. *et al.* (2014). "How much of the world's land has been urbanized, really? A hierarchical framework for avoiding confusion". *Landscape Ecology* 29(5): 763–71. doi.org/10.1007/s10980-014-0034-y.

Luo, M. & N.-C. Lau (2018). "Increasing heat stress in urban areas of eastern China: acceleration by urbanization". *Geophysical Research Letters* 45(23): 13060–9. doi.org/10.1029/2018GL080306.

Miller, J. *et al.* (2020). "Large and seasonally varying biospheric CO_2 fluxes in the Los Angeles megacity revealed by atmospheric radiocarbon". *Proceedings of the National Academy of Sciences* 117(43): 26681–7. doi.org/10.1073/pnas.2005253117.

Moran, D. *et al.* (2018). "Carbon footprints of 13,000 cities". *Environmental Research Letters* 13(6): 064041. doi.org/10.1016/j.gloenvcha.2020.102205.

Nowak, D. & E. Greenfield (2018). "Declining urban and community tree cover in the United States". *Urban Forestry & Urban Greening* 32: 32–55. doi.org/10.1016/j.ufug.2018.03.006.

Seto, K. *et al.* (2012). "Global forecasts of urban expansion to 2030 and direct impacts on biodiversity and carbon pools". *Proceedings of the National Academy of Sciences* 109(40): 16083–8. doi.org/10.1073/pnas.1211658109.

Seto, K. *et al.* (2000). "Landsat reveals China's farmland reserves, but they're vanishing fast". *Nature* 406(6792): 121. doi.org/10.1038/35018267.

Seto, K. & J. Shepherd (2009). "Global urban land-use trends and climate impacts". *Current Opinion in Environmental Sustainability* 1(1): 89–95. doi.org/10.1016/j.cosust.2009.07.012.

Teo, H. C. *et al.* (2021). Global urban reforestation can be an important natural climate solution. *Environmental Research Letters*, 16 (3): 034059. doi.org/10.1088/1748-9326/abe783

UN DESA-PD (UN Department of Economic and Social Affairs and Population Division) (2018). *World Urbanization Prospects: The 2018 Revision*. New York: United Nations. population.un.org/wup/.

Vivoni, E. *et al.* (2020). "Abiotic mechanisms drive enhanced evaporative losses under urban oasis conditions". *Geophysical Research Letters*: e2020GL090123. doi.org/10.1029/2020GL090123.

SECTION IV

HUMAN CHANGES TO THE HYDROSPHERE

The hydrosphere includes all forms of water (H_2O or dihydrogen monoxide) in the Earth system – solid, liquid and gas or, respectively, ice and snow (sleet), fresh and saline water, and water vapour. The hydrological (or water) cycle refers to the transformation of these forms of H_2O as they move through the spheres. This movement creates different amounts or stocks of H_2O on land and in the oceans and atmosphere. The greatest proportion of water is saline (96.5 per cent; Table IV.1), overwhelmingly found within the oceans. The vast majority of life on land, however, requires freshwater.

Humankind has long employed multiple ways to improve access to the stocks of freshwater for various uses. Canal construction to move water to cultivated fields and cities dates back to ~4,000 BCE in Mesopotamia (Wilkinson, Rayne & Jotheri 2015). Qanats – underground tunnels to move subterranean water over distances and to the surface – date

Table IV.1 Global stocks of water

Water bodies	Distribution area	Volume	Global reserves (per cent)	
			Total water	Freshwater
Oceans and seas	361,300	1,338,000	96.5	n/a
Groundwater	134,800	23,400	1.7	n/a
Freshwater		10,530	0.76	30.1
Soil moisture		16.5	0.001	0.05
Glaciers and permanent snow cover	16,227	24,064	1.74	68.7
Ground ice / permafrost	21,000	300	0.022	0.86
Water reserves in lakes	2,058.7	176.4	0.013	n/a
Fresh	1,236.4	91	0.007	0.26
Saline	822.3	85.4	0.006	n/a
Water in swamps	2,682,6	11.47	0.0008	0.03
Water in rivers	148,000	2.12	0.0002	0.006
Water in biomass	510,000	1.12	0.0001	0.003
Atmospheric water	510,000	12.9	0.001	0.04
Total water reserves	510,000	1,385,984	100	n/a
Total freshwater reserves	148,000	35,029	2.53	100

Area: 10^3 km²; volume: 10^3 km³; n/a: not applicable.
Source: Williams (2012).

to the early first millennium (Avni 2018). During the fourth century, Byzantium (Istanbul) constructed aqueducts stretching 250 km inland to provide water to the city and stored in subterranean cisterns (Crow 2012). The longest boat-transport canal, 1,776 km (1,104 mi) in length, was begun in the fifth century (CE) in China (Fig. IV.1) (Zhang & Lenzer 2020). By 1445, the Mexica (Aztec) built a 16 km dyke to control the salinity and flooding of lake water surrounding their island capital, as well as numerous aqueducts to supply fresh water (Torres-Alves & Morales-Nápoles 2020). Today irrigation systems reach enormous expanses: 2.9 MH in central Asia (Micklin 2016) and 3.6 MH in California. Water is even moved up 1,100 m in elevation to stabilize the weight of Mexico City and attenuate the otherwise sinking metropolis in the ancient lake bed on which it is built (Mautner *et al.* 2020) (**Q30**). In these uses and many others, humankind has not only affected the stocks of water (**Q33**), for example, by damming rivers and draining wetlands, but degraded water quality by polluting lakes, rivers and even the oceans, creating zones with insufficient oxygen to support marine life (**Q41, Q42**). In addition, human activity is increasing sea level through climate change (**Q39**), threatening the oceans' conveyer belts (**Q40**) and altering the nature of precipitation globally. Essentially, anthropogenic activities have changed the hydrological cycle itself (**Q32**). These impacts are addressed in this section, in which the term "water" is used generically to refer to all forms of H_2O unless otherwise specified in one of its states.

Ocean dynamics are critical to the states and flows of the Earth system. For the most part, however, these dynamics are intimately tied to interactions with the other spheres, such as global warming and changes in the thermohaline or meridional oceanic circulation (**Q40**). Unlike plate tectonics (Section III) and the Milankovitch cycle (Section V), the hydrosphere is not treated as possessing forces of change exogenous to the Earth system in this book.

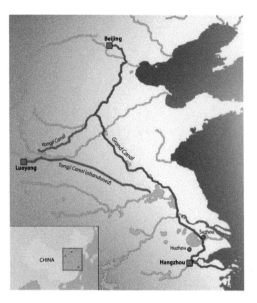

Figure IV.1 Grand Canal, China

Grand Canal (in blue) and other transport canals (in red) built in antiquity (see text); inset denotes the position of three main cities relative to the canal.
Source: Porfyriou (2019).

References

Avni, G. (2018). "Early Islamic irrigated farmsteads and the spread of qanats in Eurasia". *Water History* 10(4): 313–38. doi.org/10.1007/s12685-018-0225-6.

Crow, J. (2012). "Ruling the waters: managing the water supply of Constantinople, AD 330–1204". *Water History* 4(1): 35–55. doi.10.1007/s12685-012-0054-y.

Mautner, M. *et al.* (2020). "Urban growth and groundwater sustainability: evaluating spatially distributed recharge alternatives in the Mexico City Metropolitan Area". *Journal of Hydrology* 586: 124909. doi.org/10.1016/j.jhydrol.2020.124909.

Micklin, P. (2016). "The future Aral Sea: hope and despair". *Environmental Earth Sciences* 75(9): 844. doi.10.1007/s12665-016-5614-5.

Porfyriou, H. (2019). "Urban heritage conservation of China's historic water towns and the role of Professor Ruan Yisan: Nanxun, Tongli, and Wuzheen". *Heritage* 2(3): 2417–43. doi.org/10.3390/heritage230149.

Torres-Alves, G. & O. Morales-Nápoles (2020). "Reliability analysis of flood defenses: the case of the Nezahualcoyotl dike in the Aztec city of Tenochtitlan". *Reliability Engineering & System Safety* 203: 107057. doi.org/10.1016/j.ress.2020.107057.

Wilkinson, T., L. Rayne & J. Jotheri (2015). "Hydraulic landscapes in Mesopotamia: the role of human niche construction". *Water History* 7(4): 397–418. doi.10.1007/s12685-015-0127-9.

Williams Jr, R. (2012). "Introduction – changes in the Earth's cryosphere and global environmental change". In R. Williams, Jr & J. Ferrigno (eds), *Satellite Image Atlas of Glaciers of the World*. Denver, CO: US Geological Survey. pubs.usgs.gov/pp/p1386a/pdf/pp1386a-1-web.pdf

Zhang, M. & J. Lenzer Jr. (2020). "Mismatched canal conservation and the authorized heritage discourse in urban China: a case of the Hangzhou Section of the Grand Canal". *International Journal of Heritage Studies* 26(2): 105–19. doi.org/10.1080/13527258.2019.1608458.

Q32

HAS HUMAN ACTIVITY ALTERED THE HYDROLOGICAL (WATER) CYCLE?

Short answer: yes. Human uses of water stocks and through climate change have altered and continue to alter the amount and distribution of water in the cycle, with various consequences.

The hydrological or water cycle refers to the movement of the various forms or states of water through the processes of:

- condensation: water vapour cooling to liquid, e.g. clouds or dew;
- precipitation: condensing water vapour turns to rain, snow, hail or sleet;
- infiltration: liquid water soaks or sinks through a surface, such as soil;
- evaporation: liquid or solid water turns into water vapour in the atmosphere;
- transpiration: plants release water vapour, akin to sweating in animals; and
- sublimation: solid water moving to water vapour without passing through the stage of liquid water (e.g. vapour from dry ice); the reverse process is desublimation.

The hydrological cycle plays numerous roles in the Earth system (Oki & Kanae 2006) (Fig. 32.1). Water vapour in the atmosphere is a powerful greenhouse gas owing to its abundance (Q44). As air rises in the atmosphere, however, water vapour condenses to create clouds, which reradiate shortwave solar radiation back into space, reducing heat (energy) reaching the Earth's surface. With sufficient water vapour condensation, precipitation is released as rain, snow, hail or sleet. Since oceans cover about 71 per cent of the Earth's surface, the majority of precipitation (~78 per cent) arrives and is evaporated (~87 per cent) or sublimated, in the case of marine ice, from this stock back to the atmosphere. On land, where ~22 per cent of precipitation falls, the cycle is more complicated. Some of this water is temporarily stored in lakes, wetlands, reservoirs, snow packs and glaciers. In turn, it is evaporated or sublimated, used for human activities (domestic, industrial, agricultural uses) or flows to the oceans. Other precipitation supports vegetation and crop growth, the majority of which is transpired to the atmosphere. The bulk of precipitation reaching the land surface is retained in glaciers and snow or infiltrates the surface as groundwater (Q34),

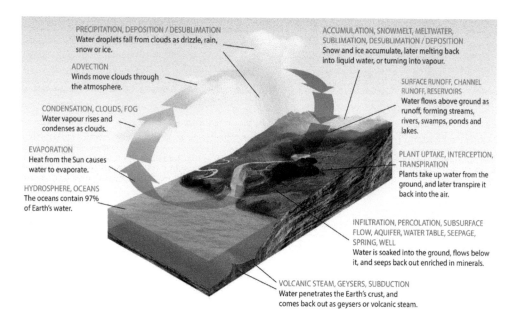

Figure 32.1 The hydrological (water) cycle
Source: Tal (2016).

either stored as permafrost or frozen soils **(Q37)** or moving through the subsurface, eventually reaching the oceans. About 13 per cent of precipitation is evaporated or transpired back to the atmosphere.

Water in the Earth system resides in different states or stocks and radically different volumes (Gleick 1993; Oki & Kanae 2006) (Fig. 32.2). More than 97 per cent of all water is saline, almost all residing in the oceans and seas. About 2.5 per cent of water is fresh. Of this small amount, almost 69 per cent is found in glaciers and ice caps (ice over polar waters), and the other sizeable portion, a bit more than 31 per cent, is groundwater, including permafrost. Of the remaining freshwater (0.4 per cent), the overwhelming majority is found in lakes and wetlands, whereas the remainder is present in soil, rivers, biota and the atmosphere.

Human activity has direct and indirect impacts on the hydrological cycle (Bosmans *et al.* 2017; Wada, de Graff & van Beek 2016). This activity does not destroy or create water but changes where in the Earth system and in what state it occurs, with important consequences. Agricultural, industrial and domestic uses directly collect and draw down freshwater stocks via water withdrawals, evapotranspiration and river discharge by altering land cover, such as creating cultivated land or through deforestation (Bagley *et al.* 2014). If land changes are of sufficient magnitude, evapotranspiration may decrease or increase regional water vapour, reducing or enhancing precipitation (i.e. land–atmosphere dynamics). For example, precipitation declines are associated with large-scale deforestation in Amazonia (Bagley *et al.* 2014).

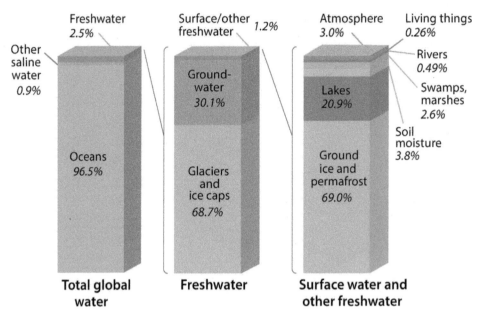

Figure 32.2 Global water stocks
Source: Gleick (1993).

Human activity indirectly affects the global hydrological cycle through the delivery of greenhouse gases, aerosols and other pollutants to the atmosphere as well as albedo changes (e.g. the cultivation of land previously forested), much of which contributes to global warming. Net warming increases the concentration of water vapour in the atmosphere (**Q43**), increasing evapotranspiration by 10 per cent between 2003 and 2019 (Pascolini-Campbell *et al.* 2021). Water vapour is the most abundant greenhouse gas by weight and volume. It is, however, short-lived and returns to the surface as precipitation in a few days, apparently with increasing intensity of rainfall events owing to atmospheric temperature increases (Madakumbura *et al.* 2021). Unlike other greenhouse gases, water vapour serves as a positive and negative feedback on atmospheric temperature; it captures longwave radiation, thereby increasing warming, but in cloud formations it blocks incoming solar radiation creating a cooling effect. Increasing average Earth temperatures affect the atmosphere's energy balance with consequences for increased droughts, intensive storms and floods (Rahmstorf 2017), water resources in general and ecosystem services (Wada *et al.* 2013; Wu, Christidis & Stott 2013), impacts discussed elsewhere in this section. Significantly, how water vapour, especially clouds, is treated in climate models is significant to the outcomes projected for global warming (Sood & Smakhtin 2015; Hu *et al.* 2020).

References

Bagley, J. *et al.* (2014). "Drought and deforestation: has land cover change influenced recent precipitation extremes in the Amazon?" *Journal of Climate* 27(1): 345–61. doi.org/10.1175/JCLI-D-12-00369.1.

Bosmans, J. *et al.* (2017). "Hydrological impacts of global land cover change and human water use". *Hydrology and Earth System Science* 21: 5603–26. doi.org/10.5194/hess-21-5603-2017.

Gleick, P. (1993). "Water reserves on the Earth". In P. Gleick (ed.), *Water in Crisis: A Guide to the World's Freshwater Resources.* Oxford: Oxford University Press.

Hu, X. *et al.* (2020). "A less cloudy picture of the inter-model spread in global warming projections". *Nature Communications* 11(1): 1–11. doi.org/10.1038/s41467-020-18227-9.

Madakumbura, G. *et al.* (2021). "Anthropogenic influence on extreme precipitation over global land areas seen in multiple observational datasets". *Nature Communications* 12: 3994. doi.org/10.1038/s41467-021-24262-x.

Oki, T. & S. Kanae (2006). "Global hydrological cycles and world water resources". *Science* 313: 1068–72. doi:10.1126/science.1128845.

Pascolini-Campbell, M. *et al.* (2021). "A 10 per cent increase in global land evapotranspiration from 2003 to 2019". *Nature* 593(7860): 43–547. doi.org/10.1038/s41586-021-03503-5.

Rahmstorf, S. (2017). "Rising hazard of storm-surge flooding". *Proceedings of the National Academy of Sciences* 114(45): 11806–08. doi.org/10.1073/pnas.1715895114.

Sood, A. & V. Smakhtin (2015). "Global hydrological models: a review". *Hydrological Sciences Journal* 60(4): 549–65. doi.org/10.1080/02626667.2014.950580.

Tal. E. (2016). "Diagram of the water cycle". commons.wikimedia.org/wiki/File:Diagram_of_the_Water_Cycle.jpg.

Wada, Y., I. de Graaf & L. van Beek (2016). "High-resolution modeling of human and climate impacts on global water resources". *Journal of Advances in Modeling Earth Systems* 8(2): 735–63. doi.org/10.1002/2015MS000618.

Wada, Y. *et al.* (2013). "Human water consumption intensifies hydrological drought worldwide". *Environmental Research Letters* 8(3): 034036. doi:10.1088/1748-9326/8/3/034036.

Wu, P., N. Christidis & P. Stott (2013). "Anthropogenic impact on Earth's hydrological cycle". *Nature Climate Change* 3(9): 807–10. doi.org/10.1038/nclimate1932.

WHAT ARE WATER WITHDRAWAL, WATER CONSUMPTION AND WATER FOOTPRINT, AND THEIR IMPLICATIONS FOR WATER AVAILABILITY?

Short answer: withdrawal and consumption refer to human uses of freshwater; the footprint refers to water associated with some unit or activity; water uses affect the global water flux, and in some cases have led to large-scale degradation of water sources.

Water withdrawal refers to the freshwater taken from the surface or groundwater for human use. Water consumption is withdrawn water that is not returned to its source but is consumed (e.g. industrial production, drinking), evaporated (e.g. pools, irrigation) or transpired by biota. The water footprint is the total water use for consumption by individuals, communities or countries per unit of time, accounting for water directly consumed and used or evaporated, polluted or fixed in production (e.g. wood products) (Hoekstra & Mekonnen 2012).

Humankind consumes about 50 per cent of all terrestrial freshwater, close to ~4 trillion $m^3 yr^{-1}$ (or 4,000 $km^3 yr^{-1}$), an eight-fold surge since the beginning of the twentieth century, and increasing (Wada, de Graff & van Beek 2016). Of this amount, ~70 per cent is taken for irrigation. Projected global increases in the use of irrigation, as well as expected increases in domestic and industrial water use, raise serious concerns about future water scarcities, which vary geographically and over time (Mekonnen & Hoekstra 2016). As noted in **Question 32**, water withdrawal and consumption affect the global water flux and hence the hydrological cycle, if perhaps modestly, but can impact local to regional runoff and precipitation significantly (Haddeland *et al.* 2014). In some instances, however, withdrawals and consumption have degraded, and even transformed, water sources. For example, the Ogallala aquifer – the once gigantic pool of fossil water developed over millennia below the High Plains of the United States, stretching from North Dakota to Texas (aka High Plains Aquifer) – has been significantly reduced by withdrawals for irrigation (Fig. 33.1). The equivalent of the annual flow of 18 Colorado rivers is pumped to the surface annually for irrigation. The aquifer recharge, in contrast, is slightly more than 1 cm per annum (Little 2009)!

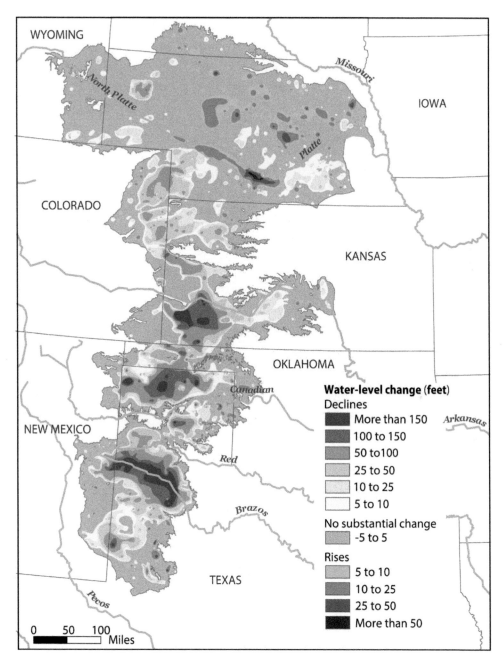

Figure 33.1 Changes in water depth of the Ogallala aquifer, Southern Great Plains, United States
Water pumped from the aquifer to the surface for irrigated agriculture has reduced the volume and increased the depth of its waters. The original figure was published with US customary measures in which one foot is the same as that in the imperial measure of one foot. To keep the legend simple, these measures were applied rather than converting them to SI units.
Source: Scott (2019).

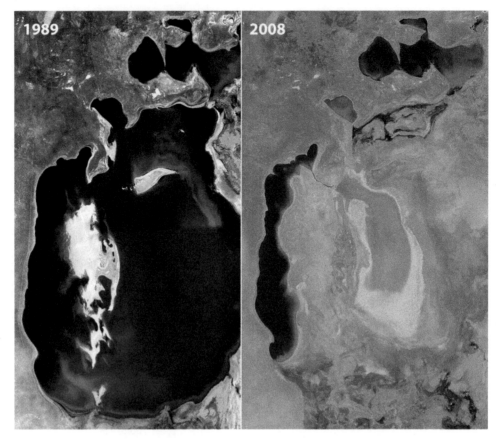

Figure 33.2 The collapse of the Aral Sea
The area of the sea (1989, left) before major irrigation networks reduced the water reaching the sea (2008, right) and after the irrigation networks were in place that continually intercepted the flow of water maintaining the sea (left). The main basin is now covered by evaporative salts (light colour, right). The remaining sea water colour reflects its shallow depth.
Source: NASA Earth Observatory (2008).

Perhaps the most dramatic impact of water withdrawal, however, is the desiccation of the Aral Sea. This lake, located in the Aralkum Desert of central Asia, is principally fed by two rivers and minimal groundwater inflow. Previous to 1960, the lake was the fourth-largest inland water body globally, supporting a large fishing industry in its brackish waters. From 1965 to 2000, irrigated agriculture increased by about 2.9 million ha, extracting fresh-water from the two rivers feeding the lake. The diminished flow of freshwater accounts for about one-half of the water lost from the Aral Sea (or ~510 km^3), suggesting that decreased regional rainfall created by the desiccated lake served as one cause of the lake's demise (Pokhrel *et al.* 2017). Desiccation decreased the water level (−23 m), area (−74 per cent) and volume (−90 per cent) (Fig. 32.2). As a result, the salinity of the remaining water increased from 10 g/l to >100 g/l (Micklin 2007), essentially killing the once prosperous fishing industry of the sea. Attempts are under way to restore the smaller, northern portion of the lake.

Surface salts (largely sodium chloride), however, cover most of the much larger portion of the southern lake. These salts blow across the irrigated land and even reach the snow and glaciers in the Hindu Kush mountains that feed the rivers of the irrigation system.

Recent attention has been given to the area footprint of human activity on ocean waters, resulting from various uses such as aquaculture or oil platforms within oceans and seas. Some 30,000 km² of this footprint is estimated, but this area accounts for only 0.0008 per cent of the vast global ocean area. Accounting for changes in coastal water flow and pollution, however, the area consumed is 2 Mkm² or 0.5 per cent of ocean area (Bugnot *et al.* 2020). Attention to the human footprint on oceans continues to increase, as noted in the questions that follow.

References

Bugnot, A. *et al.* (2020). "Current and projected global extent of marine built structures". *Nature Sustainability* 4: 33–41. doi.org/10.1038/s41893-020-00595-1.

Haddeland, I. *et al.* (2014). "Global water resources affected by human interventions and climate change". *Proceedings of the National Academy of Sciences* 111(9): 3251–6. doi.org/10.1073/pnas.1222475110.

Hoekstra, A. & M. Mekonnen (2012). "The water footprint of humanity". *Proceedings of the National Academy of Sciences* **109(9): 3232–7. doi.org/10.1073/pnas.1109936109.**

Little, J. (2009). "Saving the Ogallala aquifer". *Scientific American* 19(1): 32–9. doi:10.1038/scientificamericanearth0309-32.

Mekonnen, M. & A. Hoekstra (2016). "Four billion people facing severe water scarcity". *Science Advances* 2(2): e1500323. doi:10.1126/sciadv.1500323.

Micklin, P. (2007). "The Aral Sea disaster". *Annual Review of Earth Planetary Science* **35: 47–72. doi.org/10.1146/annurev.earth.35.031306.140120.**

NASA Earth Observatory (2008). "Aral Sea, 1988–2003". earthobservatory.nasa.gov/images/3730/aral-sea (accessed 7 February 2022).

Pokhrel, Y. *et al.* (2017). "Modeling large-scale human alteration of land surface hydrology and climate". *Geoscience Letters* 4(1): 1–13. doi.org/10.1186/s40562-017-0076-5.

Scott, M. (2019). "National climate assessment: Great Plains' Ogallala aquifer drying out". www.climate.gov/news-features/featured-images/national-climate-assessment-great-plains'-ogallala-aquifer-drying-out.

Wada, Y., I. de Graaf & L. van Beek (2016). "High-resolution modeling of human and climate impacts on global water resources". *Journal of Advances in Modeling Earth Systems* 8(2): 735–63. doi.org/10.1002/2015MS000618.

Q34

HAS HUMAN ACTIVITY ALTERED FRESHWATER SURFACE STOCKS AND WITH WHAT ENVIRONMENTAL CONSEQUENCES?

Short answer: yes. Dams and their reservoirs increase local to regional water stocks but increase river-water evaporation, decrease downstream sediments and nutrients, alter biodiversity and collapse some ecosystems.

Of the 1.2 per cent of freshwater residing on the continental surfaces, about 24 per cent is in lakes, reservoirs, wetlands and rivers (**Q32**). The geographical distribution of this surface freshwater is highly skewed (Fig. 34.1), with some areas possessing an abundance of surface stocks and others having sparse sources, foremost warm desert biomes without rivers crossing them. Indeed, one body of freshwater, Lake Baikal in Siberia, holds just under one-quarter of all freshwater in the world, owing to its depth, an astounding 1,642 m. Lakes and reservoirs worldwide hold the majority of fresh surface water.

Humankind has long impounded rivers and surface collection areas (reservoirs) by way of dams and barrages in order to control the flow and storage of water, creating stocks for human uses: potable water, agricultural irrigation and, increasingly, hydroelectric power. In 2015, 3,700 major hydroelectric dams were planned worldwide (Zarfl *et al.* 2015). These dams joined the estimated 16.7 million reservoirs with dams that existed in 2011, creating a global surface area of about 305,723 km^2 of water, which along with regulated lakes (i.e. those with controlled outflow), store about 8,069 km^3 of water (Lehner *et al.* 2011). Regulated lakes and reservoirs have numerous environmental impacts, including increasing evaporation estimated to amount to ~5 per cent of total flow of water in rivers globally (Lehner *et al.* 2011). They also serve as impediments to marine biota and downstream deliveries of sediments and nutrients to floodplains and deltas.

Several examples are insightful. Globally, an estimated 12 per cent of river phosphorus was captured in sediments behind dams in 2000 (Maavara *et al.* 2015), depleting this needed nutrient for downstream floodplain soils and plant growth. Dams have also homogenized river and stream flows in their magnitude and timing, with consequences on biotic diversity (Poff *et al.* 2007). In addition, the flow and/or fragmentation of ~48 per cent of river water volume has been moderately to strongly affected (Grill *et al.* 2015), although this percentage is debated (Hennig & Magee 2017). In extreme cases, water diversion from rivers has led to

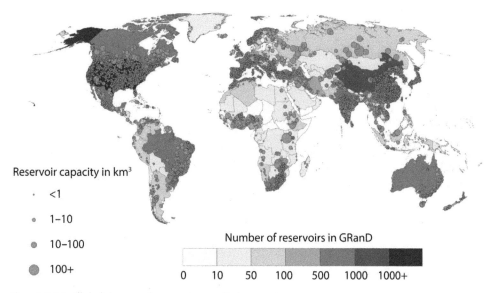

Figure 34.1 Global dams and reservoir capacity by country
Includes waters behind dams regardless of the water source damned, as generated by GRanD (Global Reservoir and Dam Database).
Source: www.gwsp.org/products/grand-database.html; basemap: Esri.

ecological collapse, such as that of the Aral Sea (**Q33**), owing to the 92 per cent water inflow loss due to irrigated agriculture (Micklin & Aladin 2008). In addition, dams disrupt the connectivity of rivers, highly threatening access of freshwater fish to feeding and spawning grounds (Barbarossa *et al.* 2020). Interestingly, the massive construction of dams by 1970 nearly brought sea-level rise to a halt (Frederikse *et al.* 2020)! Such disruption, climate change and land use have apparently increased the proportion of non-perennial rivers (i.e. river flow stops for at least one day a year) to 51 per cent of all rivers worldwide, accounting for those naturally in this condition (Messager *et al.* 2021).

Recognizing these dangers, and the imbalance of water location versus needs, a transformation in water management would be useful for much of the world (Gleick 2018), especially in light of human-induced climate change. For example, human-induced greenhouse gases and aerosols accounted for 60 per cent of the decreasing river flow and factors affecting that flow (i.e. winter temperature and mountain snowpack) for the Colorado River between 1950 and 1999 (Barnett *et al.* 2008). This river in the western United States, critical to $1 trillion of economic activity annually, loses 9.3 per cent of it discharge per 1°C in climate warming (Milly & Dunne 2020). In 2022 the reservoir-lakes on the river are already at their lowest level ever recorded, with significant consequences for water deliveries to the surrounding service area. Examination of past warming periods suggests that rainfall patterns will be altered worldwide with climate warming (Putnam & Broecker 2017), which in turn will affect freshwater hydrology. A climate modelling exercise, for example, indicates that in the 2050s climate change will severely reduce the flow of rivers in Europe, especially in the Mediterranean region (Schneider *et al.* 2013).

References

Barbarossa, V. *et al.* (2020). "Impacts of current and future large dams on the geographic range connectivity of freshwater fish worldwide". *Proceedings of the National Academy of Sciences* 117(7): 3648–3655. doi.org/10.1073/pnas.1912776117.

Barnett, T. *et al.* (2008). "Human-induced changes in the hydrology of the western United States. *Science* 319(5866): 1080–83. doi:10.1126/science.1152538.

Frederikse, T. *et al.* (2020). "The causes of sea-level rise since 1900". *Nature* 584(7821): 393–7. doi. org/10.1038/s41586-020-2591-3.

Gleick, P. (2018). "Transitions to freshwater sustainability". *Proceedings of the National Academy of Sciences* 115(36): 8863–71. doi.org/10.1073/pnas.1808893115.

Grill, G. *et al.* (2015). "An index-based framework for assessing patterns and trends in river fragmentation and flow regulation by global dams at multiple scales". *Environmental Research Letters* 10(1): 015001. doi:10.1088/1748-9326/10/1/015001.

Hennig, T. & D. Magee (2017). "Comment on 'An index-based framework for assessing patterns and trends in river fragmentation and flow regulation by global dams at multiple scales'. *Environmental Research Letters* 12(3): 038001. doi.org/10.1088/1748-9326/aa5dc6.

Lehner, B. *et al.* (2011). "High-resolution mapping of the world's reservoirs and dams for sustainable river-flow management". *Frontiers in Ecology and the Environment* 9(9): 494–502. doi.org/10.1890/100125.

Maavara, T. *et al.* (2015). "Global phosphorus retention by river damming". *Proceedings of the National Academy of Sciences* 112(51): 15603–08. doi.org/10.1073/pnas.1511797112.

Messager, M. *et al.* (2021). "Global prevalence of non-perennial rivers and streams". *Nature* 594: 391–7. doi.org/10.1038/s41586-021-03565-5.

Micklin, P. & N. Aladin (2008). "Reclaiming the Aral Sea". *Scientific American* 298(4): 64–71. doi:10.1038/scientificamerican0408-64.

Milly, P. & K. Dunne (2020). "Colorado River flow dwindles as warming-driven loss of reflective snow energizes evaporation". *Science* 367(6483): 1252–5. doi:10.1126/science.aay9187.

Poff, N. *et al.* (2007). "Homogenization of regional river dynamics by dams and global biodiversity implications". *Proceedings of the National Academy of Sciences* 104(14): 5732–7. doi.org/10.1073/pnas.0609812104.

Putnam, A. & W. Broecker (2017). "Human-induced changes in the distribution of rainfall". *Science Advances* 3(5): e1600871. doi:10.1126/sciadv.1600871.

Schneider, C. *et al.* (2013). "How will climate change modify river flow regimes in Europe?" *Hydrology and Earth System Sciences* 17(1): 325–39. doi.org/10.5194/hess-17-325-2013.

Zarfl, C. *et al.* (2015). "A global boom in hydropower dam construction". *Aquatic Sciences* 77(1): 161–70. doi.org/10.1007/s00027-014-0377-0.

Q35

HAS HUMAN ACTIVITY DEGRADED GROUNDWATER AND AQUIFER STOCKS GLOBALLY?

Short answer: yes. Depletion rates commonly exceed recharge rates globally, endangering potable and irrigation water, and contribute to sea-level rise.

Slightly more than 30 per cent of all water in the Earth system infiltrates soil and percolates through porous rocks, creating groundwater. In many cases, groundwater collects in water-bearing strata or aquifers, water that may be accessed by springs, tapped by wells or pumped to the surface (Fig. 35.1). About 33 per cent of all water withdrawals globally are taken from groundwater (Famiglietti 2014), supplying 36 per cent of domestic, 42 per cent of agricultural and 27 per cent of industrial uses worldwide (Taylor *et al.* 2012). Estimates of the potential diffuse recharge rates of groundwater globally (i.e. infiltration over large areas) constitutes ~30 per cent of the Earth's renewable freshwater, or ~13,000 to ~15,000 km^3 yr^{-1} (Famiglietti 2014; Taylor *et al.* 2012). Estimates of global groundwater depletion rates range from 113 km^3 yr^{-1} (2000–09) to 204 (±30) km^3 yr^{-1} (2000) (Aeschbach-Hertig & Gleeson 2012; de Graaf *et al.* 2017; Wada *et al.* 2012), owing to differences in the data sources and methods employed in the calculations (Wada 2016). In arid and semi-arid lands, water withdrawal from aquifers exceeds the recharge rates (Castle *et al.* 2014). Such aquifers commonly hold "fossil" water, or water that takes millennia to fill the aquifer. Globally, about 15 per cent of groundwater is taken from sources that are not renewed (Döll *et al.* (2014).

Groundwater depletion threatens water access for a significant proportion of the global population and endangers irrigated systems of agriculture. For example, the aquifer below the Southern High Plains in the United States (i.e. High Plains or Ogallala aquifer) has been pumped for 330 km^3 of fossil groundwater that took 13,000 years to accumulate and is estimated to render 35 per cent of those plains unable to support irrigation by the middle of this century (Scanlon *et al.* 2012) **(Q33)**. From 2002 to 2012, Tamil Nadu state in India, which is highly dependent on irrigated agriculture, has been losing 21.4 km^3 yr^{-1}, a rate that is 8 per cent higher than the recharge rate (Chinnasamy & Agoramoorthy 2015). In a final example, the North China Plain has a groundwater storage loss of 7.2 ± 1.1 km^3 yr^{-1} (Feng *et al.* 2018).

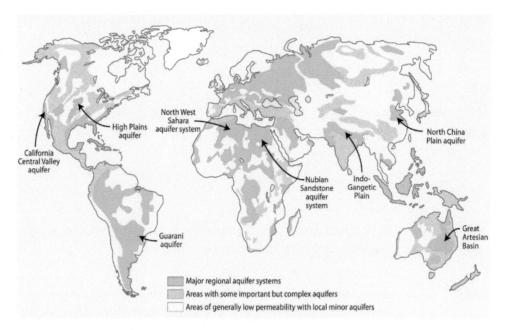

Figure 35.1 Major aquifers worldwide
The High Plains aquifer is the same as the Ogallala aquifer.
Source: Taylor *et al.* (2012). Reprinted by permission from Springer Nature. © (2012).

Surprisingly, perhaps, the global decreases in groundwater contribute to sea-level rise (Taylor *et al.* 2012). This contribution follows from the loss of groundwater storage – and its slow movement to oceans – to surface uses with more rapid means of transport to marine water. In addition, groundwater, especially at the 2–7 m depth level below the surface, affects the atmospheric boundary layer, with consequences for precipitation.

References

Aeschbach-Hertig, W. & T. Gleeson (2012). "Regional strategies for the accelerating global problem of groundwater depletion". *Nature Geoscience* 5(12): 853–61. doi.org/10.1038/ngeo1617.

Castle, S. *et al.* (2014). "Groundwater depletion during drought threatens future water security of the Colorado River Basin". *Geophysical Research Letters* 41(16): 5904–11. doi.org/10.1002/2014GL061055.

Chinnasamy, P. & G. Agoramoorthy (2015). "Groundwater storage and depletion trends in Tamil Nadu State, India". *Water Resources Management* 29(7): 2139–52. doi.org/10.1007/s11269-015-0932-z.

de Graaf, I. *et al.* (2017). "A global-scale two-layer transient groundwater model: development and application to groundwater depletion". *Advances in Water Resources* 102: 53–67. doi.org/10.1016/j.advwatres.2017.01.011.

Döll, P. *et al.* (2014). "Global-scale assessment of groundwater depletion and related groundwater abstractions: combining hydrological modeling with information from well observations and GRACE satellites". *Water Resources Research* 50(7): 5698–720. doi.org/10.1002/2014WR015595.

Famiglietti, J. (2014). "The global groundwater crisis". ***Nature Climate Change*** **4(11): 945–8. doi.org/10.1038/nclimate2425.**

Feng, W. *et al.* (2018). "Groundwater storage changes in China from satellite gravity: an overview". *Remote Sensing* 10(5): 674. doi.org/10.3390/rs10050674.

Scanlon, B. *et al.* (2012). "Groundwater depletion and sustainability of irrigation in the US High Plains and Central Valley". *Proceedings of the National Academy of Sciences* 109(24): 9320–25. doi.org/10.1073/pnas.1200311109.

Taylor, R. *et al.* (2012). "Ground water and climate change". *Nature Climate Change* 3(4): dx.doi.org/10.1038/ncliamte1744.

Wada, Y. *et al.* (2012). "Past and future contribution of global groundwater depletion to sea-level rise". *Geophysical Research Letters* 39(9): L09402. doi.org/10.1029/2012GL051230.

Wada, Y. (2016). "Modeling groundwater depletion at regional and global scales: present state and future prospects". ***Surveys in Geophysics*** **37(2): 419–51. doi:10.1007/s10712-015-9347-x.**

Q36

HAS HUMAN ACTIVITY DEPLETED WETLANDS GLOBALLY?

Short answer: yes. Significant worldwide losses of wetlands reduce the environmental services of water regulation, purification and coastal storm protection.

Wetland areas – water-saturated lands, such as swamps and marshes, including coastal saline wetlands, such as mangroves – are difficult to measure. They occupy up to an estimated 6 per cent of the Earth's surface (Junk *et al.* 2013) and hold about 2.6 per cent of the water on Earth (**Q34**). The estimates of their area coverage are large, 1.53–14.86 Mkm^2, depending on the means of calculation, although the more recent assessments coalesce around the larger estimated area (Hu *et al.* 2017; Junk *et al.* 2013). Following the large estimates, freshwater wetlands cover ~5.7 Mkm^2 and ponded rice paddies, ~1.3 Mkm^2 (Aselmann & Crutzen 1989). Wetlands maintain biodiversity, store water and recharge groundwater, regulate flooding, improve water quality, protect shorelines from erosion and generate substantial levels of annual net primary productivity (Aselmann & Crutzen 1989; Nicholls 2004). They are also major sources of methane (CH_4) emissions, providing somewhere between 40–177.2 ± 49.7 Tg yr^{-1}, mostly from tropical wetlands, including rice fields (Zhang *et al.* 2017).

Humankind has long drained wetlands for disease protection, for example against malaria-carrying mosquitos, and expanded or created wetlands anew for rice cultivation. Pre-conquest peoples of the Western hemisphere maintained extensive wetland areas for cultivation from Mexico to Bolivia (Turner & Butzer 1992). The Dutch built their country in part from wetlands. Many agricultural systems worldwide (e.g. polders, dykelands) are modifications of wetlands (Nath, van Laerhoven & Driessen 2019), as are many of the 1.3 Mkm^2 rice fields (Zhang *et al.* 2017).

Rice field expansion notwithstanding, wetlands are decreasing worldwide, perhaps as much as 33 per cent by 2009. Between 60–90 per cent of Western Europe's wetlands have been transformed to other uses; China has lost 22 per cent of its wetlands during the turn to the twenty-first century (Junk *et al.* 2013); and, as noted in **Question 28**, 35 per cent of mangrove wetlands have been lost to other human activities. One estimate indicates that the Earth would have just under 30 Mkm^2 of wetlands if humankind had not transformed them (Hu *et al.* 2017). The decline in environmental services, noted above, following the loss of wetlands, is recognized as large. The calculation of them to date, however, has been impeded by various data problems.

In addition to the wetlands noted, tidal wetlands – flats, marshes and mangroves – have a net loss of ~4,000 km^2 from 1999–2019 (Murray *et al.* 2022). These saline systems store carbon and protect coastlines, enhance fisheries, among other services (**Q28**). An estimated 27 per cent of total changes in tidal wetlands are responses to direct human activities and the remainder to indirect process, such as climate change.

References

Aselmann, I. & P. Crutzen (1989). "Global distribution of natural freshwater wetlands and rice paddies, their net primary productivity, seasonality and possible methane emissions". *Journal of Atmospheric Chemistry* 8(4): 307–58. doi.org/10.1007/BF00052709.

Hu, S. *et al.* (2017). "Global wetlands: potential distribution, wetland loss, and status". *Science of the Total Environment* 586: 319–27. doi.org/10.1016/j.scitotenv.2017.02.001.

Junk, W. *et al.* (2013). "Current state of knowledge regarding the world's wetlands and their future under global climate change: a synthesis". *Aquatic Sciences* 75(1): 151–67. doi.org/10.1007/s00027-012-0278-z.

Murray, N. J. (2022). "High-resolution mapping of losses and gains of Earth's tidal wetlands". *Science* 376(6594): 744–9. doi.10.1126/science.abm9583.

Nath, S., F. van Laerhoven & P. Driessen (2019). "Have Bangladesh's polders decreased livelihood vulnerability? A comparative case study". *Sustainability* 11(24): 7141. doi.org/10.3390/su11247141.

Nicholls, R. (2004). "Coastal flooding and wetland loss in the 21st century: changes under the SRES climate and socio-economic scenarios". *Global Climate Change* 14: 69–86. doi.org/10.1016/j.gloenvcha.2003.10.007.

Turner II, B. & K. Butzer (1992). "The Columbian encounter and land-use change". *Environment: Science and Policy for Sustainable Development* 34(8): 16–44. doi.org/10.1080/00139157.1992.9931469.

Zhang, B. *et al.* (2017). "Methane emissions from global wetlands: an assessment of the uncertainty associated with various wetland extent data sets". *Atmospheric Environment* 165: 310–21. doi.org/10.1016/j.atmosenv.2017.07.001.

Q37

HAS HUMAN ACTIVITY REDUCED FROZEN WATER – THE CRYOSPHERE – GLOBALLY?

Short answer: yes. Almost all glaciers and ice sheets, snow cover, floating/sea ice and perma-frost have been reduced by human-induced climate warming, changing the albedo of the Earth's surface, increasing sea level and increasing greenhouse gas emissions.

Frozen (solid) water on the Earth's surface – the cryosphere – is composed of glaciers, floating ice, snow cover and permafrost (Fig. 37.1). Glaciers, which include ice caps and ice sheets (combined, ice fields) are formed by snow compressed into ice. The largest areas and volume of ice sheets cover Antarctica (~13.5 Mkm^2 and ~30.1 Mkm^3) and Greenland (~1.7 Mkm^2 and ~2.6 Mkm^3) (Williams & Ferrigno 2012). Ice shelves extend from an ice field land surface in to the adjacent ocean. Glaciers, ice caps/fields and ice shelves exist in high latitudes or polar regions, much of which are climatically deserts, foremost all of Antarctica. Glaciers and ice fields develop because the small amounts of snowfall annually accumulate over millennia. Beyond the polar regions, mountain glaciers, about 160,000 in number, form at high elevations. Together, the ice cover of the Earth is ~15.9 Mkm^2 in area and holds 32.9 Mkm^3 of ice. This volume constitutes about 70 per cent of all freshwater in the Earth system (IPCC 2018; Williams & Ferrigno 2012).

Floating or sea ice differs from land-based ice in that it is frozen seawater, about 2–4 m thick, at the surface of oceans and typically covered by snow. Globally, sea ice area ranges from ~18 Mkm^2 to ~27 Mkm^2. It also varies in area by season, about ~7 km^2 in the Arctic Ocean and ~15 km^2 surrounding Antarctica (IPCC 2018; Parkinson & Cavalieri 2012).

Snow cover also varies seasonally. From 1967 to 2005, the mean annual snow cover in the Northern hemisphere was 25.6 Mkm^2, ranging from 46.9 to 3.5 Mkm^2, between winter and summer, respectively. In the Southern hemisphere, this mean is only 1 Mkm^2, mostly appearing in South America (Frei *et al.* 2012; Parkinson & Cavalieri 2012). The snow pack or accumulation of snow cover throughout the mid- to higher latitudes is critical to watersheds worldwide, providing potable and irrigation water.

Permafrost refers to soil that remains frozen, residing at high latitudes and altitudes. Frozen soil may exist at the surface, or below ground in which the upper soil freezes and unfreezes seasonally, conditions generally supporting tundra (**Q29**). In the Northern

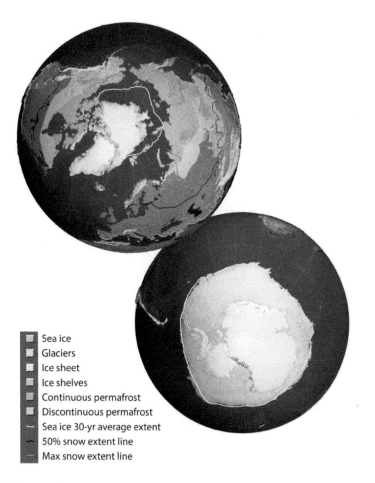

Sea ice
Glaciers
Ice sheet
Ice shelves
Continuous permafrost
Discontinuous permafrost
Sea ice 30-yr average extent
50% snow extent line
Max snow extent line

Figure 37.1 The cryosphere
The sea ice area expands in winter and retracts in summer for the oceans in the polar realms of the Northern and Southern hemispheres.
Source: IPCC (2018).

hemisphere, continuous permafrost exists in large swathes, giving way to discontinuous (dispersed) permafrost along its southern border. It also exists along the western coastal zone of Antarctica and its peninsula, and in the Andes and Himalayan mountains. Permafrost covers 12–17 Mkm^2 in the Northern hemisphere (Fig. 37.1) (Heginbottom *et al.* 2012). About one-half of permafrost (~1.85 Mkm^2) resides in boreal peatlands (Hugelius *et al.* 2020).

All forms of glaciers, floating ice and snow cover have high albedo, reflecting solar radiation back into the atmosphere and space. Sea ice insulates the ocean waters from the loss of heat (energy) that would follow from direct water-surface exposure to the extreme cold of the polar atmosphere (Parkinson & Cavalieri 2012). The area covered in sea ice used in models of climate change has a significant influence on warming projections (Hu *et al.*

2020). Permafrost holds large amounts of CO_2 and CH_4, serving as an important sink for these greenhouse gases, although the estimated amounts vary (Hugelius *et al.* 2020). At only a 3 m depth, permafrost is estimated to hold as much 1,035±150 Pg carbon, of which 800 Pg are frozen (Schuur *et al.* 2015). Recent evidence indicates that modest atmospheric warming has released massive amounts of this carbon into the Earth system, about three times more than 28 kya when warming of the Earth gave way to the Holocene (Knoblauch *et al.* 2018; Martens *et al.* 2020).

The human impacts on the cryosphere are largely indirect through global warming. This warming has generated a loss in mass balance (net change from a fixed year) in ice fields and glaciers worldwide (Table 37.1; Fig. 37.2) (IPCC 2018), including the loss of 28 trillion tonnes of ice between 1994 and 2017 (Slater *et al.* 2021). In the first two decades of the twenty-first century, the two largest ice fields, Antarctica and Greenland, had average annual rates of loss ranging from 155 to 278 Gt (Williams & Ferrigno 2012). Meanwhile, marine-terminating glaciers (land-based glaciers that end in the ocean) lost a total of 390 km^2 annually in the first two decades of the new century. Of the 1,704 such glaciers, 123 no longer reach the sea (Kochtitzky *et al.* 2002). Overall, the cryosphere area declined at a rate of ~87k km^2 yr^{-1} (Peng *et al.* 2021), increasing sea level by 34.6 ± 3.1 mm from 1994 to 2017 at an annual rate of ~1.44 mm (Slater *et al.* 2021). Beyond those years, the loss of the cryosphere may have contributed to about 50 per cent of sea-level rise (Williams & Ferrigno 2012). If Antarctica and Greenland were to lose their ice, global sea levels would rise 60–90 m (Williams & Ferrigno 2012).

The warming of the Earth has also decreased snow cover and sea ice globally, reducing their capacity to reflect solar radiation and increasing warming further through positive feedback. The ice cover of the Arctic Ocean, for example, is now 50 per cent less in area than it was in 1980 and only 75 per cent of its former volume owing to a reduction in its thickness (Voosen 2020) (Fig. 37.2). The seas between the central Arctic Ocean and the continents (i.e. Beaufort, Chukchi, East Siberian, Laptev, Kara and Barents seas) show dramatic rates in the decline of mean sea ice thickness (Mallett *et al.* 2021). Recent studies project that the Arctic Ocean may be ice-free by 2035 (Guarino *et al.* 2020). Permafrost, which holds nearly twice as much carbon as the atmosphere, has huge potential to serve as a greenhouse gas source, especially by way of CH_4, which is 28 times more powerful than CO_2 as a warming

Table 37.1 Global ice loss, 1944–2017

Ice surface	Ice type	Trillion tonnes
Arctic sea ice	Floating	7.6
Antarctic ice shelves	Grounded	6.5
Mountain glaciers	Grounded	6.1
Greenland ice sheet	Grounded	3.8
Antarctic ice sheet	Grounded	2.5
Southern Ocean ice sheet	Floating	0.9

Source: Slater *et al.* (2021).

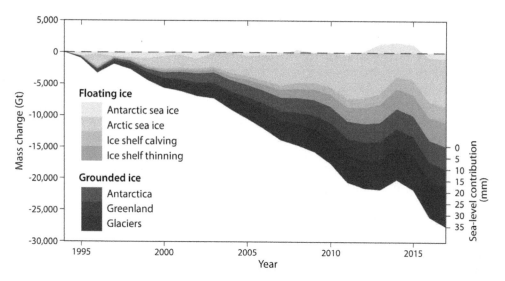

Figure 37.2 Mass change of the cryosphere and contribution to sea-level rise, 1994–2017
Source: Slater *et al.* (2021).

agent (Anthony *et al.* 2018). In addition, evidence indicates significant emissions have begun from the estimated 3.6 million lakes and ponds in the Arctic (Paltan, Dash & Edwards 2015); the low albedo of water captures heat and emits methane, apparently on a similar level as melting permafrost (Anthony *et al.* 2016).

Finally, losses in the cryosphere, foremost that of land ice or glaciers, are subtly affecting Earth's mass and its axis. The North Pole, as a result, has moved about 4 m southeast since 1980 and appears to be drifting with increasing pace owing in part to the changes in the cryosphere (Deng 2021). Such drift in the distant past disrupted climate globally.

References

Anthony, K. *et al.* (2016). "Methane emissions proportional to permafrost carbon thawed in Arctic lakes since the 1950s". *Nature Geoscience* 9(9): 679–82. doi.org/10.1038/ngeo2795.

Anthony, W. *et al.* (2018). "21st-century modeled permafrost carbon emissions accelerated by abrupt thaw beneath lakes". *Nature Communications* 9: 3626. doi:10.1038/s41467-018-05738-9.

Deng, S. (2021). "Polar drift in the 1990s explained by terrestrial water storage changes". *Geophysical Research Letters* 48(7): e2020GL092114. doi.org/10.1029/2020GL092114.

Frei, A. *et al.* (2012). "A review of global satellite-derived snow products". *Advances in Space Research* 50(8): 1007–29. doi.org/10.1016/j.asr.2011.12.021.

Guarino, M. *et al.* (2020). "Sea-ice-free Arctic during the Last Interglacial supports fast future loss". *Nature Climate Change* 10: 1–5. doi.org/10.1038/s41558-020-0865-2.

Heginbottom, J. *et al.* (2012). "Permafrost and periglacial environments". In R. Williams Jr & J. Ferrigno (eds), *Satellite Image of Glaciers of the World*. Denver, CO: US Geological Survey. pubs.usgs.gov/pp/p1386a/pdf/pp1386a-4-web.pdf.

Hu, X. *et al.* (2020). "A less cloudy picture of the inter-model spread in global warming projections". *Nature Communications* 11: 4472. doi.org/10.1038/s41467-020-18227-9.

Hugelius, G. *et al.* (2020). "Large stocks of peatland carbon and nitrogen are vulnerable to permafrost thaw". *Proceedings of the National Academy of Sciences* 117(34): 20438–46.

IPCC (Intergovernmental Panel on Climate Change) (2018). "Special Report on the Ocean and Cryosphere in a Changing Climate". www.ipcc.ch/srocc/.

Knoblauch, C. *et al.* (2018). "Methane production as key to the greenhouse gas budget of thawing permafrost". *Nature Climate Change* 8(4): 309–12. doi.org/10.1038/s41558-018-0095-z.

Kochtitzky, W. *et al.* (2022). "Retreat of Northern hemisphere marine-terminating glaciers, 2000–2020". *Geophysical Letters* 49(3): e2021GL096501. doi.org/10.1029/2021GL096501.

Mallett, R. *et al.* (2021). "Faster decline and higher variability in the sea ice thickness of the marginal Arctic seas when accounting for dynamic snow cover". *The Cryosphere* 15(5): 2429–50. doi.org/10.5194/tc-15-2429-2021.

Martens, J. *et al.* (2020). "Remobilization of dormant carbon from Siberian-Arctic permafrost during three past warming events". *Science Advances* 6(42): eabb6546. doi:10.1126/sciadv.abb6546.

Paltan, H., J. Dash & M. Edwards (2015). "A refined mapping of Arctic lakes using Landsat imagery". *International Journal of Remote Sensing* 36(23): 5970–82. doi.org/10.1080/01431161.2015.1110263.

Parkinson, C. & D. Cavalieri (2012). "Floating ice: sea ice". In R. Williams Jr & J. Ferrigno (eds), *Satellite Image of Glaciers of the World*. Denver, CO: US Geological Survey. pubs.usgs.gov/pp/p1386a/pdf/pp1386a-4-web.pdf

Peng, X. *et al.* (2021). "A holistic assessment of 1979–2016 global cryosphere extent. *Earth's Future* 9(8): e2020EF001969. doi.org/10.1029/2020EF001969.

Schuur, E. *et al.* (2015). "Climate change and the permafrost carbon feedback". *Nature* 520(7546): 171–9. doi.org/10.1038/nature14338.

Slater, T. *et al.* (2021). "Earth's ice imbalance". *The Cryosphere* 15: 233–46. doi.org/10.5194/tc-15-233-2021.

Voosen, P. (2020). "New feedbacks speed up the demise of Arctic sea ice". *Science* 369: 1043–4. doi:10.1126/science.369.6507.1043.

Williams Jr., R. & J. Ferrigno (2012). "Glaciers". In R. Williams Jr & J. Ferrigno (eds), *Satellite Image of Glaciers of the World*. Denver, CO: US Geological Survey. pubs.usgs.gov/pp/p1386a/pdf/pp1386a-4-web.pdf.

Q38

HAS HUMAN ACTIVITY INCREASED OCEAN HEAT?

Short answer: yes. Oceans have taken up the vast majority (~93 per cent) of the human-induced heat that is warming the Earth, with important consequences for global climates, sea-level rise, acidification, ecosystem change, and probable increases in tropical storm intensity and precipitation.

Oceans cover more than 75 per cent of the Earth's surface, serving as a vast sink of heat (energy) from solar radiation and atmospheric CO_2 and as a source of reradiated heat and CO_2 to the atmosphere. The ocean heat content has ebbed and flowed in the past owing to climate changes. For example, during the Medieval Warm Period (950–1250 CE or 1072–772 ya from 2022) the northern Pacific and Antarctic oceans were warmer than they have been in the past few decades (Gebbie & Huybers 2019). About 57 per cent of the global oceans, however, now exhibit extreme heat (Tanaka & van Houtan 2022). Such warming follows from human impacts on the atmosphere, foremost through added greenhouse gases (IPPC 2021) with consequences for acidification (Q47).

About 93 per cent of the warming of the Earth since 1995 has been taken up by the oceans, and about one-third of that warming was captured at a depth range of 700–2,000 m (Cheng *et al.* 2020; Gleckler *et al.* 2016; Levitus *et al.* 2012). From 1995 to 2010, the oceans to a depth of 2,000 m have had a mean increase in temperature of 0.09°C (IPCC 2021) (Fig. 38.1). These figures may be conservative, however, given recent evidence that the decadal rise in ocean temperature from 2003 to 2017 was 0.12°C, rather than the previously calculated 0.07°C (Hausfather *et al.* 2017). Estimates from 1992 to 2018 indicate that the ocean take-up of carbon has been 0.8–0.9 PgC yr^{-1} (Watson *et al.* 2020). This means that the warming of the atmosphere has been dampened by the ocean uptake of heat, generating major increases in "marine heat waves" (IPCC 2021; Cheng *et al.* 2020). Recent evidence indicates that the Atlantic Ocean is now warmer than it has been during the past three millennia (Lapointe *et al.* 2020). Importantly, however, a considerable time lag exists between the movement of marine water in the deep ocean and the surface. For example, only about a quarter of the recent heat content absorbed by the upper oceans has been offset by heat loss from the deep ocean since 1750 (Gebbie & Huybers 2019). This dynamic indicates uncertainty in various model projections and the need for further research.

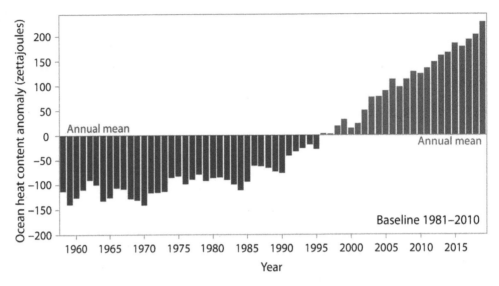

Figure 38.1 Increasing ocean heat content, 1958–2019

The anomaly is the difference between the baseline annual mean in zettajoules for 1981–2010 and a given year. Zettajoules are based on the average for the upper 2,000 ft (610 m) of the ocean. A zettajoule is 10^6 joules; 1 joule = 1 W of power 1 second. The increasing joules reflect temperature increases associated with ocean carbon uptake.

Source: Cheng *et al.* (2020). Reprinted by permission from Springer Nature. © (2019).

Nevertheless, as ocean surface water heats, it expands, contributing to sea-level rise. This expansion is projected to increase sea-level rise by 17–26 cm during the last two decades of this century (Lyu, Zhang & Church 2021). Various issues follow from this heating, other than the near-shore impacts of higher tides, some of which portend to hold serious Earth system consequences (Luo, Sasaki & Masumoto 2012; Cheng *et al.* 2019). In general, warm surface water from the tropics moves toward the poles where it mixes with freshwater from high latitude glaciers and ice sheets, becomes cooler, denser and more saline, and sinks to the deep ocean in a returning southern flow (IPCC 2021), a movement known as the meridional overturning circulation (MOC aka thermohaline circulation; **Q40**) (Li *et al.* 2020). The heat transport in the Atlantic MOC (AMOC) is expected to increase rainfall intensity associated with tropical storms and perhaps also the intensity of hurricanes (Knutson *et al.* 2019, 2020). In tandem with sea-level rise **(Q39)**, this intensity will increase the impacts of coastal storm surges. Such surges, in turn, lead to coastal overtopping, essentially flooding, if sporadically, the immediate interior lands. Over the past two decades the globally aggregated annual hours of overtopping have increased by 50 hours (Almar *et al.* 2021). Significantly, ocean warming is expected to slow the great ocean conveyor belt or MOC **(Q40)**, which will have major climate impacts and increase toxic algal blooms affecting aquatic ecosystems, fisheries, and ultimately human health from ocean food products (Gobler *et al.* 2017). In addition, warming of oceans is also leading to acidification and deoxygenation **(Q41)**, as well as degrading ecosystems, such as coral reefs **(Q59)**.

References

Almar, R. *et al.* (2021). "A global analysis of extreme coastal water levels with implications for potential coastal overtopping". *Nature Communications* 12(1): 3775. doi.org/10.1038/s41467-021-24008-9.

Cheng, L. *et al.* (2019). "How fast are the oceans warming?" *Science* 363(6423): 128–9. doi:10.1126/science.aav7619.

Cheng, L. *et al.* (2020). "Record-setting ocean warmth continued in 2019". *Advances in Atmospheric Sciences* 37: 137–42. doi.org/10.1007/s00376-020-9283-7.

Gebbie, G. & P. Huybers (2019). "The little ice age and 20th-century deep Pacific cooling". *Science* 363(6422): 70–74. doi:10.1126/science.aar8413.

Gleckler, P. *et al.* (2016). "Industrial-era global ocean heat uptake doubles in recent decades". *Nature Climate Change* 6(4): 394–8. doi.org/10.1038/nclimate2915.

Gobler, C. *et al.* (2017). "Ocean warming since 1982 has expanded the niche of toxic algal blooms in the North Atlantic and North Pacific oceans". *Proceedings of the National Academy of Sciences* 114(19): 4975–80. doi.org/10.1073/pnas.1619575114.

Hausfather, Z. *et al.* (2017). "Assessing recent warming using instrumentally homogeneous sea surface temperature records". *Science Advances* 3(1): e1601207. doi:10.1126/sciadv.1601207.

IPPC (Intergovernmental Panel on Climate Change) (2021). *IPCC Special Report on the Ocean and Cryosphere in a Changing Climate.* H.-O. Pörtner *et al.* (eds). In press. www.ipcc.ch/srocc/cite-report/.

Knutson, T. *et al.* (2019). "Tropical cyclones and climate change assessment Part I: detection and attribution". *Bulletin of the American Meteorological Society* 100(10): 1987–2007. doi.org/10.1175/BAMS-D-18-0189.1.

Knutson, T. *et al.* (2020). "Tropical cyclones and climate change assessment Part II: projected response to anthropogenic warming". *Bulletin of the American Meteorological Society* 101(3): E303–32. doi.org/10.1175/BAMS-D-18-0194.1.

Lapointe, F. *et al.* (2020). "Annually resolved Atlantic sea surface temperature variability over the past 2,900 y". *Proceedings of the National Academy of Sciences*, 117 (34): 27171–27178. doi.org/10.1073/pnas.2014166117.

Levitus, S. *et al.* (2012). "World ocean heat content and thermosteric sea level change (0–2000 m), 1955–2010". *Geophysical Research Letters* 39(10): l106003. doi.org/10.1029/2012GL051106.

Li, G. *et al.* (2020). "Increasing ocean stratification over the past half-century". *Nature Climate Change* 10: 1116–23. doi.org/10.1038/s41558-020-00918-2.

Luo, J., W. Sasaki & Y. Masumoto (2012). "Indian Ocean warming modulates Pacific climate change". *Proceedings of the National Academy of Sciences* 109(46): 18701–06. doi.org/10.1073/pnas.1210239109.

Lyu, K., X. Zhang & J. Church (2021). "Projected ocean warming constrained by ocean observational record". *Nature Climate Change* 11: 834–9. doi.org/10.1038/s41558-021-01151-1.

Tanaka, K. & K. van Houtan (2022). "The recent normalization of historical marine heat extremes". *PLOS Climate* 1(2): e0000007. doi.org/10.1371/journal.pclm.0000007.

Watson, A. *et al.* (2020). "Revised estimates of ocean-atmosphere CO_2 flux are consistent with ocean carbon inventory". *Nature Communications* 11(1): 1–6. doi.org/10.1038/s41467-020-18203-3.

Q39

HAS HUMAN ACTIVITY RAISED SEA LEVELS?

Short answer: yes. Human-induced global warming has melted land-based frozen water and thermally expanded ocean water, leading to sea-level rise with impacts on low-lying coastal zones and islands worldwide.

Sea levels vary regionally and apparently fluctuate about 1 mm on average in 64-year cycles of unknown origin (Ding *et al.* 2021). The regional variations prompt the use of global mean sea level (MSL), an area-weighted mean, as the datum to measure the level and sea-level rise (MSLR). Beyond the apparent cycle, the increase in MSLR follows from two principal sources: (1) foremost, the gain in water otherwise stored on land, largely from glaciers and ice sheets (**Q37**); and (2) secondarily, the thermal expansion of ocean water (**Q38**). Global warming is responsible for both sources (IPCC 2021) and has increased global mean sea level by about 3.3 to 3.6 mm yr^{-1} over the past several decades (Abraham *et al.* 2013; Ehlert & Zickfeld 2018) (Fig. 39.1). Even water at the depth of 4,757 m, in the South Atlantic Ocean off Argentina, has increased by ~0.2° C between 2009 and 2019 (Meinen *et al.* 2020). New analysis suggests that the rise in sea level above the longer-term background variability began in the late 1800s after large-scale fossil fuel burning of the Industrial Revolution had been under way, and between 1940 and 2000 this rise increased by 1.4 mm annually (Walker *et al.* 2022).

The warming of the Earth system is especially pronounced in the polar regions, affecting the ice sheets of Antarctica and Greenland, which hold the largest amount of frozen water on land (**Q37**). The melting of glaciers and ice sheets (i.e. reduced negative mass balance compared with a fixed year) generates meltwater which ultimately finds its way to the oceans. The mass loss of Antarctica ice sheets tripled and that for Greenland doubled from 2007 to 2016 compared with 1996 to 2006 (Ding *et al.* 2021). Beyond these two land masses, however, large-scale construction of dams over the past 50 years reduced the role of meltwater in sea level rise (Frederikse *et al.* 2020). The loss of floating or sea ice, such as that covering the Arctic Ocean, does not affect sea levels because its mass is already within the oceans. Whether this water resides as ice or melts has virtually no effect on sea level. As the meltwater heats in the oceans, however, it expands, contributing to sea-level rise.

In addition, marine heatwaves, which increase thermal expansion, are increasing in frequency, becoming more intense and more extensive in area, and lasting longer (IPCC 2021).

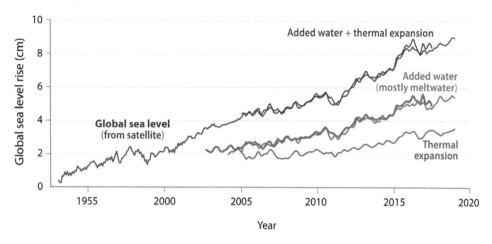

Figure 39.1 Contributions to global sea-level rise, 1993–2018
Source: NOAA (2018).

Notably, it takes the deep ocean much longer to release its heat than it does to gain it. In addition, the loss of land-based ice, foremost that in Greenland, has the potential to shut down global ocean circulations that help maintain the global climates **(Q40)**.

If Antarctica and Greenland were to lose all their ice, global sea levels would rise between 60 and 90 m (Williams & Ferrigno 2021). This drastic scenario notwithstanding, near-future sea-level increases for the year 2100 range from as low as 17 cm to 84 cm, depending on future greenhouse emissions and the assessment calculations applied (IPCC 2021). Recent fine-tuned assessments suggest the high range for the last two decades of this century should be lowered to 26 cm (Lyu, Zhang & Church 2021).

This rise has significant consequences for coastal ecosystems and people (Cohen & Small 1998; Joshi *et al.* 2016) as more than one-third of the global population lives between the sea and 100 m (Joshi *et al.* 2016). Estimates of the total global population affected by MSLR vary owing to the projected year in question, the increase in temperature explored, the type of projection (e.g. permanently flooded, storm surge area) and population dynamics in general (McMichael *et al.* 2020). In addition, satellite lidar data indicates that the land which is less than 2 m above mean sea level that is most vulnerable to MSLR accounts for roughly 1.05 Mkm^2 globally. Of this amount, 62 per cent resides in the tropics (Hooijer & Vernimmen 2021). Islands worldwide are especially affected, including both their people and biodiversity (Courchamp *et al.* 2014). This vulnerability has led the small island nations of the world to cooperate in climate change activities (Alliance of Small Island States or AOSIS).

While there is scientific consensus that sea levels are rising globally, the extent is difficult to predict given different degrees of warming and the amount of ice loss from Antarctica and Greenland. Predictions will become more accurate as the evidence improves over time. Such improvements involve the amount of land ice lost at high latitudes at different levels of global warming and consideration of isostatic rebound, the rise in land with the loss of the weight of ice sheets (Dyer *et al.* 2021).

References

Abraham, J. *et al.* (2013). "A review of global ocean temperature observations: implications for ocean heat content estimates and climate change". *Reviews of Geophysics* 51(3): 450–83. doi.org/10.1002/rog.20022.

Cohen, J. & C. Small (1998). "Hypsographic demography: the distribution of human population by altitude". *Proceedings of the National Academy of Sciences* 95(24): 14009–14. doi.org/10.1073/pnas.95.24.14009.

Courchamp, F. *et al.* (2014). "Climate change, sea-level rise, and conservation: keeping island biodiversity afloat". *Trends in Ecology & Evolution* 29(3): 127–30. doi.org/10.1016/j.tree.2014.01.001.

Ding, H. *et al.* (2021). "The contribution of a newly unraveled 64 years common oscillation on the estimate of present-day global mean sea level rise". *JGR Solid Earth* 126(8): e2021JB022147. doi.org/10.1029/2021JB022147.

Dyer, B. *et al.* (2021). "Sea-level trends across The Bahamas constrain peak last interglacial ice melt". *Proceedings of the National Academy of Sciences* 118(33): 23026839118. doi.org/10.1073/pnas.2026839118.

Ehlert, D. & K. Zickfeld (2018). "Irreversible ocean thermal expansion under carbon dioxide removal". *Earth System Dynamics* 9(1): 197–210. doi.org/10.5194/esd-9-197-2018.

Frederikse, T. *et al.* (2020). "The causes of sea-level rise since 1900". *Nature* 584(7821): 393–7. doi.org/10.1038/s41586-020-2591-3.

Hooijer, A. & R. Vernimmen (2021). "Global LiDAR land elevation data reveal greatest sea-level rise vulnerability in the tropics". *Nature Communication* 12(3592). doi.org/10.1038/s41467-021-23810-9.

Joshi, S. *et al.* (2016). "Physical and economic consequences of sea-level rise: a coupled GIS and CGE analysis under uncertainties". *Environmental and Resource Economics* 65(4): 813–39. doi.org/10.1007/s10640-015-9927-8.

IPPC (Intergovernmental Panel on Climate Change) (2021). *IPCC Special Report on the Ocean and Cryosphere in a Changing Climate.* H.-O. Pörtner *et al.* (eds). In press. www.ipcc.ch/srocc/cite-report/.

Lyu, K., X. Zhang & J. Church (2021). "Projected ocean warming constrained by the ocean observational record". *Nature Climate Change* 11: 834–9. doi.org/10.1038/s41558-021-01151.

McMichael, C. *et al.* (2020). "A review of estimating population exposure to sea-level rise and the relevance for migration". *Environmental Research Letters* 15: 123005. doi.org.10.1088/1748-9326/abb398.

Meinen, C. *et al.* (2020). "Observed ocean bottom temperature variability at four sites in the Northwestern Argentine basin: evidence of decadal deep/abyssal warming amidst hourly to interannual variability during 2009–2019". *Geophysical Research Letters* 47(18): e2020GL089093. doi.org/10.1029/2020GL089093.

NOAA (National Oceanic and Atmospheric Administration) (2018). "State of Climate". www.climate.gov/news-features/understanding-climate/climate-change-global-sea-level.

Walker, J. *et al.* (2022). "Timing of emergence of modern rates of sea-level rise by 1863". *Nature Communications* 13: 966. doi.org/10.1038/s41467-022-28564-6.

Williams Jr., R. & J. Ferrigno (2012). "Glaciers". In R. Williams Jr & J. Ferrigno (eds), *Satellite Image of Glaciers of the World.* Denver, CO: US Geological Survey. pubs.usgs.gov/pp/p1386a/pdf/pp1386a-4-web.pdf.

HAS HUMAN ACTIVITY DISRUPTED THE THERMOHALINE CIRCULATION/MERIDIONAL OVERTURNING CIRCULATION, AND WHAT ARE THE CONSEQUENCES?

Short answer: yes. The warming of the oceans and melting of land-based ice reduces ocean temperature stratification, which slows ocean circulation with the potential to change global climates.

In tandem with wind-blown surface circulation (or gyres), the oceans are connected by a global flow of surface and deep waters creating a worldwide ocean conveyor belt that distributes energy (heat) and dissolved gases (e.g. CO_2) throughout the oceans (Fig. 40.1). This flow, or the thermohaline circulation (THC) noting the roles of temperature and salinity density in its generation, is more accurately labelled the meridional overturning circulation (MOC) to account for other factors creating the flow, such as winds.

This circulation takes warm, less dense and less saline surface water from the tropics and moves it toward the poles. The wind-driven Gulf Stream in the northern Atlantic Ocean is part of the surface flow. As it moves northward, evaporation increases salinity, added to by salt ejected from sea ice, and the water cools, also increasing its salinity and density. In this condition, the water sinks to the ocean's depths from where it moves southward (Huang *et al.* 2020; Semper *et al.* 2020). It is this surface-to-depth movement in the North Atlantic that gives rise to the global MOC. From its northern "dive" in the Norwegian Sea, the circulation crosses the entire north–south length of the Atlantic to the Southern Ocean (part of the Atlantic MOC or AMOC) where it circles Antarctica. Owing to complex interactions involving the density of water, the MOC takes pathways into the warm Indian and Pacific Oceans where it surfaces and returns warm, less dense and less saline water to the Southern Ocean and back into the Atlantic (Ramstorf 2003).

This circulation is pivotal for the Earth system and climatic conditions because it transfers energy globally and serves as a carbon sink. In the North Atlantic, it reduces sea ice in the Arctic Ocean and moderates the climate of northwestern Europe. Through climate warming, human activity indirectly challenges the functioning of the THC/MOC, such as changes in ocean heat loss in the Arctic (Smedsrud *et al.* 2021). Increasing the warmth of

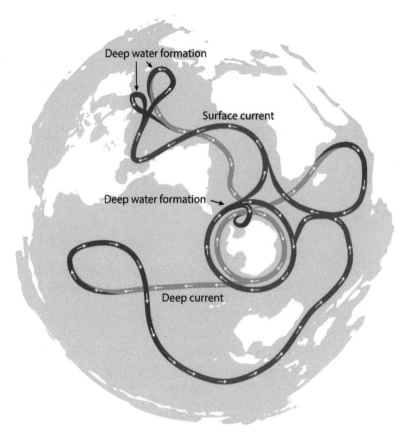

Figure 40.1 Thermohaline or meridional overturning circulation
Surface (red) and deep (blue) water circulation in oceans and formation zones (purple) where surface waters sink to deep circulation or deep waters rise to surface circulation.
Source: Avsa (2009).

the ocean at various depths reduces stratification of the density and salinity of water, major factors that keep the circulation in motion. Significantly, atmospheric warming increases transport of heat in the AMOC, melts glaciers and ice sheets, and perhaps increases precipitation (Gierz 2015). The less dense freshwater decreases salinity and the capacity of the surface water to sink, disrupting the circulation with potential major consequences for climate. Recent work demonstrates that 20–55 per cent of the Atlantic, Pacific and Indian Oceans now have different temperatures and salinities owing to climate change, figures expected to rise to 40–60 per cent by 2050 (Silvy *et al.* 2020). Significantly, a rapid decline of the strength of the AMOC has been recorded since the middle of the twentieth century (Caesar *et al.* 2021).

It has been proposed that such a decline created the Younger Dryas period (ca. 12,900–11,700 BP), in which the climate of the Northern Hemisphere, especially from Greenland to Northern Europe, ceased its warming trend and returned to glacial conditions (Fig. 40.2).

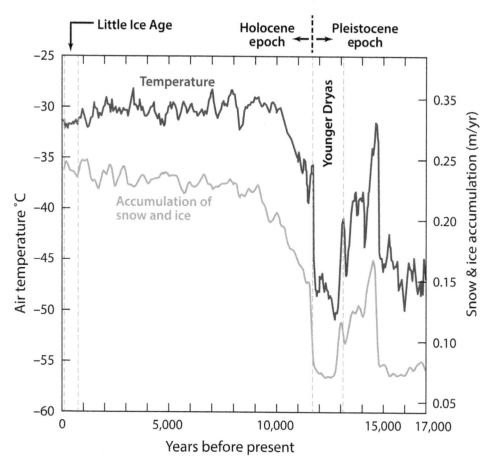

Figure 40.2 Thermohaline circulation shutdown and Younger Dryas cooling
Ice core data. Note the warming of the temperature toward the end of the last glacial period until the Gulf Stream (part of the thermohaline circulation) shut down, creating significant declines in temperature in the Northern hemisphere, followed by rebound to a warming trend once the circulation redeveloped. Also note the huge decrease in temperature compared with that of the Little Ice Age.
Source: USDOI & USGS (2016).

A long-noted thesis maintains that ice sheets holding the huge, former Lake Agassiz in central Canada gave way, generating a gigantic flood of freshwater in to the North Atlantic, shutting down the THC/AMOC (Broecker 2006). The subsequent loss or reduction of warm surface waters to the North Atlantic regenerated cold climates. It took nearly a millennium for the Earth system to reconstruct the circulation and its warming impacts. Current evidence indicates that the AMOC is slowing (Bryden, Longworth & Cunningham 2005; Caesar *et al.* 2018; Caesar *et al.* 2021; Thornalley *et al.* 2018), signalling a possible shutdown in the circulation (Latif *et al.* 2006). This observation is contested (Renssen *et al.* 2015), however, and modelling suggests that the Younger Dryas involved multiple causes, of which the slowing of the circulation was only one (Link & Tol 2009). Yet another explanation involves the Younger Dryas impact hypothesis in which a meteorite penetrating the surface of the Earth may have been the cause (Powell 2022), although the impact appears to have taken place after the initial onset of the cooling event (Cheng *et al.* 2020).

Should the THC/MOC shut down or slow considerably, conditions that appear to be in motion (Boers 2021; Dakos *et al.* 2008), the environmental consequences would be significant and involve much more than colder climates in the northern Atlantic and adjacent continents, including global telecommunications (Orihuela-Pinta, England & Taschetto 2022). It would affect the lag time in returning dissolved greenhouse gases back to the atmosphere, as well as have negative impacts on fisheries (Link & Tol 2009) and marine ecosystems. These environmental and economic impacts are questioned, however, given the potential offset of cooling in the face of current global warming (Anthoff, Estrada & Tol 2016).

References

Anthoff, D., F. Estrada & R. Tol (2016). "Shutting down the thermohaline circulation". *American Economic Review* 106(5): 602–06. doi.10.1257/aer.p20161102.

Avsa (2009). "Conveyor belt". commons.wikimedia.org/wiki/File:Conveyor_belt.svg.

Boers, N. (2021). "Observation-based early-warning signals for a collapse of the Atlantic meridional overturning circulation". *Nature Climate Change* 11: 680–8. doi.org/101038/s41558-021-01097-4.

Broecker, W. (2006). "Was the Younger Dryas triggered by a flood?" *Science* 312(5777): 1146–48. doi:10.1126/science.1123253.

Bryden, H., H. Longworth & S. Cunningham (2005). "Slowing of the Atlantic meridional overturning circulation at 25° N". *Nature* 438(7068): 655–7. doi.org/10.1038/nature04385.

Caesar, L. *et al.* (2018). "Observed fingerprint of a weakening Atlantic Ocean overturning circulation". *Nature* 556(7700): 191–6. doi.org/10.1038/s41586-018-0006-5.

Caesar, L. *et al.* (2021). "Current Atlantic meridional overturning circulation weakest in last millennium". *Nature Geoscience* 14: 118–20. doi.org/10.1038/s41561-021-00699-z.

Cheng, H. *et al.* (2020). "Timing and structure of the Younger Dryas event and its underlying climate dynamics". *Proceedings of the National Academy of Sciences* 117(38): 23408–17. doi.org/10.1073/pnas.2007869117.

Dakos, V. *et al.* (2008). "Slowing down as an early warning signal for abrupt climate change". *Proceedings of the National Academy of Sciences* 105(38): 14308–12. doi10.1073pnas.0802430105.

Gierz, P. (2015). "Response of Atlantic overturning to future warming in a coupled atmosphere–ocean-ice sheet model". *Geophysical Research Letters* 42(16): 6811–18. doi.org/10.1002/2015GL065276.

Huang, J. *et al.* (2020). "Sources and upstream pathways of the densest overflow water in the Nordic Seas". *Nature Communications* 11(1): 5389. doi.10.1038/s41467-020-19050-y.

Latif, M. *et al.* (2006). "Is the thermohaline circulation changing?" *Journal of Climate* 19(18): 4631–7. doi.org/10.1175/JCLI3876.1.

Link, P. & R. Tol (2009). "Economic impacts on key Barents Sea fisheries arising from changes in the strength of the Atlantic thermohaline circulation". *Global Environmental Change* 19(4): 422–33. doi.org/10.1016/j.gloenvcha.2009.07.007.

Orihuela-Pinto, B., England, M. & Taschetto, A. (2022). "Interbasin and interhemispheric impacts of a collapsed Atlantic Overturning Circulation." *Nature Climate Change,* 12: 558–65. doi.org/10.1038/s41558-022-01380-y

Powell, J. (2022). "Premature rejection in science: the case of the Younger Dryas impact hypothesis". *Science Progress* 105(1): 1–43. doi:10.1177/00368504211064272.

Ramstorf, S. (2003). "The concept of the thermohaline circulation". *Nature* 412: 699. doi.org/10.1038/421699a.

Renssen, H. *et al.* (2015). "Multiple causes of the Younger Dryas cold period". *Nature Geoscience* 8(12): 946–9. doi.org/10.1038/ngeo2557.

Semper, S. *et al.* (2020). "The Iceland-Faroe Slope Jet: a conduit for dense water toward the Faroe Bank Channel overflow". *Nature Communications* 11(1): 5390. doi.10.1038/s41467-020-19049-5.

Silvy, Y. *et al.* (2020). "Human-induced changes to the global ocean water masses and their time of emergence". *Nature Climate Change* 10: 1030–36. doi.org/10.1038/s41558-020-0878-x.

Smedsrud, L. *et al.* (2021). "Nordic seas heat loss, Atlantic inflow, and Arctic Sea ice cover over the last century". *Reviews of Geophysics* 60(1): e2020RG000725. doi.org/10.1029/2020RG000725.

Thornalley, D. *et al.* (2018). "Anomalously weak Labrador Sea convection and Atlantic overturning during the past 150 years". *Nature* 556(7700): 227–30. doi.org/10.1038/s41586-018-0007-4.

USDOI & USGS (US Department of Interior and US Geological Survey) (2016). "Temperature variations during the late Pleistocene Epoch and near the beginning of the Holocene Epoch, determined as proxy temperatures from ice cores extracted from the central part of the Greenland ice sheet". pubs.usgs.gov/pp/p1386a/gallery2-fig35.html.

Q41

HAS HUMAN ACTIVITY ACIDIFIED AND DEOXYGENATED THE OCEANS?

Short answer: yes. Carbon emissions acidify the ocean and, with nutrients from land-based activities, decrease marine oxygen, which degrades marine ecosystems and biodiversity.

The natural processes of the exchange of CO_2 at the sea surface and production of calcium carbonate ($CaCO_3$) (e.g. shells) by marine life give rise to variations in pH (measure of acid or basic water-based solution: 0 = most acidic; 14 = most basic or alkaline) by ocean depth. Increases in atmospheric CO_2 lead to an uptake of this compound in the oceans, which decreases the pH, creating acidification (OSB/NRC 2010).

Over the past 200 years, oceans have become about 30 per cent more acidic from a neutral pH (7) condition, assisted by a 30–40 per cent increase of human-induced atmospheric CO_2 (Doney *et al.* 2009; Mora *et al.* 2013), and they continue to acidify by 0.017–0.027 pH units per decade (IPCC 2021) (Fig. 41.1). Acidification involves decreased carbonate ions. It has significant impacts on marine biota – carbonate is essential for skeleton and shell development – food chains and coral growth (Anthony *et al.* 2008; Service 2012). Naturally induced acidification 56 Mya, for example, led to a marine mass extinction (Zachos 2005). The projected increases in ocean acidification have several effects: acidification bleaches (kills) coral, an essential marine ecosystem with high biodiversity (Q59), and affects various algae and shellfish (D'Olivo *et al.* 2019; Radford *et al.* 2021; Raven 2005). Recent research suggests that build-up of CO_2 in the oceans can reach thresholds that generate chemical feedbacks leading to extreme acidification, with large-scale Earth system consequences (Rothman 2019).

In contrast to the natural process of ocean carbon uptake, various human-induced compound pollutants in the ocean are largely novel to the oceans. Together, carbon uptake and nitrogen and phosphorous pollutants deoxygenate (oxygen-deprive) the oceans. Carbon-induced warming of the oceans causes deoxygenation, owing to reduced solubility of oxygen in warm water (Breitburg *et al.* 2018; Oschlies *et al.* 2018). Ocean stratification inhibits the upwelling of nutrients from the ocean's depth, which reduces phytoplankton (microscopic algae) and, hence, photosynthesis, which is estimated to produce about 50 per cent of oxygen globally (Cermeño *et al.* 2008; Limburg *et al.* 2020) (Fig. 41.2). In addition, pollutants deoxygenate coastal zones through the process of eutrophication, a condition created by

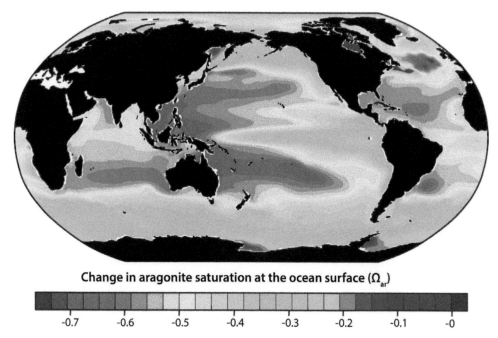

Change in aragonite saturation at the ocean surface (Ω_{ar})

-0.7   -0.6   -0.5   -0.4   -0.3   -0.2   -0.1   -0

Figure 41.1 Acidification of ocean surface waters, 1880–2015

Aragonite is essential for marine organisms to build and maintain shells and skeletons. As aragonite saturation (Ω_{ar}) levels decrease, associated with acidification or decreasing pH, the more difficult for shell building among the organisms. Note that aragonite saturation levels have decreased globally, especially in tropical waters.

Source: WHOI (2016).

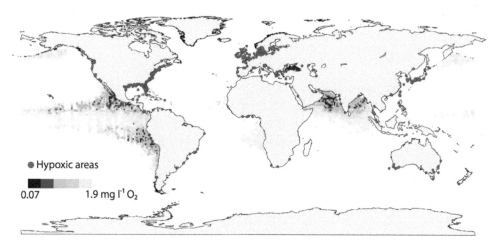

● Hypoxic areas

0.07   1.9 mg l^{-1} O$_2$

Figure 41.2 Ocean areas of deoxygenation and hypoxia

≤2 mg l^{-1} O$_2$ = insufficient oxygen. Hypoxic zones (red dots) are below this level. Blue shading reflects annual rates of loss in O$_2$ levels at 300 m depth in oceans.

Source: Breitburg *et al.* (2018).

Figure 41.3 Nitrogen source areas and hypoxia in the Gulf of Mexico
Nitrogen and phosphorous deliveries through the watersheds, much of them from agricultural fertilizer use, move through the Mississippi River Delta to create a "dead zone" in a major fishery of the Gulf of Mexico.
Source: top: USGS (2014); bottom: The Oil Drum (2013).

excessive nutrients generating dense plant growth that deprives animals of oxygen and may cause hypoxia (Breitburg *et al.* 2018; Rabalais *et al.* 2014), as illustrated in Figure 41.3. Evidence suggests that coastal zone eutrophication may be linked to open ocean deoxygenation (Melzner *et al.* 2013) and that ocean warming may be linked to coastal deoxygenation (Laurent *et al.* 2018). Such impacts add to oxygen-deficient zones that occur through natural processes in some parts of the oceans (Kwiecinski & Babbin 2021).

References

Anthony, K. *et al.* (2008). "Ocean acidification causes bleaching and productivity loss in coral reef builders". *Proceedings of the National Academy of Sciences* 105(45): 17442–6. doi.org/10.1073/pnas.0804478105.

Breitburg, D. *et al.* (2018). "Declining oxygen in the global ocean and coastal waters". *Science* 359(6371). doi:10.1126/science.aam7240.

Cermeño, P. *et al.* (2008). "The role of nutricline depth in regulating the ocean carbon cycle". *Proceedings of the National Academy of Sciences* 105(51): 20344–2034. doi.org/10.1073/pnas.0811302106.

D'Olivo, J. *et al.* (2019). "Deconvolving the long-term impacts of ocean acidification and warming on coral biomineralisation". *Earth and Planetary Science Letters* 526: 115785. doi.org/10.1016/j.epsl.2019.115785.

Doney, S. *et al.* (2009). "Ocean acidification: the other CO_2 problem". *Annual Review of Marine Science* 1: 169–92. doi.org/10.1146/annurev.marine.010908.163834.

IPCC (Intergovernmental Panel on Climate Change) (2021). *IPCC Special Report on the Ocean and Cryosphere in a Changing Climate.* H.-O. Pörtner *et al.* (eds). In press. www.ipcc.ch/srocc/cite-report/.

Kwiecinski, J. & A. Babbin (2021). "A high-resolution atlas of the eastern tropical Pacific oxygen deficient zone". *Global Biogeochemical Cycles* 35(12): e2021GB007001. doi.org/10.1029/2021GB007001.

Laurent, A. *et al.* (2018). "Climate change projected to exacerbate impacts of coastal eutrophication in the northern Gulf of Mexico". *Journal of Geophysical Research: Oceans* 123(5): 3408–26. doi.org/10.1002/2017JC013583.

Limburg, K. *et al.* (2020). "Ocean deoxygenation: a primer". *One Earth* 2(1): 24–9. doi.org/10.1016/j.oneear.2020.01.001.

Melzner, F. *et al.* (2013). "Future ocean acidification will be amplified by hypoxia in coastal habitats". *Marine Biology* 160(8): 1875–88. doi.org/10.1007/s00227-012-1954-1.

Mora, C. *et al.* (2013). "Biotic and human vulnerability to projected changes in ocean biogeochemistry over the 21st century". *PLoS Biology* 11(10): e1001682. doi.org/10.1371/journal.pbio.1001682.

OSB/NRC (Ocean Science Board and National Research Council) (2010). *Ocean Acidification: A National Strategy to Meet the Challenges of a Changing Ocean.* Washington, DC: National Academies Press.

Oschlies, A. *et al.* (2018). "Drivers and mechanisms of ocean deoxygenation". *Nature Geoscience* 11(7): 467–73. doi.org/10.1038/s41561-018-0152-2.

Rabalais, N. *et al.* (2014). "Eutrophication-driven deoxygenation in the coastal ocean". *Oceanography* 27(1): 172–83. doi.org/10.5670/oceanog.2014.21.

Radford, C. *et al.* (2021). "Ocean acidification effects on fish hearing". *Proceedings of the Royal Society B* 288(1946): 20202754. doi.org/10.1098/rspb.2020.2754.

Raven, J. (2005). "Ocean acidification due to increasing atmospheric carbon dioxide". Royal Society Policy Document. London: The Royal Society. royalsociety.org/topics-policy/publications/2005/ocean-acidification/.

Rothman, D. (2019). "Characteristic disruptions of an excitable carbon cycle". *Proceedings of the National Academy of Sciences* 116(30): 14813–22. doi.org/10.1073/pnas.1905164116.

Service, R. (2012). "Marine ecology: rising acidity brings an ocean of trouble". *Science* 337(6091): 146–8. doi:10.1126/science.337.6091.146.

The Oil Drum (2013). "Gulf of hypoxia". theoildrum.com/files/gulfhypoxia.png.

USGS (US Geological Survey) (2014). "Differences in phosphorus and nitrogen delivery to the Gulf of Mexico from the Mississippi River Basin". water.usgs.gov/nawqa/sparrow/gulf_findings/hypoxia.html.

WHOI (Woods Hole Oceanographic Institution) (2016). "Climate change indicators: ocean acidity". www.epa.gov/climate-indicators/climate-change-indicators-ocean-acidity#ref11.

Zachos, J. (2005). "Rapid acidification of the ocean during the Paleocene–Eocene thermal maximum". *Science* 308(5728): 1611–15. doi:10.1126/science.1109004.

Q42

HAS HUMAN ACTIVITY POLLUTED THE OCEANS GLOBALLY IN NEW WAYS?

Short answer: yes. Oceans have become a dumping ground for human litter, of which plastics and noise are escalating in scale, with major impacts on marine biota.

Nitrogen, phosphorus and other elements associated with fertilizer use and industrial wastes delivered to surface and groundwater have long been examined in terms of their polluting effects on oceans **(Q41)**. Human-induced oil spills and seepages have long been noted but not necessarily established in global amount. New studies identified 450,000 oil slicks between 2014 and 2019, of which 94 per cent were anthropogenic in origin (Dong *et al.* 2022). Added to these pollutants, plastics have become a more recent threat. Noise pollution, in contrast, emanates from direct marine activities, foremost massive ship transport, sonar use, seismic surveys and marine mining, such as oil and gas production (Nowacek *et al.* 2015; Simmonds *et al.* 2014).

While the magnitude and spatial extent of all pollutants are increasing throughout the world's oceans, the threat from plastics is a growing concern. This includes intact products, such as plastic waste items (e.g. bottles), and various-sized plastic particles, including micro or nano plastics transmitted through the atmosphere to the oceans (Allen *et al.* 2022). This pollution follows from the explosive use of disposable plastic products worldwide delivered to the oceans largely by rivers, or plastic particles delivered similarly, or through the atmosphere,

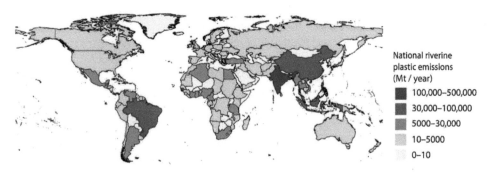

National riverine
plastic emissions
(Mt / year)

- 100,000–500,000
- 30,000–100,000
- 5000–30,000
- 10–5000
- 0–10

Figure 42.1 National plastic emissions delivered to oceans through rivers
Mt = million metric tonnes or megaton.
Source: Meijer *et al.* (2021).

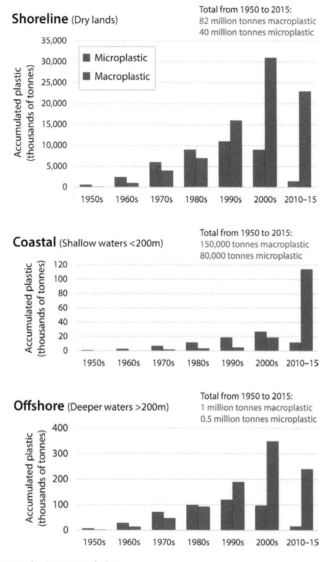

Figure 42.2 Ocean plastic accumulation
Source: Ritchie (2018).

to the oceans (Fig. 42.1) (Meijer *et al.* 2021). The accumulation of plastics in the Earth system is 0.26 PgC yr^{-1} (Stubbins *et al.* 2021) of which the majority appears to accumulate in the oceans. Plastics constitute an estimated 60–80 per cent of marine litter (Xanthos & Walker 2017), often concentrated with other litter by ocean gyres (surface wind circulations), creating ocean garbage patches or trash vortexes (Evans-Pughe 2017). The estimated range in the amount and rate of plastic pollution is large, but strongly indicative of major increases. In 2010, 4.8–12.7 million tonnes of plastic found its way to the oceans (Jambeck *et al.* 2015), contributing to an estimated 5.25 trillion plastic particles weighing 269,000 tons (Xanthos & Walker 2017) (Fig. 42.2).

Another estimate for 2013 enlarged the range from 10,000 to 40,000 tonnes of debris from plastics floating on the ocean surface (Cózar *et al.* 2014). About 14 million tonnes of microplastics are projected to reside on the ocean floor (Barrett *et al.* 2020). The 2017 rate of plastic emissions ranged from 1.15 to 2.41 million tonnes yr^{-1} (Lenreton *et al.* 2017) and 0.44 to 2.75 million tonnes yr^{-1} (Schmidt, Krauth & Wagner 2018). The overwhelming amount of all plastic particles in the oceans are degraded to less than 1 cm – microplastics – which are circulated globally and find their way into marine biota and seabirds, commonly leading to internal suffocation. Fish catch, in turn, appears to lead to human consumption of plastic through ingesting fish (MacLeod *et al.* 2021; Miranda & de Carvalho-Souza 2016).

Noise pollution affects marine faunae by the alteration of their acoustic environment (Duarte *et al.* 2021). Sound travels much farther in water than on land and affects the pressure and particle motion on marine fauna, which are extremely sensitive to noise pollution (Fig. 42.3). Between 1950 and 2000, the overall sounds travelling through oceans

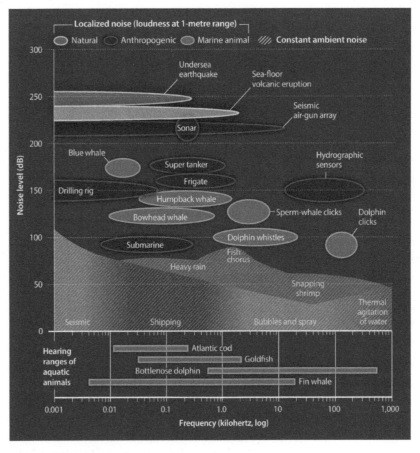

Figure 42.3 Under ocean sound

dB = decibel.

Source: Jones (2019). Reprinted by permission from Springer Nature. © (2019).

are estimated to have increased by about 3 dB per decade, creating a ten-year period to double (Jones 2019). These impacts may affect fauna anatomy, physiology and behaviour (Kunc, McLaughlin & Schmidt 2016).

References

Allen, D. *et al.* (2022). "Microplastics and nanoplastics in the marine–atmosphere environment". *Nature Reviews Earth and Environment.* doi.org/10.1038/s43017-022-00292-x.

Barrett, J. *et al.* (2020). "Microplastic pollution in deep-sea sediments from the Great Australian Bight". *Frontiers in Marine Science* 7: 576170. doi.org/10.3389/fmars.2020.576170.

Cózar, A. *et al.* (2014). "Plastic debris in the open ocean". *Proceedings of the National Academy of Sciences* 11(28): 10239–44. doi.org/10.1073/pnas.1314705111.

Dong, Y. *et al.* (2022). "Chronic oiling in global oceans". *Science* 376(6599): 1300–4. doi10.1126/science.abms5940

Duarte, C. *et al.* (2021). "The soundscape of the Anthropocene ocean". *Science* 371(6529): eaba4658. doi:10.1126/science.aba4658.

Evans-Pughe, C. (2017). "All at sea: cleaning up the great Pacific garbage patch". *Engineering & Technology* 12(1): 52–5. doi:10.1049/et.2017.0105.

Jambeck, J. *et al.* (2015). "Plastic waste inputs from land into the ocean". *Science* 347(6223): 768–71. doi:10.1126/science.1260352.

Jones, N. (2019). "Ocean uproar: saving marine life from a barrage of noise". *Nature* 568: 158–61. doi:10.1038/d41586-019-01098-6.

Kunc, H., K. McLaughlin & R. Schmidt (2016). "Aquatic noise pollution: implications for individuals, populations, and ecosystems". *Proceedings of the Royal Society B: Biological Sciences* 283(1836): 20160839. doi.org/10.1098/rspb.2016.0839.

Lenreton, L. *et al.* (2017). "River plastic emission to world's oceans". *Nature Communications* 8: 15611. doi.org/10.1038/ncomms15611.

MacLeod, M. *et al.* (2021). "The global threat from plastic pollution". *Science* 373: 61–5. doi:10.1126/science.abg5433.

Meijer, L. *et al.* (2021). "More than 1000 rivers account for 80 per cent of global riverine plastic emissions into the ocean". *Science Advances* 7(18): eaaz5803. doiI:10.1126/sciadv.aaz5803.

Miranda, D. & G. de Carvalho-Souza (2016). "Are we eating plastic-ingesting fish?" *Marine Pollution Bulletin* 103(1/2): 109–14. doi.org/10.1016/j.marpolbul.2015.12.035.

Nowacek, D. *et al.* (2015). "Marine seismic surveys and ocean noise: time for coordinated and prudent planning". *Frontiers in Ecology and the Environment* 13(7): 378–86. doi.org/10.1890/130286.

Ritchie, H. (2018). "Plastic pollution". ourworldindata.org/plastic-pollution.

Schmidt, C., T. Krauth & S. Wagner (2018). "Correction to export of plastic debris by rivers into the sea". *Environmental Science and Technology* 52(2): 927. doi.org/10.1021/acs.est.7b06377.

Simmonds, M. *et al.* (2014). "Marine noise pollution-increasing recognition but need for more practical action". *Journal of Ocean Technology* 9(1): 71–90. www.wellbeingintlstudiesrepository.org/acwp_ehlm/9/.

Stubbins, A. *et al.* (2021). "Plastics in the Earth system". *Science* 373: 51–5. doi:10.1126/science.abb0354.

Xanthos, D. & T. Walker (2017). "International policies to reduce plastic marine pollution from single-use plastics (plastic bags and microbeads): a review". *Marine Pollution Bulletin* 118(1/2): 17–26. doi.org/10.1016/j.marpolbul.2017.02.048.

HUMAN CHANGES TO THE ATMOSPHERE

The atmosphere constitutes the gases (N = 78 per cent; O_2 = ~21 per cent) surrounding the Earth, held in place by the gravitational forces of the planet. Its many layers serve several Earth system functions critical to the biosphere (Fig. V.1). These layers range from dense levels of gases held within the troposphere to the sparse but important gases approaching the boundary with "space" – the Kármán line about 100 km from the Earth's surface that protects our planet from most meteoroids entering the atmosphere. Human activity has altered the conditions or states of the troposphere (surface to 12 km) and stratosphere

(12–50 km), with significant consequences for the Earth system. Our emissions have not only increased greenhouse gases in the lower layers of the atmosphere, they are actually shrinking the thickness of the stratosphere by ~400 m since the 1980s as the troposphere moves upward. By 2080, the stratosphere may contract by an additional kilometre (Pisoft *et al.* 2021).

These and other consequences have long been ignored in human history because the atmosphere could not be owned or controlled, and – like the oceans – prompted the view that whatever we delivered to it could be absorbed and diffused by its vastness, with minimal, if any, impacts on the functioning of the environment. While some debate exists over the history of human impacts on the atmosphere, it is clear that the industrialized era, especially from the middle of the twentieth century forward, has strongly amplified atmospheric pollution and emissions of chemical compounds. These emissions, in turn, challenge the functioning of the Earth system to maintain the climatic conditions on which the global economy has been built (Steffen *et al.* 2011).

Human impacts notwithstanding, long-term cycles in Earth–Sun relationships and episodic, big bang events generate atmospheric conditions that change climates. Over a 400-million-year cycle, for example, the Earth experiences coolhouse (or icehouse) and warmhouse (or hothouse) conditions (Hoffmann & Schrag 2000), in which greenhouse

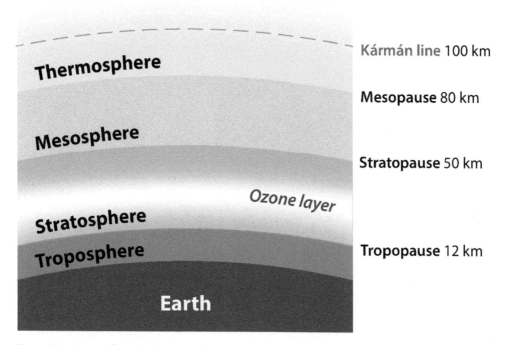

Figure V.1 Layers of the Earth's atmosphere
Source: author.

gases decrease and increase, respectively, owing to various dynamics (Smith & Pickering 2003). A brief review of some of these dynamics helps us to understand the current human impacts on the atmosphere and climate.

The 100,000-year cycle of ice ages during the Quaternary Period (**Q2**) is somewhat less dramatic than coolhouse–warmhouse Earth but holds profound Earth system consequences regarding glacial and interglacial periods. This Milankovitch cycle follows from Earth–Sun relations in which solar radiation is subtly changed by the position of our planet in its elliptical orbit (eccentricity), the shifting of the tilt of its axis (obliquity) and the wobble in its spin (precession) (Fig. VI.2) (Spiegel *et al.* 2010). The cycle varies the amount of solar irradiance by ~3.4 W/m² (Lean & Rind 2001; Raymo & Huybers 2008), sufficient to generate the cooler–warmer conditions of the glacial and interglacial periods. Importantly, the recent climate warming under way may involve a more chaotic cycle and even constitute a threshold in the Earth system in which the Milankovitch cycle disappears (Caccamo & Magazù 2021).

The amount of solar energy entering the Earth's atmosphere – the solar constant – averages 1,267 W/m² and also varies subtly by an 11-year cycle of the Sun's magnetic activity. This activity creates sunspots, or areas of cooler temperatures (ca. 3,526°C vs 5,527°C), appearing as dark spots on the surface of the Sun. Sunspots generate solar flares (or solar storms), intense outbursts of energy released as electrically charged particles, increasing the solar constant by ~0.1 per cent (Bard & Frank 2006; Lean & Rind 2001). This short-term cycle and the bursts of energy, however, are not associated with current climate warming

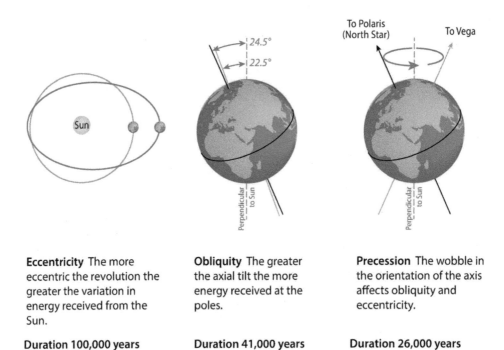

Eccentricity The more eccentric the revolution the greater the variation in energy received from the Sun.

Duration 100,000 years

Obliquity The greater the axial tilt the more energy received at the poles.

Duration 41,000 years

Precession The wobble in the orientation of the axis affects obliquity and eccentricity.

Duration 26,000 years

Figure V.2 The Milankovitch cycle
Source: adapted from socratic.org/questions/what-are-milankovic-cycles-and-how-do-they-contribute-to-climate-change.

as evidenced by the non-association between the variations in solar energy and the climate warming in our atmosphere (Fig. V.3) (NAS/RS 2020).

Finally, the atmosphere protects the planet from meteors, more often than not by evaporating them before they reach the Earth's surface. Sufficiently large meteors, however, may penetrate to the surface, becoming a meteorite. Large meteorites may hit or explode near the surface with multi-megaton impacts, creating firestorms over large areas and spraying enough surface particles (asteroid dust) in the atmosphere to block large amounts of solar radiation for extended periods, cooling the Earth and affecting biota (Borovička *et al.* 2013). The Cretaceous–Paleogene (*K–Pg*; formerly Cretaceous–Tertiary, *K–T*) extinction 65 Mya, for example, in which non-avian dinosaurs disappeared, followed from a large meteorite

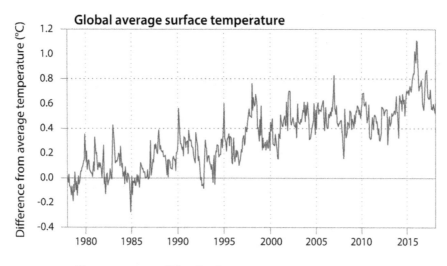

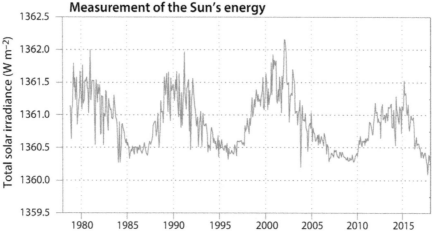

Figure V.3 Earth's warming and solar energy received, 1975–2020
Over the past 45 years the average temperature of Earth has increased, whereas the amount of irradiance (energy) from the Sun has waxed and waned.
Source: NAS/RS (2020).

hitting northwestern Yucatán (Mexico) and creating a crater about 180 km across (Schulte *et al.* 2010) (Section VI). Volcanic aerosols may have also played a role in this event, but recent work indicates that asteroid dust (aerosols) provoking significant cooling was the main driver of extinction (Chiarenza *et al.* 2020).

With the exception of multiple nuclear explosions, human impacts on the atmosphere do not have big bang consequences like those of large meteorites, the cycling of Earth–Sun relationships, or the 400-million-year coolhouse–warmhouse cycle. Our changes in states and flows of the Earth system, however, have created conditions in the atmosphere, in tandem with land and ocean feedbacks, that are changing global climates. The intensity and long-term persistence of the changes constitute a challenge for humanity.

References

Bard, E. & M. Frank (2006). "Climate change and solar variability: what's new under the sun?" *Earth and Planetary Science Letters* 248(1/2): 1–14. doi.org/10.1016/j.epsl.2006.06.016.

Borovička, J. *et al.* (2013). "The trajectory, structure and origin of the Chelyabinsk asteroidal impactor". *Nature* 503(7475): 235–7. doi.org/10.1038/nature12671.

Caccamo, M. & S. Magazù (2021). "On the breaking of the Milankovitch Cycles triggered by temperature increase: the stochastic resonance response". *Climate* 9(4): 67. doi.org/10.3390/cli9040067.

Chiarenza, A. *et al.* (2020). "Asteroid impact, not volcanism, caused the end-Cretaceous dinosaur extinction". *Proceedings of the National Academy of Sciences* 117(29): 17084–93. doi.org/10.1073/pnas.2006087117.

Hoffmann, P. & D. Schrag (2000). "Snowball Earth". *Scientific American* 282(1): 68–75. doi:10.1038/scientificamerican0100-68.

Lean, J. & D. Rind (2001). "Earth's response to a variable sun". *Science* 292(5515): 234–6. doi:10.1126/science.1060082.

NAS/RS (National Academy of Sciences/Royal Society) (2020). *Climate Change: Evidence and Causes, Update 2020.* Washington, DC: National Academy Press. doi.org/10.17226/25733.

Pisoft, P. *et al.* (2021). "Stratospheric contraction caused by increasing greenhouse gases". *Environmental Research Letters* 16(6): 064038. doi.org/10.1088/1748-9326/abfe2b.

Raymo, M. & P. Huybers (2008). "Unlocking the mysteries of the ice ages". *Nature* 451: 284–5. doi.org/10.1038/nature06589.

Schulte P. *et al.* (2010). "The Chicxulub asteroid impact and mass extinction at the Cretaceous-Paleogene boundary". *Science* 327(5970): 1214–18. doi:10.1126/science.1177265.

Smith, A. & K. Pickering (2003). "Oceanic gateways as a critical factor to initiate icehouse Earth". *Journal of the Geological Society* 160(3): 337–40. doi.org/10.1144/0016-764902-115.

Spiegel, D. *et al.* (2010). "Generalized Milankovitch cycles and long-term climatic habitability". *Astrophysical Journal* 721(2): 1308. doi:10.1088/0004-637X/721/2/1308.

Steffen, W. *et al.* (2011). "The Anthropocene: from global change to planetary stewardship". *Ambio* 40(7): 739–61. doi:10.1007/s13280-011-0185-x.

HAS HUMAN ACTIVITY ALTERED THE FUNCTIONING OF THE TROPOSPHERE AND STRATOSPHERE?

Short answer: yes. Emissions of gaseous molecules and compounds and aerosols challenge various functions of the atmosphere, warming the troposphere, degrading the ozone layer in the stratosphere, and changing the operation of the Earth system with significant health implications for the biosphere.

The troposphere (Earth's surface to ~12 km above it) is strongly affected by numerous human activities: near-surface pollution by wind-blown dust from agriculture and arid land degradation; longwave radiation from changes in albedo owing to land-cover changes; ozone and other pollution from fossil fuel use; acid rain from energy and industrial production (Fig. 43.1); and greenhouse gases (GHGs) from fossil fuel burning at large (Steffen *et al.* 2011).

In contrast to its role in the stratosphere **(Q52)**, ozone (O_3) in the troposphere not only serves as a GHG, but is a health-risk pollutant, generating smog, especially in cities with dense vehicular traffic (Cooper *et al.* 2014). While only about 10 per cent of atmospheric O_3 resides in the troposphere, it is harmful for individuals with respiratory problems and to plant growth (Fann *et al.* 2015). Fossil fuel burning and agriculture also pollute the troposphere via particulate matter (PM). Fine $PM_{2.5}$ is especially problematic for respiratory health among humans. Commonly recognized as a city issue owing to vehicular traffic, agricultural $PM_{2.5}$ in the United States is calculated to be involved in 17,900 deaths ascribed to air-quality issues, 88 per cent of which emanate from food production alone (Domingo *et al.* 2021). In China, a staggering 30.8 million people died prematurely owing to air pollution, foremost $PM_{2.5}$, between 2000 and 2016, according to one assessment (Liang *et al.* 2020).

Acidic precipitation (low pH) is created by the reaction of sulphur dioxide (SO_2) and nitrogen oxide (NO_x) emitted from industrial activities with water molecules (Fig. 43.1). Acid rain damages vegetation, surface water, including ocean acidification **(Q41)**, soils and insects, corrodes steel structures, weathers stone, and has human health impacts (Menz & Seip 2004). Coal-burning power plants have historically emitted significant amounts of SO_2 and NO_x, generating major downwind forest damage, such as in the northeastern United States, northeastern Europe and elsewhere worldwide where significant coal burning existed

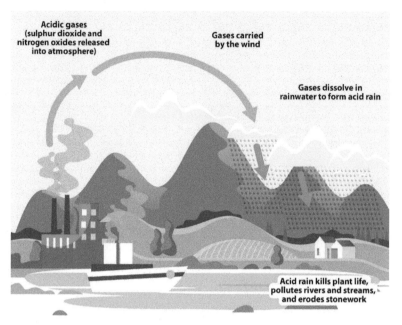

Figure 43.1 Acid rain

Industrial activity, especially at power plants, emits acidic gases into the atmosphere. These gases are carried downwind, dissolving in precipitation and creating acid rain with various consequences, including forest destruction.

Source: photo: Wikimedia Commons (2006); graphic: adapted from sites.google.com/site/bio151theteam/artifact-5.

before the use of smokestack scrubbers that reduced the emissions in question (Srivastava, Jozewicz & Singer 2001).

GHGs affect atmospheric layers in different ways (Santer 2013). These gases, for example, concentrate in the troposphere where they capture reradiated energy (longwave radiation) and emit it within the troposphere and to the Earth's surface (**Q44**). Overall, the emitted compounds and other factors differentially heat and cool the troposphere, captured in the IPCC's assessment of the different radiative forcings (Fig. 43.2), which are detailed in other questions in this section (Myhre & Shindell 2013). In the troposphere, GHGs overwhelmingly warm, whereas aerosols commonly cool (**Q45**); the energy from the Sun warms. Combined human activities have probably increased the effective radiated forcing (or energy transfer) by 2.3 W/m² (Myhre & Shindell 2013), increasing the Earth's average temperature (**Q46, Q47**). This forcing has increased rapidly since 1970.

In the stratosphere, complex interactions of radiation with CO_2, perhaps in the context of the presence of aerosols and ozone, leads to cooling at this high elevation (Goessling & Bathiany 2016). In addition, synthetic compounds, in this case chlorofluorocarbons (CFCs) and halocarbons, threaten the ozone layer (20–30 km), which protects life from ultraviolet radiation (**Q53**). Importantly, ozone is also a GHG and its reduction or addition to the stratosphere presents a cooling or warming effect, respectively.

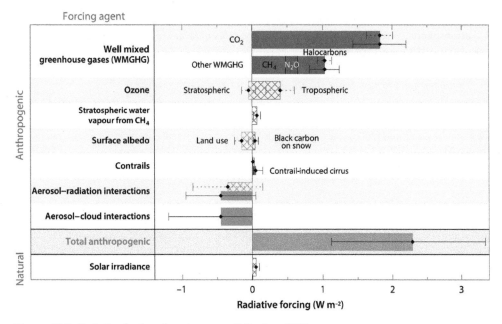

Figure 43.2 Radiative forcings from human activity since 1750

Radiative forcing refers to energy (W/m²) generated by forcing agents that heat (+) or cool (−) the atmosphere. Hatching means the agent heats or cools by conditions. The black lines are error bars. Note the small and large radiative forcing of solar irradiance (from the Sun) and the various agents from human activity, respectively.

Source: Myhre & Shindell (2013).

References

Cooper, O. *et al.* (2014). "Global distribution and trends of tropospheric ozone: an observation-based review". *Elementa: Science of the Anthropocene* 2: 29. doi:10.12952/journal.elementa.000029.

Domingo, N. *et al.* (2021). "Air quality-related health damages of food". *Proceedings of the National Academy of Sciences* 118(20): e2013637118. doi.org/10.1073/pnas.2013637118.

Fann, N. *et al.* (2015). "The geographic distribution and economic value of climate change-related ozone health impacts in the United States in 2030". *Journal of the Air & Waste Management Association* 65(5): 570–80. doi.org/10.1080/10962247.2014.996270.

Goessling, H. & S. Bathiany (2016). "Why CO_2 cools the middle atmosphere – a consolidating model perspective". *Earth System Dynamics* 7(3): 697–715. doi.org/10.5194/esd-7-697-2016.

Liang, F. *et al.* (2020). "The 17-y spatiotemporal trend of PM 2.5 and its mortality burden in China". *Proceedings of the National Academy of Sciences* 117(41): 25601–08. doi.org/10.1073/pnas.1919641117.

Menz, F. & H. Seip (2004). "Acid rain in Europe and the United States: an update". *Environmental Science & Policy* 7(4): 253–65. doi.org/10.1016/j.envsci.2004.05.005.

Myhre, G. & D. Shindell (2013). "Anthropogenic and natural radiative forcing". In T. Stocker *et al.* (eds), *Climate Change 2013: The Physical Science Basis. Contribution of Working Group I to the Fifth Assessment Report of the Intergovernmental Panel on Climate Change.* Cambridge: Cambridge University Press.

Santer, B. (2013). "Human and natural influences on the changing thermal structure of the atmosphere". *Proceedings of the National Academy of Sciences* 110(43): 17235–40. doi.org/10.1073/pnas.1305332110.

Srivastava, R., W. Jozewicz & C. Singer (2001). "SO_2 scrubbing technologies: a review". *Environmental Progress* 20(4): 219–28. doi.org/10.1002/ep.670200410.

Steffen, W. *et al.* (2011). "The Anthropocene: from global change to planetary stewardship". *Ambio* 40(7): 739–61. doi.org/10.1007/s13280-011-0185-x.

Wikimedia Commons (2006). "Acid rain woods". commons.wikimedia.org/wiki/File:Acid_rain_woods1.JPG.

HAS HUMAN ACTIVITY INCREASED GREENHOUSE GASES IN THE ATMOSPHERE?

Short answer: yes. Human-added greenhouse gases to the Earth system increase the capacity of the atmosphere and oceans to absorb and reradiate (longwave) energy, warming the troposphere and oceans.

A greenhouse permits solar (shortwave) radiation to penetrate its glass cover to be absorbed by materials inside it. These materials reradiate longwave radiation that cannot penetrate the glass, warming the air within the structure. Greenhouse gases (GHGs) – molecules and compounds – operate differently. They permit shortwave radiation to penetrate to the Earth's surface, but capture the returning longwave radiation before reradiating it back into the atmosphere and planet's surface, warming the Earth (Fig. 44.1). Without their natural presence, the average temperature of the Earth (**Q46**) would be around −18°C (~0°F). Biogeochemical cycling of elements and compounds through the Earth system (**Q4**) has maintained a relatively consistent balance of GHGs during the Holocene, until the Anthropocene.

Human activity has changed the surface of Earth, it has mined and burned fossil fuels, and invented synthetic compounds that add significant amounts of GHGs to the Earth system. These gases, in turn, increase the capacity of the atmosphere to trap longwave radiation and warm the Earth – the principal process of human-induced climate change. The characteristics of the major GHGs are listed in Tables 44.1 and 44.2. Other GHGs exist, such as halogens, but the overwhelming majority of current global warming is caused by the gases listed. Their overall contribution to human-induced global warming is created by their individual abundance, warming capacity or global warming potential (GWP), and atmospheric lifetime (emission to removal) (Text box 44.1). The GWP is measured as the amount of energy absorbed from 1 tonne of a GHG relative to 1 tonne of carbon dioxide (CO_2) over a period of time, 20 years in Table 44.1. For example, a GWP of 200 means that 1 tonne of a GHG is equal to 200 tonnes of CO_2 over a 20-year lifespan (Blasing 2016; Kiehl & Trenberth 1997; Myhre & Shindell 2013).

Water vapour (H_2O) is a major part of the hydrological cycle (**Q32**), largely entering the atmosphere from surface water evaporation and transpiration from vegetation. It remains in

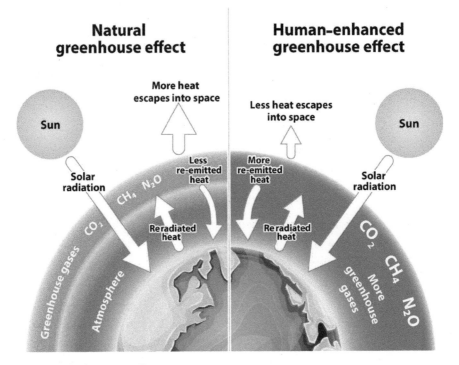

Figure 44.1 Greenhouse gas effect

Solar radiation – shortwave radiation – passes through greenhouse gases. It is absorbed by different phenomena (e.g. land and oceans) and transformed to longwave radiation which is reradiated into the atmosphere. Greenhouse gases absorb and reradiate this radiation through the atmosphere and to the Earth's surface, although some gases escape to space. The more greenhouse gases in the atmosphere (right panel), the more longwave radiation is captured, retaining more heat in the Earth system.

Source: adapted from NPS (2020).

the atmosphere for only a short time before being precipitated to the Earth's surface, but is overwhelmingly the most abundant GHG. Its GWP has yet to be endorsed by international science conventions, although at least one estimate provides a 20-year GWP of −0.004 to +0.002 (Sherwood, Dixit & Salomez 2018). This low GWP is affected by the "cooling" role of water droplets and ice in clouds serving as aerosols (**Q45**) that reflect solar (shortwave) radiation. The amount of water vapour in the troposphere is directly affected by various human activities, foremost directly through irrigation and storage (reservoirs and human-made lakes) and indirectly by warming the atmosphere and aerosols. Increasing temperature allows the troposphere to hold more water vapour, but also generates more clouds.

Carbon dioxide (CO_2) is the most abundant of the GHGs other than H_2O, with atmospheric lifespans in the hundreds of years. It is part of the carbon cycle, naturally emitted from marine and terrestrial biota and plant decomposition. To these emissions, the human activities involving the use of fossil fuels, deforestation and other land uses add large amounts of CO_2 to the troposphere and stratosphere, where they are more and less dense, respectively. Fossil fuels constitute carbon mined from subterranean storage (sinks), and are emitted by

Text box 44.1 Global warming potential metrics

Global warming potential (GWP) is a metric employed to compare the warming capacity of different greenhouse gases owing to their different power to heat the atmosphere. The original GWP calculation does not account for the lifetime of the gas in the atmosphere, however. A new metric GWP* attempts to add this lifetime component to the calculation, improving the comparisons among the gases (Lynch *et al.* 2020). It does so by comparing the pulse emissions of a long-lived gas with the continuous emissions of shorter-lived gases. The GWP* results illustrate how powerful gases, such as CH_4, with short lives in the atmosphere evaluated over long periods of time compare to different levels of less powerful but long-lived gases, such as CO_2. Such information provides insights about mitigating strategies for global warming.

Table 44.1 Major greenhouse gases, sources and global warming potential

GHG (rank)	Sources (N = natural; H = human-generated)	GWP (20 yr; see text for explanation)
H_2O (1)	N = part of hydrologic cycle	To be determined
CO_2 (2)	N = part of carbon cycle	1
	H = burning fossil fuels and deforestation	
CH_4 (3)	N = part of methane cycle	872–960
	H = agriculture-livestock, energy production, landfills	
N_2O (4)	N = part of nitrogen cycle	264–289
	H = agricultural fertilizers; fossil fuel burning	
O_3 (5)	N = continuously regenerated in stratosphere	62–69
	H = mining and industrial production; transportation	
CFCs (6)	H = industrial uses, refrigerants	6,100–10,200
HCFCs (7)	H = industrial uses, refrigerants	3,300–1,200

Source: Myhre & Shindell (2013).

burning them as an energy source, amplifying the amount of carbon in the cycle. CO_2 is also released from the removal of vegetation (e.g. deforestation) and via cultivation, including that of soil carbon, constituting a significant proportion of human-induced emissions of the gas. About 127 ppm of CO_2 has been added to the atmosphere since pre-industrial times, with sustained, rapid increases since measurements began in the 1960s (Fig. 44.2). In May 2020, the amount of CO_2 recorded at the Mauna Loa Observatory on the island of Hawaii was 414.7 ppm. CO_2 lasts in the atmosphere for hundreds of years, and has resulted in an increase of one m² of the atmosphere holding almost 2 W more of energy (heat) (Table 44.1). It must be recognized that the oceans take up most of the CO_2 emitted by human activities, reducing the amount in the atmosphere, but ultimately returning it after an extended stay in the marine environment.

Historically, the United States and western Europe, those entities that entered the Industrial Revolution early, have contributed the most CO_2 to the atmosphere, or about 47 per cent of the global cumulative total (Fig. 44.3). Throughout the twentieth century and today, the United States has been the largest emitter of CO_2 per capita, although its emission

Table 44.2 Change in major greenhouse gas concentrations, atmospheric lifetime and increased radiative forcing

Gas	Pre-1750 concentration*	Recent tropospheric concentration*	Atmospheric lifetime (yrs)	Increased radiative forcing (W/m²)**
Water vapour (H₂O)				
Carbon dioxide (CO₂)	280 ppm	407.4 ppm (2020)	100–300	1.94
Methane (CH₄)	722 ppb	1866 ppb (2019)***	12.4	0.50
Nitrous oxide (N₂O)	270 ppb	331.1 (2018)	121	0.20
Ozone (O₃)	237 ppb	337 (2016)	hours–days	0.40
Chlorofluorocarbons (CFCs)	0	72–516 ppt (2016)	45–100	0.022–0.166
Hydrochlorofluorocarbons (HCFCs)	0	22–233 ppt (2016)	92.–17.2	0.0039–0.049

* Concentration noted by year from sources cited in text.
** Radiative forcing = rate per m² of energy absorbed in atmosphere; increase = change in the rate from pre-1750
*** Reported at 1875 ppb for 2020

Source: Blasing (2016).

rate has dropped in the twenty-first century, as has that of Europe (Fig. 44.4). Currently China is the largest producer of CO_2 emissions and has the most rapid rate of emission increases, followed by India and several other Asian countries.

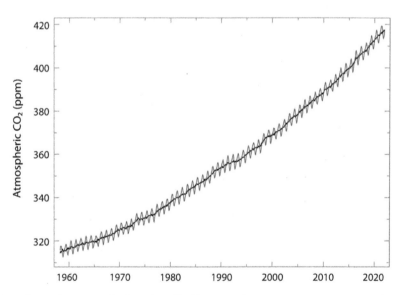

Figure 44.2 Increasing atmospheric carbon dioxide
The iconic measurement of CO_2 from the Mauna Loa Observatory, island of Hawaii, indicating the significant increase of this gas in the atmosphere. The red line is the annual seasonal variance. Note that the observatory, near the peak of Mauna Loa (4,169 m), avoids immediate vicinity emissions of carbon dioxide, capturing the global diffusion of this gas as it spreads across the troposphere.
Source: GML (2020).

Methane (CH_4) cycles naturally in the Earth system and is derived from soil, wetland plant decay, termites, ruminants (multiple-stomach animals), forest fires and oceans (Fig. 44.5). A much more powerful GHG than CO_2 (28 times more powerful over 100 years), it is less abundant and maintains a short lifespan in the atmosphere, of about 12 years. Human activities directly adding CH_4 to the cycle are rice paddies (essentially anthropogenic wetlands), domestic livestock (foremost cattle) and fossil fuel extraction and uses. Indirect human emissions come from melting permafrost and forest fires triggered by climate warming. These activities have increased CH_4 concentrations by 1,144 ppb with an increased radiative forcing of 0.50 W/m^2 (Table 44.2) based on 2019 data (and 1,153 ppb based on 2020 data) (Jackson *et al.* 2020; Myhre & Shindell 2013). Various experiments indicate that warming waters caused by climate change increase the amount of methane emitted much more than previously expected, perhaps by 50–80 per cent (Koffi *et al.* 2020; Zhu *et al.* 2020), which should have consequences for climate model projections. Indeed, data released in 2021 indicate methane reached 1907.2 ppb in October of that year, the majority apparently associated with natural sources, foremost from wetlands (NOAA 2021).

Nitrogen is the most abundant molecule in the Earth system, and it cycles (in various compounds) throughout the Earth system. Nitrogen fixation takes place owing to lightning, legume-soil bacteria and other natural processes. Part of this cycle is nitrous oxide (N_2O), the fourth most important GHG, also more powerful than CO_2 and with atmospheric lifespans in excess of 100 years. About 10.3.3 TgN yr^{-1} is emitted to the atmosphere, of which ~20 per cent is from human activities (Kanter *et al.* 2013). The overwhelming human source is agricultural soil management, linked to ammonia emissions from fertilizer, made abundant by the Haber–Bosch process (Liu *et al.* 2022). About 60 per cent of anthropogenic N_2O emissions come from agricultural lands, of which the total among from these lands may be underestimated, not accounting for pulses during the Spring in cold climates (Del Grasso *et al.* 2022). Other sources include livestock waste, motor vehicles, industrial production,

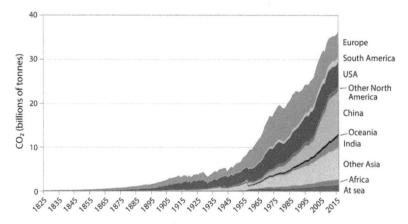

Figure 44.3 Annual carbon dioxide (CO_2) emissions by country/area since 1850
Source: GCP (2020).

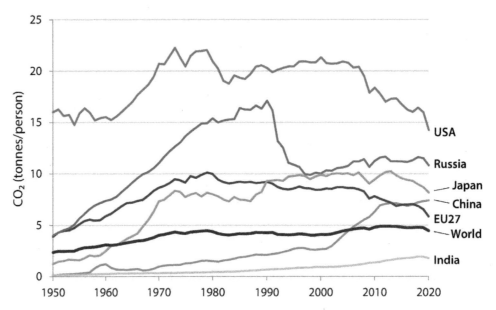

Figure 44.4 Per capita carbon emissions by country/area
Source: GCP (2021a).

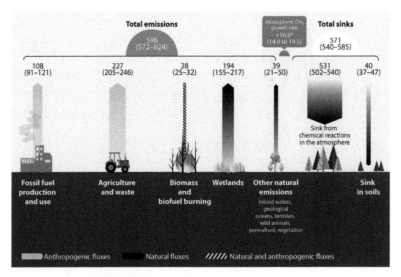

Figure 44.5 Global methane budget, 2007
Source: GCP (2021b).

biomass burning, sewage treatment and landfills (Fig. 44.6) (Bouwman *et al.* 2013). N_2O has increased from 275 to 325 ppb since 1960, increasing the energy capacity of the atmosphere by 0.20 W/m^2 (Table 44.2) (Kanter *et al.* 2013). It also reacts in the stratosphere to reduce ozone. In October 2021 atmospheric N_2O reached 334.6 ppb (NOAA 2021).

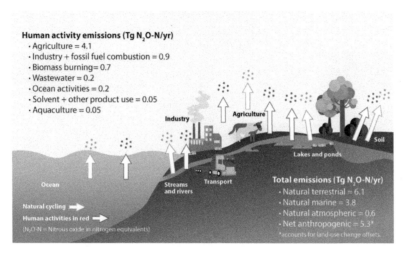

Figure 44.6 Nitrogen emission estimates, 2015
N₂O-N = Nitrous oxide in nitrogen equivalents.
Source: Bouwman *et al.* (2013).

Ozone (O_3) is generated in the stratosphere (**Q53**) and by lightning. It serves as a pollutant in the troposphere, and is also a GHG with a powerful GWP (in the 1,000 range), but has such a short lifetime in the atmosphere that its 20-year GWP is reduced to the 60s. Human activities adding O_3 include motor vehicles, utilities and industrial production in which oxides of nitrogen combine with volatile organic compounds and heat-sunlight. O_3 has increased by 100 ppb, leading to an increased radiative forcing of 0.40 W/m² (Table 44.2).

Chlorofluorocarbons (CFCs) and halocarbons (e.g. hydrochlorofluorocarbons or HCFCs) and others do not occur in nature but are invented, synthetic compounds used for a wide assortment of industrial uses, especially as refrigerants. They are extremely powerful GHGs with significantly different lifetimes (45–100 years). Their concentration varies across the troposphere and stratosphere, raising radiative forcing from 0.0039 to 0.166 W/m², depending on the compound (Table 44.2). Owing to their role in ozone depletion, they were expected to be phased out of use between 2020–30 (but see **Q98**).

Greenhouse gases have feedbacks within the Earth system that have important impacts, although they are often left out of climate models because of their complexity. Those for the carbon cycle exemplify these impacts (Fig. 44.7) (Lade *et al.* 2018). Increasing CO_2 in the atmosphere creates a fertilization impact on plant growth (increased primary productivity on land), whereas in the oceans various processes lead to carbon sinking to depth for extended periods. Either case reduces potential carbon uptake from land and water. Warming climates, for example, depress carbon uptake on land and oceans owing to increased respiration among vegetation and to increased solubility of CO_2. Accounting for these feedbacks lends insights about options to address mitigation strategies (**Q92**).

If human emissions of GHGs were to stop or be reduced in a major way, the Earth system would not immediately cool. Why? As noted above, many GHGs have long lives, remaining in the atmosphere for up to a thousand years. In addition, the process of ocean uptake of CO_2

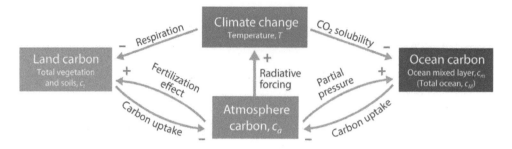

Figure 44.7 Simplified feedbacks in the carbon cycle
Key: c_t – aggregated carbon stock of vegetation and soil; c_a – carbon stock in atmosphere; c_m and c_M – ocean mixed layer and total ocean carbon stocks.
Source: Lade *et al.* (2018).

and storing it in the deep ocean (sink) before it is released to the atmosphere is a slow process. The larger this uptake, however, the larger the ultimate emission down the road. As a result, without some means to reduce atmospheric GHGs, foremost CO_2, the Earth system will remain in a human-induced warming state for the long-term future (NAS/RS, 2020). In addition, modelling experiments indicate that the addition and subtraction of CO_2 emissions is asymmetrical. It takes a larger amount of CO_2 uptake to offset that emitted, an amount that escalates as the CO_2 emissions increase (Zickfeld *et al.* 2021). For this and other reasons, much attention is given to geoengineering, such as new technology to sequester CO_2 from the atmosphere and place it, for example, in deep, subterranean storage (Boyd 2008). To date, the technological capacity to undertake such efforts is insufficient and the costs are high. The Paris Agreement of 2016 was an international effort consistent with the United Nations Framework Convention on Climate Change to reduce greenhouse gases and keep the increase in global mean temperature to no more than 2°C above the pre-industrial level. This effort does not appear to have worked. The expert community believes that the 2°C goal will not be achieved.

References

Blasing, T. (2016). "Recent greenhouse gas concentrations". Carbon Dioxide Information Analysis Center. doi:10.3334/CDIAC/atg.032.

Bouwman, L. *et al.* (2013). *Drawing down N₂O to protect climate and the ozone layer. A UNEP Synthesis Report*. United Nations Environment Programme (UNEP). nora.nerc.ac.uk/id/eprint/505848.

Boyd, P. (2008). "Ranking geo-engineering schemes". *Nature Geoscience* 1(11): 722–4. doi.org/10.1038/ngeo348.

Del Grosso, S. (2022). "A gap in nitrous oxide emissions reporting complicates long term climate mitigation". *Proceedings of the National Academy of Sciences* 119(31): e2200354119 doi.org/10/1073/pnas.2200354119

GCP (Global Carbon Project) (2020). "Energy and climate change: a story in 11 charts". www.morganstanley.com/ideas/energy-economies-climate-change-infographic.

GCP (Global Carbon Project) (2021a). "Figures from the Global Carbon Budget 2021". robbieandrew.github.io/GCB2021/.

GCP (Global Carbon Project) (2021b). "Global methane budget". www.globalcarbonproject.org/methanebudget/.gcp.

GML (Global Monitoring Laboratory) (2020). "Trends in atmospheric carbon dioxide". gml.noaa.gov/ccgg/trends/.

Jackson, R. _et al._ (2020). "Increasing anthropogenic methane emissions arise equally from agricultural and fossil fuel sources". _Environmental Research Letters_ 15(7): 071002. doi.org/10.1088/1748-9326/ab9ed2.

Kanter, D. _et al._ (2013). "A post-Kyoto partner: considering the stratospheric ozone regime as a tool to manage nitrous oxide". _Proceedings of the National Academy of Sciences_ 110(12): 4451-7. doi.10.1073/pnas.1222231110.

Kiehl, J. & K. Trenberth (1997). "Earth's annual global mean energy budget". _Bulletin of the American Meteorological Society_ 78(2): 197-208. doi.org/10.1175/1520-0477(1997)078<0197:EAGMEB>2.0.CO;2.

Koffi, E. _et al._ (2020). "An observation-constrained assessment of the climate sensitivity and future trajectories of wetland methane emissions". _Science Advances_ 6(15): eaay4444. doi:10.1126/sciadv.aay4444.

Lade, S. _et al._ (2018). "Analytically tractable climate–carbon cycle feedbacks under 21st century anthropogenic forcing". _Earth System Dynamics_ 9(2): 507-23. doi.org/10.5194/esd-9-507-2018.

Liu, L. _et al._ (2022). "Exploring global changes in agricultural ammonia emissions and their contribution to nitrogen deposition since 1980". _Proceedings of the National Academy of Sciences_ 119(14): e2121998119 doi.org/10.1073/pnas.2121998119.

Lynch, J. _et al._ (2020). "Demonstrating GWP*: a means of reporting warming-equivalent emissions that captures the contrasting impacts of short-and long-lived climate pollutants". _Environmental Research Letters_ 15(4): 044023. doi.org/10.1088/1748-9326/ab6d7e.

Myhre, G. & D. Shindell (2013). "Anthropogenic and natural radiative forcing". In T. Stocker _et al._ (eds), _Climate Change 2013: The Physical Science Basis. Contribution of Working Group I to the Fifth Assessment Report of the Intergovernmental Panel on Climate Change_, 659-740. Cambridge: Cambridge University Press.

NAS/RS (National Academy of Sciences/Royal Society) (2020). _Climate Change: Evidence and Causes: Update 2020_. Washington, DC: National Academies Press. doi.org/10.17226/25733.

NOAA (National Oceanic and Atmospheric Administration) (2021). Global Monitoring Laboratory. Earth System Research Laboratories. https://gml.noaa.gov/webdata/ccgg/trends/n2o/n2o_mm_gl.txt.

NPS (National Park Service, US) (2020). "What is climate change?" www.nps.gov/goga/learn/nature/climate-change-causes.htm.

Sherwood, S., V. Dixit & C. Salomez (2018). "The global warming potential of near-surface emitted water vapour". _Environmental Research Letters_ 13(10): 104006. doi.org/10.1088/1748-9326/aae018.

Zhu, Y. _et al._ (2020). "Disproportionate increase in freshwater methane emissions induced by experimental warming". _Nature Climate Change_ 10(7): 685-90. doi.org/10.1038/s41558-020-0824-y.

Zickfeld, K. _et al._ (2021). "Asymmetry in the climate–carbon cycle response to positive and negative CO_2 emissions". _Nature Climate Change_ 11: 613–617. doi:10.1038/s41558-021-01061-2.

Q45

HAS HUMAN ACTIVITY INCREASED AEROSOLS IN THE ATMOSPHERE?

Short answer: yes, but different aerosols warm or cool the Earth, creating uncertainties in relation to their various impacts.

Aerosols are tiny (nanometres to microns) solid or liquid particles suspended in the atmosphere, largely in the troposphere and stratosphere. Those affecting the Earth system emanate from natural and human causes, such as volcanoes, dust, fires, vegetation and sea spray, fossil fuel burning and land change (Carslaw *et al.* 2010; Unger *et al.* 2010). These sources emit aerosol particles into the atmosphere or generate them in the atmosphere by chemical reactions. Human activities account for about 10 per cent of atmospheric aerosols and their precursors, SO_2, black carbon and organic carbon.

Different aerosols may cool or warm the Earth system, depending on their qualities, foremost their albedo impacts. Dust, volcanic sulphate and sea spray in the atmosphere reflect solar radiation, cooling the atmosphere, and increase cloud droplets, reducing precipitation efficiency with unclear impacts on cloud reflectivity of solar radiation (Boucher *et al.* 2013). The Mt Pinatubo eruption in 1991 emitted so many sulphate particles into the stratosphere that the global average temperature **(Q47)** dropped by 0.6°C for two years (McCormick, Thomason & Trepte 1995) (Fig. 45.1). In contrast, black carbon and dark soot, largely from fossil fuel burning and human burning of landscapes as well as desert dust, absorb solar radiation and reradiate it within the Earth system. About 7,500 Gg yr^{-1} of black carbon (high uncertainty) is emitted to the atmosphere, of which about 13 per cent is attributed to landscape burning, especially tropical forests (Bond *et al.* 2013). For the most part, intensive industrial burning of fossil fuels increases emissions, whereas a decrease in this intensity reduces them.

The role of aerosols in the atmosphere maintains several uncertainties, in part because of their direct and indirect complexities. For example, black carbon aerosols warm the atmosphere as noted, but also shade the land surface from shortwave radiation, providing a cooling effect (Fig. 45.2). In addition, recent work indicates that aerosol-induced cloud-water downwind of anthropogenic sources decreases cloud-water density, offsetting by 29 per cent the cooling effect of clouds, an offset not yet included in climate models (Toll *et al.* 2019). Various synthesis estimates project that, overall, aerosols cool the Earth, perhaps by

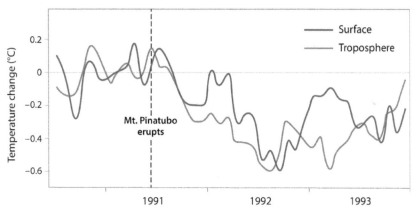

Figure 45.1 Mt Pinatubo impacts on global temperature

Mt Pinatubo, on the island of Luzon in the Philippines, is an active volcano. It erupted in June 1991 emitting 15–20 million tonnes of sulphur dioxide – an aerosol – into the atmosphere, which spread globally throughout the lower latitudes reflecting solar radiation. The result was a significant drop in the global mean surface temperature over a year, as depicted in the graphic.

Source: photo: Harlow (1991); graph: Self *et al.* (1999).

–0.9 W m², reducing the impact of greenhouse gases (Boucher *et al.* 2013; Unger *et al.* 2010). In contrast, estimates of the radiative forcing of black carbon aerosols, now proposed to be 1.1 W m² (Bond *et al.* 2013), a warming rate exceeded by only one greenhouse gas, CO_2, may be overestimated as a mitigator of warming owing to its short atmospheric life (Smith & Mizrahi 2013; Toll *et al.* 2019). Nevertheless, geoengineering proposals include the use of black carbon and sulphate aerosols to reduce global warming, seeking to create impacts akin to the Mt Pinatubo climate results (Rasch *et al.* 2008). The consequences of this proposed

Atmospheric heating

Surface cooling

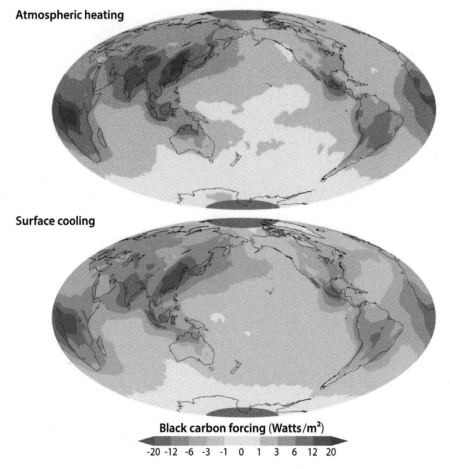

Black carbon forcing (Watts/m²)

-20 -12 -6 -3 -1 0 1 3 6 12 20

Figure 45.2 Black carbon aerosol impacts
Black carbon (e.g. soot) demonstrates the complexity of the agents of radiative forcing. The top graphic indicates the warming of this carbon in the atmosphere as it absorbs solar radiation and reradiates it. As this absorption takes place, less solar radiation finds its way to the surface of the Earth, creating a cloud-like cooling on the surface.
Source: NASA Earth Observatory (2010).

mitigation solution are not sufficiently understood, however. Such action could affect the ozone layer **(Q52)** and increase ocean acidification, among other impacts (Robock 2008).

Beyond the global climatic impacts, aerosols in the near-surface troposphere constitute pollutants with significant human health consequences. Here, especially in cities and indoors where wood fuels or coal are used for cooking and heating, particulate matter (PM) of various sizes – coarse (PM_{10}, <2.5 μm) and fine ($PM_{2.5}$, <10 μm) – concentrate in the air, leading to respiratory and related problems (Kim, Kabir & Kabir, 2015). About 800,000 premature deaths per year are attributed to PM air pollution according to the World Health Organization (Anderson, Thundiyil & Stolbach 2012).

References

Anderson, J., J. Thundiyil & A. Stolbach (2012). "Clearing the air: a review of the effects of particulate matter air pollution on human health". *Journal of Medical Toxicology* 8(2): 166–75. doi.org/10.1007/s13181-011-0203-1.

Bond, T. *et al.* (2013). "Bounding the role of black carbon in the climate system: a scientific assessment". *Journal of Geophysical Research: Atmospheres* 118(11): 5380–552. doi.org/10.1002/jgrd.50171.

Boucher, O. *et al.* (2013). "Clouds and aerosols". In T. Stocker *et al.* (eds), *Climate Change 2013: The Physical Science Basis. Contribution of Working Group I to the Fifth Assessment Report of the Intergovernmental Panel on Climate Change*, 571–658. Cambridge: Cambridge University Press.

Carslaw, K. *et al.* (2010). "A review of natural aerosol interactions and feedbacks within the Earth system". *Atmospheric Chemistry and Physics* 10(4): 1701–37. doi.org/10.5194/acp-10-1701-2010.

Harlow, D. (1991). "The June 12, 1991 eruption column from Mount Pinatubo taken from the east side of Clark Air Base". US Geological Survey. CVO Photo Archives. commons.wikimedia.org/wiki/File:Pinatubo91eruption_plume.jpg.

Kim, K., E. Kabir & S. Kabir (2015). "A review on the human health impact of airborne particulate matter". *Environment International* 74: 136–43. doi.org/10.1016/j.envint.2014.10.005.

McCormick, M., L. Thomason & C. Trepte (1995). "Atmospheric effects of the Mt Pinatubo eruption". *Nature* 373(6513): 99–404. doi.org/10.1038/373399a0.

NASA Earth Observatory (2010). "Aerosols and incoming sunlight (direct effects)". earthobservatory.nasa.gov/features/Aerosols/page3.php (accessed 12 January 2022).

Rasch, P. *et al.* (2008). "An overview of geoengineering of climate using stratospheric sulphate aerosols". *Philosophical Transactions of the Royal Society A: Mathematical, Physical and Engineering Sciences* 366(1882): 4007–37. doi.org/10.1098/rsta.2008.0131.

Robock, A. (2008). "20 reasons why geoengineering may be a bad idea". *Bulletin of the Atomic Scientists* 64(2): 14–18. doi.org/10.2968/064002006.

Self, S. *et al.* (1999). "The atmospheric impact of the 1991 Mount Pinatubo eruption". US Geological Survey. pubs.usgs.gov/pinatubo/self/.

Smith, S. & A. Mizrahi (2013). "Near-term climate mitigation by short-lived forcers". *Proceedings of the National Academy of Sciences* 110(35): 14202–06. doi.org/10.1073/pnas.1308470110.

Toll, V. *et al.* (2019). "Weak average liquid-cloud-water response to anthropogenic aerosols". *Nature* 572(7767): 51–55. doi.org/10.1038/s41586-019-1423-9.

Unger, N. *et al.* (2010). "Attribution of climate forcing to economic sectors". *Proceedings of the National Academy of Sciences* 107(8): 3382–7. doi.org/10.1073/pnas.0906548107.

IS THE EARTH SYSTEM WARMING – IS CLIMATE CHANGE REAL – AND HOW DO WE KNOW?

Short answer: yes. The average temperature of the Earth is increasing as registered in the atmosphere, oceans and various other biophysical indicators of increasing heat.

Climate change refers to long-term changes – multiple decades to centuries or more – in the averages of the components comprising climate, such as temperature and precipitation. Since 1900, the global mean surface temperature (GMST) has increased by ~1°C (1.8°F) (**Q47**) (NAS/RS 2020) as illustrated by the noted "hockey stick" graph of temperature change (Fig. 46.1). The graph denotes the average Earth temperature anomaly (i.e. the yearly global average temperature difference) from the 1850–1990 global average temperature over two millennia. This anomaly was below the baseline (0.0°C) for most of the last 500 years until the accelerated rise above the baseline in the early 1900s which continues to this day and will do so into the immediate future (Fig. 46.1) (IPCC 2021; Mann 2007).

The original hockey-stick graph and interpretations were debated within the science community on the grounds of the data employed, assumptions, methods of analysis and uncertainties involved in its creation (Mann 2021). It was severely challenged by some individuals (e.g. von Storch & Zorita 2005), think tanks and governments that were uncertain of its validity or antithetical to its implications that human activity is the cause of the increasing temperature (Oreskes *et al.* 2018). As the data and analyses improved, however, the basic claims of the graph have prevailed; dozens of independent reconstructions generate highly similar results (Brumfield 2006), as indicated in Figure 46.1 (IPCC 2021). The trends noted and their causes guide climate change science. The public and political controversy surrounding the reconstruction is addressed in **Question 79**.

The spatial and temporal dimensions of climate change are global and long-term, as in the Earth's average temperature over centuries. A polar vortex (**Q52**) – hypothesized to be made more frequent because of climate warming – may reach winter records for cold temperatures in parts of the Northern hemisphere, lowering regional average temperatures for a year or so, while the global average temperature continues to increase over a longer period. Also, what may appear to be small, incremental increases in global warming (or cooling) have significant impacts on climatic conditions. For example, the global GMST

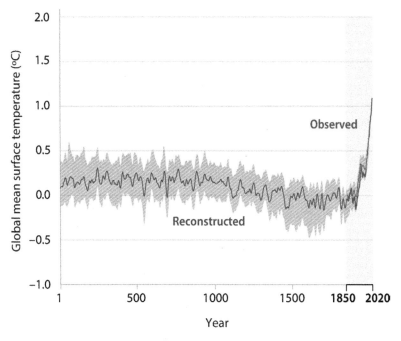

Figure 46.1 The "hockey stick" formulation of current warming
The global surface temperature (decadal average) anomaly from the 1850–1990 average (0.0°C line) combines reconstructed (1–2000) and observed (1850–2020) results showing the dramatic increase in the temperature anomaly from the early 1900s. The grey shading constitutes the uncertainty range.
Source: IPCC (2021).

during the last Ice Age was only 4°C below the warmer temperatures of the Holocene in general. It is in this perspective that the approximate human-induced 1°C increase in global temperatures since 1880 and the projected 1.5–2.0°C or higher increase soon to be reached (Masson-Delmotte *et al.* 2018) should be viewed. The consequences for the global climate and the Earth system will become profound.

Climate warming is consistent with the processes operating within the atmosphere, foremost that of the greenhouse gases **(Q44)**. The role of CO_2 in warming the atmosphere was advanced almost 200 years ago (Jones & Henderson-Sellers 1990), followed by the identification of other greenhouse gases. The measurements of these gases in the atmosphere continuously improve, demonstrating their rapid increases, as do the models used to analyse the gases' interactions within and impacts on the atmosphere (Flato *et al.* 2014). Care is required, however, in model applications owing to their sensitivities (Hausfather *et al.* 2022). For example, day and night temperatures continue to warm globally, on average, although an asymmetry exists between them, apparently generated primarily by cloud cover. About twice as much land area has experienced a >0.25°C increase in night-time temperature over daytime, and has become wetter than previously. Hot and dry climatic areas, however, experience greater daytime temperature increases (Cox *et al.* 2020).

The amount of ice and snow on the Earth's surface, both on land and water (**Q37**), is decreasing. The extent of winter and summer ice over the Arctic Ocean continues to decline, opening a northwest passage between the Atlantic to Pacific Oceans (Dawson *et al.* 2013; Swart *et al.* 2015). The Northern hemisphere snow cover for June has declined; it has not topped the long-term average in cover extent for more than a quarter of a century and is consistently 2 Mkm² or more below that average (Estilow, Young & Robinson 2015). Land ice – mountain glaciers, ice sheets – and permafrost are losing their bulk rapidly (**Q37**). The first decade of the twenty-first century witnessed the largest mass balance loss (ice loss) of mountain glaciers from their recordings in the middle of the nineteenth century (Marshall 2014). The two largest ice sheets globally, Antarctica and Greenland (the elephants in the ice-world room), lost more than 95 ±25 and 265 ± 25 Gt yr⁻¹ of ice, respectively, in the first 15 years of this century (Forsberg, Sørensen & Simonsen 2017). The twentieth-century loss of permafrost is estimated to cover 3.4 Mkm², although relatively large uncertainties are involved (Chadburn *et al.* 2017).

The loss of land ice due to warming ultimately finds its way to the oceans, leading to sea-level rise (**Q39**). Since the late nineteenth century sea level has risen ~16 cm, a response to both the interglacial warming of the Holocene and human-induced warming. The upper oceans have soaked up about 93 per cent of warming of the Earth system since 1995 (Levitus *et al.* 2012), expanding water molecules, which along with the loss of continental ice, have created recent sea-level rise increasing at ~3.6 mm yr⁻¹ (Domingues *et al.* 2008; NAS/RS 2020). This warming is also increasing ocean acidification (**Q41**) and bleaching of coral reefs (i.e. the loss of algae exposing the corals' skeleton). The overwhelming number of coral reefs show signs of heat and acidification stress (Heron *et al.* 2016) (**Q59**).

Land biota are also responding to the ~1°C (1.8°F) increase over the 1961–90 global average Earth temperature (NAS/RS, 2020). Land flora, for example, have altered their phenology. The end of the growing season in the Northern hemisphere has been increasing at a mean rate of ~0.18 days yr⁻¹, and the spring "green up" time appears to be earlier (Liu *et al.* 2016). Indeed, tree cover is moving poleward in the Arctic area, reducing the spatio-temporal coverage of ice and snow and, as such, warming the Earth through the albedo changes (Pearce 2022) (**Q5**). Such changes have a series of impacts on regulating environmental services (**Q7**) and ecosystem carbon dynamics and feedbacks to climate (Zhang *et al.* 2022).

References

Brumfield, G. (2006). "Academy affirm hockey stick graphic". *Nature* 44: 1032–33. doi.org/10.1038/4411032a.

Chadburn, S. *et al.* (2017). "An observation-based constraint on permafrost loss as a function of global warming". *Nature Climate Change* 7(5): 340–44. doi.org/10.1038/nclimate3262.

Cox, D. *et al.* (2020). "Global variation in diurnal asymmetry in temperature, cloud cover, specific humidity and precipitation and its association with leaf area index". *Global Change Biology* 26: 7099–111. doi.org/10.1111/gcb.15326.

Dawson, J. *et al.* (2013). "Local-level responses to sea ice change and cruise tourism in Arctic Canada's Northwest Passage". *Polar Geography* 36(1/2): 142–62. doi.org/10.1080/1088937X.2012.705352.

Domingues, C. *et al.* (2008). "Improved estimates of upper-ocean warming and multi-decadal sea-level rise". *Nature* 453(7198): 1090–93. doi.org/10.1038/nature07080.

Estilow, T., A. Young & D. Robinson (2015). "A long-term Northern hemisphere snow cover extent data record for climate studies and monitoring". *Earth System Science Data* 7(1): 137–42. doi.org/10.5194/essd-7-137-2015.

Flato, G. *et al.* (2014). "Evaluation of climate models". In T. Stocker *et al.* (eds), *Climate Change 2013: The Physical Science Basis. Contribution of Working Group I to the Fifth Assessment Report of the Intergovernmental Panel on Climate Change.* Cambridge: Cambridge University Press.

Forsberg, R., L. Sørensen & S. Simonsen (2017). "Greenland and Antarctica ice sheet mass changes and effects on global sea level". *Survey in Geophysics* 38: 89–104. doi.10.1007/s10712-016-9398-7.

Hausfather, Z. *et al.* (2022). "Climate simulations: recognize the 'hot model' problem". *Nature* 605: 26–29. www.nature.com/articles/d41586-022-01192-2.

Heron, S. *et al.* (2016). "Warming trends and bleaching stress of the world's coral reefs 1985–2012". *Scientific Reports* 6: 38402. doi.org/10.1038/srep38402.

IPCC (Intergovernmental Panel on Climate Change) (2021). *Climate Change (2021): The Physical Science Basis. Summary for Policy Makers.* www.ipcc.ch/report/ar6/wg1/downloads/report/IPCC_AR6_WGI_SPM.pdf.

Jones, M. & A. Henderson-Sellers (1990). "History of the greenhouse effect". *Progress in Physical Geography* 14(1): 1–18. doi.org/10.1177/030913339001400101.

Levitus, S. *et al.* (2012). "World ocean heat content and thermosteric sea level change (0–2000 m), 1955–2010". *Geophysical Research Letters* 39: L10603. doi.org/10.1029/2012GL051106.

Liu, Q. *et al.* (2016). "Delayed autumn phenology in the Northern hemisphere is related to change in both climate and spring phenology". *Global Change Biology* 22(11): 3702–11. doi.org/10.1111/gcb.13311.

Mann, M. (2007). "Climate over the past two millennia". *Annual Review of Earth Planetary Science* 35: 111–36. doi.10.1146/annurev.earth.35.031306.140042.

Mann, M. (2021). "Beyond the hockey stick: climate lessons from the Common Era". *Proceedings of the National Academy of Sciences* 118(39): e2112797118. doi.org/10.1073/pnas.2112797118.

Marshall S. (2014). "Glacier retreat crosses a line". *Science* 345(6199): 872. doi10.1126/science.1258584.

Masson-Delmotte, V. *et al.* (eds) (2018). *Global Warming of 1.5°C.* Geneva: World Meteorological Organization. www.ipcc.ch/sr15/.

NAS/RS (National Academy of Sciences/Royal Society) (2020). *Climate Change: Evidence and Causes: Update 2020.* Washington, DC: National Academies Press. doi.org/10.17226/25733.

Oreskes, N. *et al.* (2018). "The denial of global warming". In S. White *et al.* (eds), *The Palgrave Handbook of Climate History*, 149–71. London: Palgrave Macmillan.

Pearce, F. (2022). "The forest forecast". *Science* 316(6595): 789–91. doi:10.1126/science.adc9867.

Swart, N. *et al.* (2015). "Influence of internal variability on Arctic sea-ice trends". *Nature Climate Change* 5(2): 86–9. doi.org/10.1038/nclimate2483.

von Storch, H. & E. Zorita (2005). "Comment on 'Hockey sticks, principal components, and spurious significance' by S. McIntyre and R. McKitrick". *Geophysical Research Letters* 32: L20701. doi:10.1029/2005GL022753.

Q47

WHAT IS THE EARTH'S AVERAGE TEMPERATURE AND HOW IS IT DETERMINED?

Short answer: the average of land and ocean near-surface temperature – 14.6°C in 2019 – continues to increase, determined by land and ocean measurements globally.

The global mean surface temperature (GMST) or average temperature of the Earth combines the averages of the min–max readings near the land surface and submerged but near the surface in oceans. Various measurement instruments – from buoy thermometers to satellite microwaves – provide grid-based data that are averaged and weighted by the area of the grid. Past differences between satellite and ground measures have been overcome; the two sources are now in sync (Susskind *et al.* 2019). The twentieth-century average temperature was 13.9°C (57.0°F). That for 2019 was 14.6°C (58.3°F), which was 0.7°C above the twentieth-century average and 1.71°C above that pre-industrial period average (Fig. 47.1). Indeed, 2019 and 2020 were, respectively, the third and second warmest in the history of calculations by the National Oceanic and Atmospheric Administration (NOAA), with the five warmest years all occurring since 2015 (NOAA 2020). The certainty in the temperature calculations continues to rise, accurate within 0.05 of one degree Celsius. Average temperatures of the past are estimated from proxies, such as tree rings and ice cores. These data are central to long-term reconstructions of climate change, such as the "hockey stick" reconstruction (Q46). The long-term trends in average temperature are critical to assess climate change. New efforts aggregate the past 12,000 years of Holocene temperature anomalies using a baseline calculated from the GMST of 1850–1900. They indicate the warmest 200-year-long period during the Holocene to have occurred 6,500 ya (Fig. 47.2), averaging 0.7°C above the nineteenth-century GMST. Subsequently, the Earth's atmosphere has cooled at an average rate of –0.08°C per 1,000 years, until, of course, the current warming trend (Kaufman *et al.* 2020).

Focusing on the recent past (1850–2019), the Earth's average temperature has increased based on 60-, 30-, and 10-year averages. Significantly, small changes in the global mean surface temperature, can create large changes in climate. The last ice age, for example, was sustained by only 4–5°C (7–9°F) cooler average Earth temperatures (NAS/RS 2020). The average temperature is not the same worldwide, and for any given year, some areas may

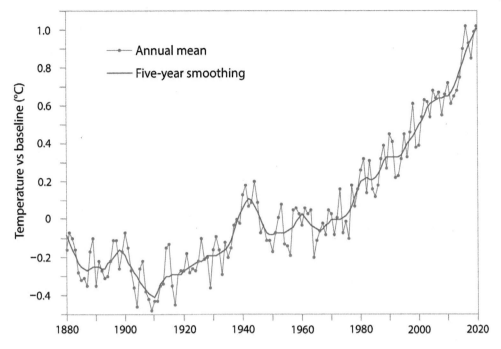

Figure 47.1 Global mean surface temperatures (GMST), 1880–2020

The temperature represents anomalies from the global average (0°C, the baseline, is 14°C). Note the continual increases in the anomalies in the early 1900s and increasing rate since the 1960s.

Source: NASA GISS.

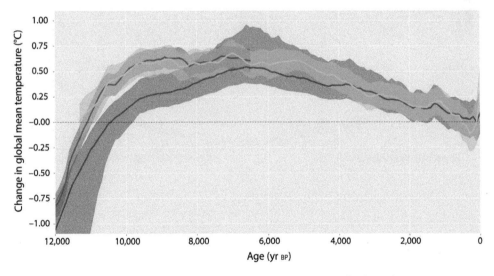

Figure 47.2 Global mean surface temperature (GMST) reconstructions for the Holocene

Three reconstructions compared with the GMST anomaly for 1800–1900 (0.00 baseline). The results indicate warming from the last glacial period until 6.5 ka, followed by three millennia of cooling until the warming beginning in the 1900s. See Kaufman *et al.* (2020) for each reconstruction.

Source: Kaufman *et al.* (2020).

experience a negative temperature anomaly (e.g. below the baseline averages in Fig. 47.3 registered in blue that range from –0.5°C to –1.0°C). Overall, most of the world has had a positive anomaly with each decade since 1980 witnessing increases in global average temperatures (Blunden & Arndt 2020). The higher latitudes and polar regions have had much greater warming than the Earth's average, as depicted in Figure 47.3 of temperature anomalies for 2019 (°C above or below the 1951–80 average). This increase has been attributed to albedo changes (less annual coverage of ice and snow on land and ocean) and feedbacks from greater reradiation to space in the lower latitudes (Pithan & Mauritsen 2014; Screen & Simmonds 2010), which may also facilitate significant energy transfers from low to high latitudes. Recent work has also indicated that that the larger warming of Antarctica in its western, compared to the eastern, parts is owing to ocean–atmosphere feedbacks (Jun *et al.* 2020).

Various projections for the year 2100, largely following the different emission scenarios from the Intergovernmental Panel on Climate Change (IPCC), indicate an average Earth temperature increasing from 1.7°C to 4.6°C (averages of various models by scenario) (BI 2019).

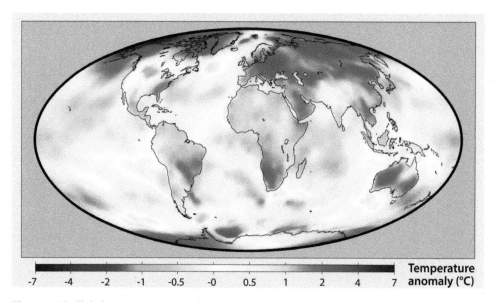

Figure 47.3 Global temperature anomaly, June 2019

The anomaly is the difference between the 1951 and 1980 averages across the Earth. Note the few regions/areas that remained the same or cooled (white to blue), whereas the majority of Earth warmed (yellow to red), with the significant warming in the high latitudes.

Source: Berkeley Earth (2021).

References

Berkeley Earth (2021). "June 2021 temperature report". berkeleyearth.org/june-2021-temperature-update/.

Blunden, J. & D. Arndt (eds) (2020). "State of the climate 2019". *Bulletin of the American Meteorological Association* **101(8): S1–429. doi.org/10.1175/2020BAMSStateoftheClimate.1.**

BI (Breakthrough Institute) (2019). "A 3C world is 'business as usual'". thebreakthrough.org/issues/energy/3c-world.

Jun, S. *et al.* (2020). "The internal origin of the west–east asymmetry of Antarctic climate change". *Science Advances* 6(24): eaaz1490. doi:10.1126/sciadv.aaz1490.

Kaufman, D. *et al.* (2020). "Holocene global mean surface temperature, a multi-method reconstruction approach". *Scientific Data* 7(1): 1–13. doi.org/10.1038/s41597-020-0530-7.

NAS/RS (National Academy of Sciences/Royal Society) (2020). *Climate Change: Evidence and Causes: Update 2020.* **Washington, DC: National Academies Press. doi.org/10.17226/25733.**

NOAA (National Oceanic and Atmospheric Administration) (2020). "2019 was 2nd hottest year on record for Earth say NOAA, NASA". www.noaa.gov/news/2019-was-2nd-hottest-year-on-record-for-earth-say-noaa-nasa.

Pithan, F. & T. Mauritsen (2014). "Arctic amplification dominated by temperature feedbacks in contemporary climate models". *Nature Geoscience* 7(3): 181–4. doi.org/10.1038/ngeo2071.

Screen, J. & I. Simmonds (2010). "The central role of diminishing sea ice in recent Arctic temperature amplification". *Nature* 464(7293): 1334–7. doi.org/10.1038/nature09051.

Susskind, J. *et al.* (2019). "Recent global warming as confirmed by AIRS". *Environmental Research Letters* 14(4): 044030. doi.org/10.1088/1748-9326/aafd4e.

WHAT IS THE EVIDENCE FOR HUMAN-INDUCED GLOBAL CLIMATE WARMING?

Short answer: the observed amount and pace of climate warming can only be reconciled by accounting for the various human changes to the Earth system, foremost greenhouse gas emissions.

Virtually all studies demonstrate a 97 per cent or higher consensus among climate scientists that global warming is human-induced; human activity plays an important role in the increases in the Earth's average temperature (Anderegg 2010; Cook *et al.* 2013). This consensus is also strongly supported by international and national science organizations worldwide (Table 48.1) and by special publications from these organizations, such as those by the US Global Change Research Program (USGCRP 2017) and by the National Academy of Sciences, Engineering and Medicine, and the Royal Society (UK) (NAS/RS 2020).

The Earth system is in an interglacial period in which natural forces warm the planet. The rates of warming are unusually fast-paced, however, about ten times faster than warming at the end of last ice age. This pace is exceptional but consistent with the emissions of greenhouse gases from human activity (**Q44**) (e.g. CO_2, CH_4, N_2O), especially since the industrial era began. These emissions are well documented and linked to numerous human activities (e.g. energy production, automotive travel, irrigated agriculture). No emission is more obviously "human-induced" than the large amount of CO_2 increasing in the atmosphere and oceans. The majority of these emissions are due to the extraction of fossil fuels (e.g. coal, oil) from subterranean sources where the carbon compounds have been held in sinks for hundreds of millions of years.

Possible natural causes – warming sources and agents – have proven insufficient to account for the amount and rate of climate warming. Solar energy (irradiance) is increasing over the very long term, but the incremental energy increases cannot account for the pace of rapid warming. Irradiance also varies through 11-year cycles in which sun spots increase and decline in number, affecting solar radiation arriving at Earth by ~0.1 per cent, largely affecting the stratosphere (NAS/RS 2020). The trajectory of irradiance and global mean surface temperature do not match, however (Fig. VI.1). Since 1980, for example, four solar energy cycles have taken place with no significant increases in energy per cycle, whereas the surface temperature continues its warming trajectory (NAS/RS, 2020). None of the natural

Table 48.1 Examples of science organizations supporting climate warming owing to human activities

Organizations, programmes, agencies: international	Programmes' scientific reports on the subject
European Climate Change Programme	Second ECCP Progress Report. Can we meet our Kyoto targets? ec.europa.eu/clima/document/download/bcf73ad9-3311-473d-902b-432cf2f41cd8_en
Intergovernmental Panel on Climate Change	IPCC Fifth Assessment Report, Summary for Policymakers (2014)
Joint Science Academies	Joint academies' statement: Global response to climate change (Brazil, Canada, China, France, Germany, Indian, Italy, Japan, Russia, United Kingdom, United States) www.academie-sciences.fr/archivage_site/activite/rapport/avis0605a_gb.pdf
World Meteorological Organization	State of the Global Climate 2020 public.wmo.int/en/our-mandate/climate/wmo-statement-state-of-global-climate
World Climate Research Programme	WCRP Grand Challenge on Near-term Climate Predictions www.wcrp-climate.org/gc-near-term-climate-prediction
17 Global Science Academies	The Science of Climate Change (2001). *Science*, 292: 1261. (Academies of Science from Australia, Belgium, Brazil, Canada, the Caribbean, China, France, Germany, India, Indonesia, Ireland, Italy, Malaysia, New Zealand, Sweden, Turkey, and the United Kingdom) doi:10.1126/science.292.5520.1261
Organizations, programs, agencies: United States	
American Association for the Advancement of Science	AAAS Board Statement on Climate Change (2014) whatweknow.aaas.org/get-the-facts/
American Chemical Society	ACS Public Policy Statement: Climate Change (2016–2019) www.acs.org/content/acs/en/policy/publicpolicies/sustainability/globalclimatechange.html
American Geophysical Union	Society Must Address the Growing Climate Crisis Now (2019) www.agu.org/Share-and-Advocate/Share/Policymakers/Position-Statements/Position_Climate
American Meteorological Society	Climate Change: An Information Statement of the American Meteorological Society (2019) www.ametsoc.org/index.cfm/ams/about-ams/ams-statements/statements-of-the-ams-in-force/climate-change1/
American Physical Society	APS National Policy 07.1 Climate Change (2015) www.aps.org/policy/statements/15_3.cfm
National Academy of Sciences (US)	Royal Academy & National Research Council 2020. Climate Change: Evidence and Causes: Update 2020. Washington, DC: National Academies Press. doi.org/10.17226/25733; sites.nationalacademies.org/sites/climate/index.htm
US Global Change Research Program	Fourth National Climate Assessment: Volume II (2018) nca2018.globalchange.gov/

Source: author.

causes or their combinations can account for the observable evidence of greenhouse gas concentrations in the atmosphere and the pace of warming of our planet (NAS/RS 2020).

Perhaps the strongest indication of the human input to global warming is the inability to simulate global surface temperature anomalies using natural forcings alone (Fig. 48.1). Such simulations strongly underestimate observed anomalies. Anthropogenic models alone

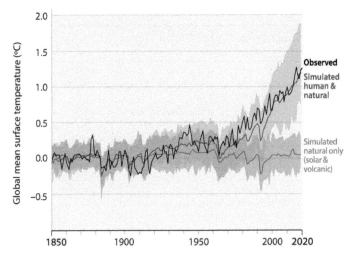

Figure 48.1 Simulated versus observed global average temperature, 1850–2020

Global mean surface temperature (annual average) anomaly (1850–2020 average = baseline). The black line is the observed or measured temperature anomaly. The green line is simulated average temperature anomaly applying only natural forcings. The brown line is simulated average temperature anomaly applying natural and human forcings. The shaded area constitutes the simulated range of temperatures.

Source: IPCC (2021).

come much closer to the observed temperatures, but combining the natural and anthropogenic agents means a very close match is simulated (IPCC 2021; Watson 2001). The most sophisticated method of estimating the range of possible global warming pegs the range between 2.6–3.9°C; a doubling of CO_2 in the atmosphere almost assures an increase of 2°C, if not higher (Voosen 2020).

References

Anderegg, W. (2010). "Expert credibility in climate change". *Proceedings of the National Academy of Sciences* 107(27): 12107–09. doi:10.1073/pnas.1003187107.

Cook, J. *et al.* (2013). "Quantifying the consensus on anthropogenic global warming in the scientific literature". *Environmental Research Letters* 8(2): 024024. doi:10.1088/1748-9326/8/2/024024.

IPCC (Intergovernmental Panel on Climate Change) (2021). *Climate Change (2021) The Physical Science Basis. Summary for Policy Makers.* www.ipcc.ch/report/ar6/wg1/downloads/report/IPCC_AR6_WGI_SPM.pdf

NAS/RS (National Academy of Sciences/Royal Society) (2020). *Climate Change: Evidence and Causes: Update 2020.* Washington, DC: National Academies Press. doi.org/10.17226/25733.

US Global Change Research Program (USGCRP) (2017). *Climate Science Special Report: Fourth National Climate Assessment.* K. Hibbard *et al.* (eds). Washington, DC: USGCRP.

Voosen, P. (2020). "After 40 years, researchers finally see Earth's climate destiny more clearly". *Science* 369(6502): 345–55. doi:10.1126/science.abd9184.

Watson, R. (ed.) (2001). *Climate Change 2001: Synthesis Report.* Cambridge: Cambridge University Press.

DO EL NIÑO AND LA NIÑA (ENSO) EVENTS CAUSE WARMING OF THE EARTH SYSTEM?

Short answer: during an El Niño or La Niña event, the average Earth temperature may slightly increase or decrease, but any such change is not part of the overall trajectory of warming.

El Niño (Spanish for "Christ Child" or "little boy") and La Niña ("little girl") refer to special conditions in temperature and winds occurring sporadically every several years (~3–7 years) in the equatorial waters of the Pacific Ocean that can have short-term (9–12 months) global impacts on climate. Together, and combined with wind conditions, they create the El Niño–Southern Oscillation (ENSO) phenomenon.

In neutral conditions, trade winds move warm equatorial waters westward across the Pacific Ocean toward Oceania, permitting cool, nutrient rich water to upwell or rise toward the surface along the equatorial coastline of South America. In El Niño events, the trade winds break down and the warm surface waters move eastward toward northern South America (Fig. 49.1), reducing the upwelling (NOAA 2021; Philander & Fedorov 2003). The result is warm, wet weather in much of the Western hemisphere and beyond, commonly associated with floods but also drought, depending on location. The 2015 super El Niño was sufficiently intense to have increased the global mean surface temperature by adding unusual warming for the last three months of the year. Such effects, however, are for the El Niño period alone. In contrast to El Niño, La Niña events involve trade winds strengthening, pushing the cool, upwelled water near South America eastward into the central Pacific. The result is cooler temperatures in the eastern equatorial Pacific.

El Niño and La Niña, respectively, warm and cool the tropical Pacific, disrupting atmospheric circulation with temperature and precipitation impacts in the mid-latitudes. What is yet to be determined is the impact of global warming on the frequency and intensity of future ENSO events (Yeh, Kug & An 2014). It appears, however, that El Niño events may be wetter and La Niña events dryer (IPCC 2021).

Figure 49.1 El Niño sea surface temperature, 2016
The orange to red colouring shows increasing temperatures from the "normal" temperature. The dark red band extending from South America westward marks the warm ocean surface temperature during the 2016 El Niño event. Note that much of the tropics north of the equator also warmed as well. The precise warming by ocean area differs by El Niño events.
Source: NOAA-NESDIS (2017).

References

IPCC (Intergovernmental Panel on Climate Change) (2021). "Summary for policymakers".
    In T. Stocker *et al.* (eds), *Climate Change 2021: The Physical Science Basis. Contribution of Working Group I to the Sixth Assessment Report of the Intergovernmental Panel on Climate Change.* Cambridge: Cambridge University Press.

NOAA (National Oceanic and Atmospheric Administration) (2021). "El Niño and La Niña: frequently asked questions". www.climate.gov/news-features/understanding-climate/el-ni%C3%B1o-and-la-ni%C3%B1a-frequently-asked-questions.

NOAA-NESDIS (National Oceanic and Atmospheric Administration-National Environmental Satellite Data and Information Services) (2017). "El Nino". www.nesdis.noaa.gov/news/el-nino.

Philander, S. & A. Fedorov (2003). "Is El Niño sporadic or cyclic?" *Annual Review of Earth and Planetary Sciences* 31(1): 579–94. doi.org/10.1146/annurev.earth.31.100901.141255.

Yeh, S., J. Kug & S. An (2014). "Recent progress on two types of El Niño: observations, dynamics, and future changes". *Asia-Pacific Journal of Atmospheric Sciences* 50(1): 69–81. doi.org/10.1007/s13143-014-0028-3.

WILL HUMAN ACTIVITY INCREASE THE OCCURRENCE OF DROUGHTS GLOBALLY?

Short answer: yes. If modelling projects are accurate, droughts will increase worldwide, even if they vary in scale and intensity by region.

Droughts are prolonged periods with a shortfall of precipitation, relative to the climatic zone in question and its long-term averages in precipitation. Droughts adversely affect plants, animals and people, and are commonly categorized by their consequences (Text box 50.1). Despite increased atmospheric water vapour and the projected intensity of tropical storm precipitation (**Q51**) associated with human-induced climate warming, worldwide increases in drought impacts follow from various modelling assessments. These assessments have been hindered by the paucity of standardized historical data, but a new catalogue of drought indicates, perhaps surprisingly, that short-term meteorological droughts significantly decreased between 1950 and 2016 (He *et al.* 2020). This result stands in contrast to future forecasts, however, which point to increasing drought problems (Liu *et al.* 2018; Xu, Chen & Zhang 2019; Zhao & Dai 2015). These various exercises, using increases of 1.5°C and 2.0°C, indicate: (1) increasing drought duration from 2.9 to 3.2 months, with hotspots in the Amazon, northeastern Brazil, southern Africa and central Europe (Liu *et al.* 2018); (2) probable ~36 per cent and ~62 per cent increases in the frequency and ~15 per cent and ~20 per cent increases in the duration of low and high temperature conditions, respectively, with hotspots in North America, South America, southern Africa, Australia and Europe (Xu, Chen & Zhang 2019); and (3) moderate to severe agricultural drought increases as high as 50–100 per cent by 2090 over most of the Americas, Europe, southern Africa and parts of East and West Asia and Australia (Zhao & Dai 2015).

Recognizing the various uncertainties in forecasting drought in the context of climate change, the IPCC concludes that human-induced warming will increase agricultural and ecological drought in various regions of the world, especially in western North America and the Mediterranean region. Other areas identified are consistent with the hotspots noted above, but are subject to disagreement at this time among the expert community (IPCC 2021).

Text box 50.1 Categories of drought

Meteorological drought: precipitation significantly below long-term average.
Agricultural drought: precipitation endangers crop production.
Ecological drought: precipitation degrades ecosystem functions and environmental services.

References

He, X. *et al.* (2020). "A global drought and flood catalogue from 1950 to 2016". *Bulletin of the American Meteorological Society* 101(5): E508–35. doi.org/10.1175/BAMS-D-18-0269.1.

IPCC (Intergovernmental Panel on Climate Change) (2021). "Summary for policymakers". In T. Stocker *et al.* (eds), *Climate Change 2021: The Physical Science Basis. Contribution of Working Group I to the Sixth Assessment Report of the Intergovernmental Panel on Climate Change*. Cambridge: Cambridge University Press.

Liu, W. *et al.* (2018). "Global drought and severe drought-affected populations in 1.5 and 2°C warmer worlds". *Earth System Dynamics* 9(1): 267–83. doi.org/10.5194/esd-9-267-2018.

Xu, L., N. Chen & X. Zhang (2019). "Global drought trends under 1.5 and 2°C warming". *International Journal of Climatology* 39(4): 2375–85. doi.org/10.1002/joc.5958.

Zhao, T. & A. Dai (2015). "The magnitude and causes of global drought changes in the twenty-first century under a low–moderate emissions scenario". *Journal of Climate* 28(11): 4490–512. doi.org/10.1175/JCLI-D-14-00363.1.

Q51

HAS HUMAN ACTIVITY INCREASED TROPICAL STORMS?

Short answer: in some ways. The intensity of tropical cyclones poleward with increased precipitation, storm surges and flooding are probably the result of human-induced warming, recognizing uncertainties in modelling results.

Tropical storms or cyclones (aka hurricanes and typhoons), designated by a minimum wind speed of 119 km/h (74 mph), are generated by the build-up of atmospheric heat and water vapour (Murakami *et al.* 2020). This intensity may also be associated with slower moving storms, owing to the movement of the westerlies poleward (Zhang *et al.* 2020), followed by storms reaching higher latitudes. An increase in temperature of 1°C yields ~7 per cent increase in water vapour (Kossin, Emanuel & Vecchi 2014), suggesting that climate warming has and will continue to increase the intensity of tropical cyclones, with various levels of support about their frequency (Fig. 51.1). Those with winds ≥250 km per hour have tripled since 1980 (Bhatia *et al.* 2019). As this warming moves poleward, especially ocean heat, the latitude at which these cyclones achieve their strongest intensity apparently does so as well (Kossin, Emanuel & Vecchi 2014; Zhang *et al.* 2020). This latitudinal movement has proceeded at a rate of 53 km/decade and 62 km/decade poleward in the Northern and Southern hemispheres, respectively (Kossin, Emanuel & Vecchi 2014). These conditions with sea-level rise (**Q39**), perhaps together with the decline in the strength of the AMOC (**Q40**) (Yan, Zhang & Knutson 2017), mean that the severity of storm surges and flooding along coastlines is expected to increase (Murakami *et al.* 2020).

It is important to recognize that the observed changes and modelled projections of tropical storms owing to human causes have yet to be confirmed (Lee 2020). Indeed, various observations and models disagree on changes in intensity with climate change (e.g. Emanuel 2013; Walsh 2016; Wright, Knutson & Smith 2015) and on the proportion of tropical cyclones projected to reach categories 4 and 5, estimated in one study as an increase by 25 per cent to 30 per cent for each 1°C increase in temperature (Knutson *et al.* 2019). Models also vary about the frequency of tropical storms due to global warming. The changes in intensity observed, however, are not a product of natural variability (Bhatia *et al.* 2019; Yan, Zhang & Knutson 2017). Synthesis analysis of medium to high confidence results indicates

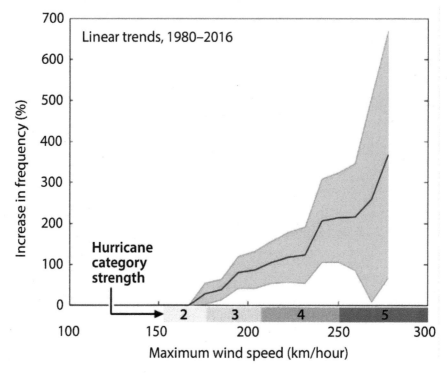

Figure 51.1 Increasing frequency of intense tropical storms
The blue area represents the range in uncertainty from the projected trend line.
Source: Emanuel (2013).

that a 2°C warming will lead to a median 14 per cent increase in precipitation, and 5 per cent increase in wind speed per storm, a 13 per cent increase in the proportion of category 4–5 storms and in storm surges, and a decrease in the frequency of tropical storms (Knutson *et al.* 2019).

References

Bhatia, K. *et al.* (2019). "Recent increases in tropical cyclone intensification rates". *Nature Communications* 10(1): 349–52. doi.org/10.1038/nature13278.

Emanuel, K. (2013). "Downscaling CMIP5 climate models shows increased tropical cyclone activity over the 21st century". *Proceedings of the National Academy of Sciences* 110(30): 12219–24. doi.org/10.1073/pnas.1301293110.

Knutson, T. *et al.* (2019). "Tropical cyclones and climate change assessment: Part II. Projected response to anthropogenic warming". *Bulletin of the American Meteorological Society* 101(3): E303–22. doi.org/10.1175/BAMS-D-18-0194.1.

Kossin, J., K. Emanuel & G. Vecchi (2014). "The poleward migration of the location of tropical cyclone maximum intensity". *Nature* 509(7500): 349–52.

Lee, T. (2020). "Third assessment on impacts of climate change on tropical cyclones in typhoon committee region – Part 1: observed changes, detection and attribution". *Tropical Cyclone Research and Review* 9(1): 1–22. doi.org/10.1016/j.tcrr.2020.03.001.

Murakami, H. *et al.* (2020). "Detected climatic change in global distribution of tropical cyclones". *Proceedings of the National Academy of Sciences* 117(20): 10706–14. doi.org/10.1073/pnas.1922500117.

Walsh, K. (2016). "Tropical cyclones and climate change: unresolved issues". *Climate Research* 27(1): 77–83. doi:10.3354/cr027077.

Wright, D., T. Knutson & J. Smith (2015). "Regional climate model projections of rainfall from US landfalling tropical cyclones". *Climate Dynamics* 45(11/12): 3365–79. doi.org/10.1007/s00382-015-2544-y.

Yan, X., R. Zhang & T. Knutson (2017). "The role of Atlantic overturning circulation in the recent decline of Atlantic major hurricane frequency". *Nature Communications* 8(1): 1695. doi.org/10.1038/s41467-017-01377-8.

Zhang, G. *et al.* (2020). "Tropical cyclone motion in a changing climate". *Science Advances* 6(17): eaaz7610. doi:10.1126/sciadv.aaz7610.

Q52

IS THE EXTREME COLD OF THE POLAR VORTEX HUMAN-INDUCED AND DOES IT NEGATE THE TRENDS IN GLOBAL WARMING?

Short answer: perhaps and no. It is hypothesized that global warming increases the frequency of the polar vortex reaching the mid-latitudes; the vortex does not negate the trajectory of global warming.

The polar vortex is a low-pressure area (about 1,000 km in diameter) extending from the middle and upper troposphere into the stratosphere above the North and South Poles, generating winds that move, respectively, eastward and westward. Each pole has two vortices, creating a seasonal (autumn to spring) but relatively stable air movement in the stratosphere and a permanent but often spatially variable air movement in the troposphere (Fig. 52.1) (Waugh, Sobel & Polvani 2017). Cold and dense air masses reside below the vortices.

Affected by the difference in temperature between the poles and equator, the vortices strengthen in the winter and weaken in the summer at either pole. The vortices also wax and wane by year. When they weaken, in some cases in both vortices simultaneously (Waugh, Sobel & Polvani 2017), the tropospheric vortex (registered by the jet stream or the westerlies in the Northern hemisphere) meander as indicated in Figure 52.1, creating north–south waves in their movement around the globe. These meanders permit cold, polar air to reach the mid-latitudes. Models indicate that weakened vortices increase these movements of cold air toward the mid-latitudes by 50 per cent (Wang & Chen 2010).

Global climate warming may be decreasing the strength of the polar vortex in the Northern hemisphere, perhaps increasing the frequency of cold polar air masses reaching the mid-latitudes – this while the average Earth temperature increases! This proposition was advanced at the turn of this century and subsequently supported by a large number of studies (Kretschmer *et al.* 2018; Zhang *et al.* 2016). It holds that the reduced snow and ice cover associated with global warming (**Q37**) weakens the vortex and increases the meandering of the jet stream. The relationship is also affected by ENSO events (**Q49**). In addition, outbreaks of cold winter air appear to be shifting away from North America and toward Eurasia (Zhang *et al.* 2016). Given the relatively brief time frame of the research in question, this proposition remains uncertain, constituting a hypothesis awaiting more

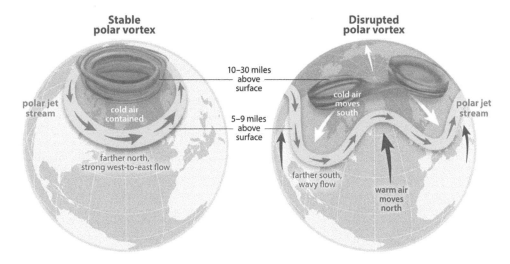

Figure 52.1 Polar vortex variations, Northern hemisphere
With the northern polar vortex in its stable state (left), the jet stream (westerlies) in the troposphere has a non-sinuous route. The build-up of heat in the tropics pushes energy poleward, generating waves in the jet steam which permit the cold polar air to penetrate into the mid-latitudes (right).
Source: NOAA (2021).

data and testing. Importantly, these bursts of cold air are not sufficient to change the global warming trends associated with the increasing greenhouse gas concentrations in the atmosphere.

References

Kretschmer, M. *et al.* (2018). "More-persistent weak stratospheric polar vortex states linked to cold extremes". *Bulletin of the American Meteorological Society* 99(1): 49–60. doi.org/10.1175/BAMS-D-16-0259.1.

NOAA (National Oceanic and Atmospheric Administration) (2021). "Understanding the polar vortex". www.climate.gov/media/11999.

Wang, L. & W. Chen (2010). "Downward Arctic oscillation signal associated with moderate weak stratospheric polar vortex and the cold December 2009". *Geophysical Research Letters* 37(9). doi.org/10.1029/2010GL042659.

Waugh, D., A. Sobel & L. Polvani (2017). "What is the polar vortex and how does it influence weather?" *Bulletin of the American Meteorological Society* **98(1): 37–44. doi.org/10.1175/BAMS-D-15-00212.1.**

Zhang, J. *et al.* (2016). "Persistent shift of the Arctic polar vortex towards the Eurasian continent in recent decades". *Nature Climate Change* 6(12): 1094–99. doi.org/10.1038/nclimate3136.

HAS HUMAN ACTIVITY DAMAGED
THE OZONE LAYER?

Short answer: yes. Emissions of chlorofluorocarbons and hydrochlorofluorocarbons thin the ozone layer, especially over polar areas, reducing the layer's capacity to protect life from ultraviolet radiation.

Human-induced emissions of chlorofluorocarbons (CFCs) and hydrochlorofluorocarbons (HCFCs) that reach the stratosphere thin the protective ozone layer. That layer, with a density of about 10 ppm of ozone (O_3), exists between 15 km to 35 km in the lower stratosphere (Fig. 53.1). It is created by an interconversion of solar radiation breaking up oxygen molecules (O_2), which recombine as O_3 in a constant cycle. In total there is about 3 billion tonnes of ozone in the ozone layer, a minuscule proportion (0.00006 per cent) of the gases comprising the atmosphere. This thin layer is essential for the biosphere because it protects life on Earth from excessive exposure to ultraviolet radiation, which can cause cataracts, genetic damage and skin cancer as well as impair photosynthesis of vegetation (Matsumi & Kawasaki 2003).

CFCs and HCFCs do not occur in nature. They (and bromofluorocarbons) are synthetic compounds invented to overcome the dangers of explosive-prone ammonia-based refrigerants, but are also used in a wide assortment of industrial products. Large-scale use of CFCs in the middle of the twentieth century increased their emissions, and their stability in the atmosphere allowed them to rise into the stratosphere. There, ultraviolet radiation breaks down the CFCs, releasing chlorine, foremost via nitrous oxide, depleting O_3 (Chipperfield *et al.* 2010; Solomon *et al.* 2016). The resulting thinning of the ozone layer is especially significant over the polar regions as extreme cold and certain cloud formations stimulate this depletion.

The identification of this process in Antarctica was announced in 1985 (Farman, Gardiner & Shanklin 1985), drawing rapid global attention. An international convention, the Montreal Protocol (1987), to rid the use of ozone-depleting substances followed, substituting the use of HCFCs which were planned to be phased out as well (Chipperfield *et al.* 2010). The protocol appeared to be working, leading to a thickening of the ozone layer over Antarctica. In 2019 this thinning was the smallest in area since the detection of the "ozone hole", varying between 10 Mkm² and 16.4 Mkm² (NASA 2020) (Fig. 53.2). In 2018 and 2020, however, owing to natural factors, the hole increased in area, but was smaller than it would have been "naturally" had not ozone-depleting substances been reduced.

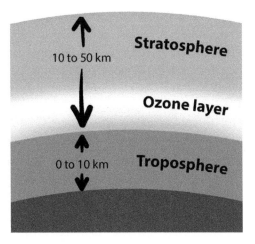

Figure 53.1 The ozone layer
The ozone layer protects life on Earth from ultraviolet radiation by reducing the amount of that radiation penetrating the layer. This layer is "ultra" thin by atmospheric standards, averaging about 300 Dobson units or 3 mm thick.
Source: author.

Unfortunately, recent observations indicate that the use of ozone-depleting substances may be on the rise with major increases in HFC-23 in the atmosphere, a by-product in the production of HCFCs (Park *et al.* 2021; Stanley *et al.* 2020). This increase is associated with

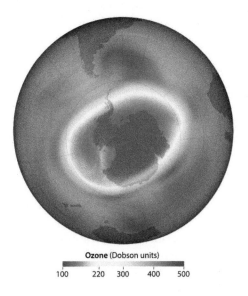

Ozone (Dobson units)
100   220   300   400   500

Figure 53.2 The ozone "hole" 2020: ozone thinning over Antarctica
The yellow to red colouring indicates decreasing thickness of the ozone layer over virtually all of Antarctica in September 2020. 100 Dobson units measure about 1 mm. The area, 24.8 Mkm², was about three times the size of the continental United States.
Source: NASA (2020).

a thinning of lower stratosphere ozone between 60° N and S and the poles (Ball *et al.* 2018). Another issue is the release of CFCs that have been taken up by the oceans (as a sink): by the mid-2100s it is expected CFC-11 will be released from the oceans to the atmosphere at a rate of ~0.5 Gt yr^{-1}, a source not accounted for in the Montreal Protocol (Wang *et al.* 2021). Finally, recall as well that CFCs and HCFCs are also powerful greenhouse gases (**Q44**).

The recent increases in HFC-23 notwithstanding, the success of the Montreal Protocol in reducing CFCs/HCFCs worldwide has shifted attention to other ODS (ozone-depleting substances). Today, N_2O (nitrous oxide) emissions, which serve as a greenhouse gas (**Q44**), appear to be the most important ODS (Bouwman *et al.* 2013; Ravishankara, Daniel & Portmann 2009). Human activity has increased natural emissions of N by ~20 per cent or 5.5–8.2 TgN yr^{-1} (Kanter *et al.* 2013).

References

Ball, W. *et al.* (2018). "Evidence for a continuous decline in lower stratospheric ozone offsets ozone layer recovery". *Atmospheric Chemistry and Physics* 18: 1379–94. doi.org/10.5194/acp-2017-862.

Bouwman, L. *et al.* (2013). *Drawing Down N2O to Protect Climate and the Ozone Layer: A UNEP Synthesis Report.* United Nations Environment Programme (UNEP). nora.nerc.ac.uk/id/eprint/505848.

Chipperfield, M. *et al.* (2010). "Detecting recovery of the stratospheric ozone layer". *Nature* 549(7671): 211–18. doi.org/10.1038/nature23681.

Farman, J., B. Gardiner & J. Shanklin (1985). "Large losses of total ozone in Antarctica reveal seasonal ClO_x/NO_x interaction". *Nature* 315(6016): 290–98. doi.org/10.7208/9780226284163-039.

Kanter, D. *et al.* (2013). "A post-Kyoto partner: considering the stratospheric ozone regime as a tool to manage nitrous oxide". *Proceedings of the National Academy of Sciences* 110(12): 4451–7. doi.10.1073/pnas.1222231110.

Matsumi, Y. & M. Kawasaki (2003). "Photolysis of atmospheric ozone in the ultraviolet region". *Chemical Reviews* 103(12): 4767–82. doi.org/10.1021/cr0205255.

NASA (National Aeronautics and Space Administration) (2020). "Large, deep Antarctic ozone hole in 2020". earthobservatory.nasa.gov/images/147465/large-deep-antarctic-ozone-hole-in-2020#.

Park, S., *et al.* (2021). "A decline in emissions of CFC-11 and related chemicals from eastern China". *Nature* 590: 433–7. doi.org/10.1038/s41586-021-03277-w.

Ravishankara, A., J. Daniel & R. Portmann (2009). "Nitrous oxide (N_2O): the dominant ozone-depleting substance emitted in the 21st century". *Science* 326(5949): 123–5. doi:10.1126/science.1176985.

Solomon, S. *et al.* (2016). "Emergence of healing in the Antarctic ozone layer". *Science* 353(6296): 269–74. doi:10.1126/science.aae0061.

Stanley, K. *et al.* (2020). "Increase in global emissions of HFC-23 despite near-total expected reductions". *Nature Communications* 11(1): 1–6. doi.org/10.1038/s41467-019-13899-4.

Wang, P. *et al.* (2021). "On the effects of the ocean on atmospheric CFC-11 lifetimes and emissions". *Proceedings of the National Academy of Sciences* 118(12). e2021528118. doi.org/10.1073/pnas.2021528118.

DO CITIES AMPLIFY LOCAL TEMPERATURES AND PRECIPITATION AND CONTRIBUTE TO GLOBAL WARMING?

Short answer: yes. Cities create urban heat islands that amplify temperatures and increase precipitation compared with the ambient atmospheric conditions, generate waste heat with long-distant consequences, and are the largest source of greenhouse gas emissions and other pollutants affecting global climate warming.

Cities and especially large metro-areas and megalopolises (i.e. spatial concentration of metro-areas) affect local and regional climates in various ways. The concentration of impervious surfaces (i.e. buildings, roadways, pavements, car parks) in cities, the energy used to maintain urban populations and functions (e.g. heating and cooling, transportation) and, in many cases, the decrease in trees and vegetation, have the effect of reradiating longwave radiation, emitting waste heat and reducing evapotranspiration. In the process, a bubble of air develops over the city that is warmer than the surrounding ambient (regional) air, a condition referred to as the urban heat island (UHI) effect (Fig. 54.1). The difference between the rural ambient and UHI tends to be larger at night than during the day (Deilami, Kamruzzaman & Liu 2018; Martilli, Krayenhoff & Nazarian 2020) owing to the amount of longwave radiation emitted by the city contrasted with the cooling of the rural, ambient air (Kalnay & Cai 2003). The intensity of the UHI is related to building configurations, urban landscaping, and the geographic–climatic conditions of city locations. Large metropolitan areas can create "regional" heat islands (Georgescu *et al.* 2014; Kalnay & Cai 2003). Overall, climate change in tandem with the UHI effect are expected to amplify cities' temperatures and, hence, environmental and human health problems (Gago *et al.* 2013; Ward *et al.* 2016). Research indicates that by 2100, cities on average will have increased in temperatures by as much as 4.4°C (7.9°F) (Zhao *et al.* 2020).

The UHI effect also generates unstable air over the city, affecting rainfall. A global meta-analysis reveals that precipitation is enhanced by 16 per cent over cities and by 18 per cent 20–50 km downwind from the city centres (Liu & Niyogi 2019). The expanse of Houston, Texas, for example, amplified the amount of precipitation from Hurricane Harvey, contributing to the extensive flooding of that city in 2017 (Zhang *et al.* 2018). In the context of global climate change, however, variations in temperature and precipitation are evident in

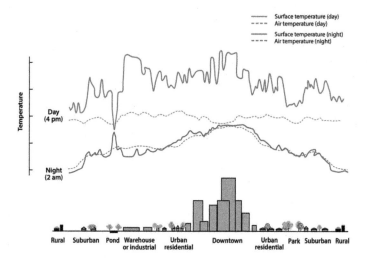

Figure 54.1 The urban heat island (UHI) effect
The UHI effect is created by land-cover and waste heat emitted in cities. Impervious surfaces and structures capture shortwave radiation during the day, convert it to longwave radiation, and reradiate it into the city. This process wanes during the night, as does waste heat from transportation.
Source: USGS-EROSC (2019).

observational and modelling assessments worldwide. For example, long-term observational data for 139 cities in India indicates significant variability in temperature and precipitation as the global climate has warmed (Mohammed & Goswami, 2019). Likewise, climate change models of several African cities indicate variability in precipitation frequency and intensity (Abiodun *et al.* 2017).

The various attributes of urbanization (e.g. scale of transportation, industry, energy use) that give rise to the UHI effect, make cities the loci of greenhouse gas emissions (GHG) (Du & Xia, 2018) and waste heat (Zhang, Cai & Hu 2013). Furthermore, significant amounts of these GHG emissions originate from a few large cities worldwide (Fig. 54.2) (Marcotullio *et al.* 2013; Moran *et al.* 2018). One assessment calculates that while all cities contribute 48.6 per cent of all GHG emissions globally, 38.8 per cent is emitted by the 50 largest urban areas (Marcotullio *et al.* 2013). Another assessment of 13,000 cities globally indicates that the largest 100 emitted 50 per cent of all carbon dioxide emissions (Moran *et al.* 2018). As such, urban activities are major drivers of climate warming. These and other emissions also generate local to regional air pollution, especially particulate matter and ozone, which have become especially dangerous in many cities within developing economies (Becken *et al.* 2017). Cities are increasingly concerned with sustainability and human health, directing attention to improvements in emissions. They vary in their sources of different emissions owing to the density of occupation, transportation structure and other such characteristics (Croci, Melandri & Molteni 2011). Nevertheless, research addressing several cities in North America and Europe indicates that per capita emissions per year in terms of carbon dioxide equivalents are being reduced annually by 0.27 t (Kennedy, Demoullin & Mohareb 2012), owing largely to improved efficiency in stationary combustion (e.g. fuels used for electricity).

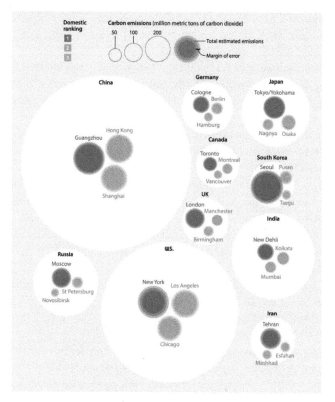

Figure 54.2 Cities with high carbon dioxide emissions

The ten highest carbon footprint countries in the world are shown, along with the three urban clusters that contribute the most carbon emissions to each national total. Researcher-calculated carbon emissions are based on household consumption only, so the data do not include emissions related to building infrastructure or transportation, for example. *Source*: Miller (2018) (adapted from Moran *et al.* 2018, crediting A. Montañez). © (2018) *Scientific American*, a division of Nature America, Inc. All rights reserved.

In addition, recent research indicates that waste heat from cities (e.g. transportation, heat-cooling of buildings and longwave radiation from impervious surfaces), especially in the mid-latitudes, alters the jet stream and other atmospheric processes, which in turn affects temperatures at distances of thousands of miles from the cities. Remote areas in northern North America and northern Asia, for example, may be increased by 1°C (1.8°F) owing to this effect. Overall, however, this waste heat only increases the global average temperature by 0.01°C because it constitutes a minuscule 0.3 per cent of heat transported at altitude. Nevertheless, its "place-based" impacts are significant.

References

Abiodun, B. *et al.* (2017). "Potential impacts of climate change on extreme precipitation over four African coastal cities". *Climatic Change* 143: 399–413. doi.org/10.1007/s10584-017-2001-5.

Becken, S. *et al.* (2017). "Urban air pollution in China: destination image and risk perceptions". *Journal of Sustainable Tourism* 25(1): 130–47. doi.org/10.1080/09669582.2016.1177067.

Croci, E., S. Melandri & T. Molteni (2011). "Determinants of cities' GHG emissions: a comparison of seven global cities". *International Journal of Climate Change Strategies and Management* 3(3): 275–300. doi:10.1108/17568691111153429.

Deilami, K., M. Kamruzzaman & Y. Liu (2018). "Urban heat island effect: a systematic review of spatio-temporal factors, data, methods, and mitigation measures". *International Journal of Applied Earth Observation and Geoinformation* **67: 30–42. doi.org/10.1016/ j.jag.2017.12.009.**

Du, W. & X. Xia (2018). "How does urbanization affect GHG emissions? A cross-country panel threshold data analysis". *Applied Energy* **229: 872–83. doi.org/10.1016/ j.apenergy.2018.08.050.**

Gago, E. *et al.* (2013). "The city and urban heat islands: a review of strategies to mitigate adverse effects". *Renewable and Sustainable Energy Reviews* 25: 749–58. doi.org/10.1016/ j.rser.2013.05.057.

Georgescu, M. *et al.* (2014). "Urban adaptation can roll back warming of emerging megapolitan regions". *Proceedings of the National Academy of Sciences* 111(8): 2909–14. doi.org/10.1073/ pnas.1322280111.

Kalnay, E. & M. Cai (2003). "Impact of urbanization and land-use change on climate". *Nature* 423(6939): 528–31. doi.org/10.1038/nature01675.

Kennedy, C., S. Demoullin & E. Mohareb (2012). "Cities reducing their greenhouse gas emissions". *Energy Policy* 49: 774–7. doi.org/10.1016/j.enpol.2012.07.030.

Liu, J. & D. Niyogi (2019). "Meta-analysis of urbanization impact on rainfall modification". *Scientific Reports* **9(1): 1–14. doi.org/10.1038/s41598-019-42494-2.**

Marcotullio, P. *et al.* (2013). "The geography of global urban greenhouse gas emissions: An exploratory analysis". *Climatic Change* 121(4): 621–34. doi.org/10.1007/s10584-013-0977-z.

Martilli, A., E. Krayenhoff & N. Nazarian (2020). "Is the urban heat island intensity relevant for heat mitigation studies?" *Urban Climate* 31: 100541. doi.org/10.1016/j.uclim.2019.100541.

Miller, M. (2018). "Here's how much cities contribute to the world's carbon footprint". *Scientific American*. www.scientificamerican.

Mohammed, P. & A. Goswami (2019). "Temperature and precipitation trend over 139 major Indian cities: an assessment over a century". *Modeling Earth Systems and Environments* 5: 1481–93. doi. org/10.1007/s40808-019-00642-7.

Moran, D. *et al.* (2018). "Carbon footprints of 13,000 cities". *Environmental Research Letters* 13: 064041. doi.org/10.1088/1748-9326/aac72a.

USGS–EROSC (US Geological Survey–Earth Resources Observation and Science Center) (2019). "Urban heat island". www.usgu.gov/media/images/urban-heat-island.

Ward, K. *et al.* (2016). "Heat waves and urban heat islands in Europe: a review of relevant drivers". *Science of the Total Environment* 569–570: 527–39. doi.org/10.1016/j.scitotenv.2016.06.119.

Zhang, W. *et al.* (2018). "Urbanization exacerbated the rainfall and flooding caused by hurricane Harvey in Houston". *Nature* 563(7731): 384–8. doi.org/10.1038/s41586-018-0676-z.

Zhang, G., M. Cai & A. Hu (2013). "Energy consumption and the unexplained winter warming over northern Asia and North America". *Nature Climate Change* 3(5): 466–70. doi.org/10.1038/ nclimate1803.

Zhao, L. *et al.* (2020). "Global multi-model projections of local urban climates". *Nature Climate Change* **11: 152–7. doi.org/10.1038/s41558-020-00958-8.**

HUMAN CHANGES TO LIFE IN THE BIOSPHERE

Coined by Eduard Suess in 1875, the biosphere (aka ecosphere) constitutes that part of the Earth system in which life exists, from microorganisms in the stratosphere to those deep below the surface of continents and near hydrothermal vents on the ocean floors (Olson *et al.* 2001) (Fig. VI.1). Regarding our species, however, the liveable biosphere is reduced to the zone from the Earth's pedosphere (soil) surface to the lower levels of the troposphere (the highest human settlement is La Rinconada, Peru, at 5,100 m), although we draw on resources from deep aquifers to glacial water melting from elevations in excess of 7,000 m.

The biosphere supports three special, interactive parts: abundant water, ample solar energy, and interactive liquid, solid and gaseous matter (Hutchinson 1970). The interaction of these parts maintains the living organisms on Earth. Humankind has not been content to abide by the vagaries of the services that nature provides, however. Rather, our history is one of increasing our capacity to obtain the availability and reliability of environmental services and reduce disservices (Section II). In this process, which is estimated to support about 9.8 billion people by 2050, humankind has not only fundamentally challenged the capacity of the Earth system to maintain the biosphere in the past (Section II), but increasingly continues today to challenge much of the life – the flora and fauna – within it.

Beyond this source of change in the biosphere are the subtle changes in ecosystems from background extinction and speciation and, more so, from mass extinctions which can change

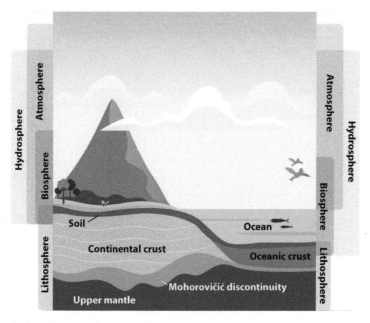

Figure VI.1 The biosphere in relation to other Earth system spheres
Biosphere for land–atmosphere (left) and oceans–atmosphere (right).
Source: adapted by author.

biogeochemical flows and evolution. Mass extinction events involve the loss of a majority of species worldwide, well beyond the rate of speciation (Table VI.1). These extinctions tend to be associated with the impacts of various forcings on the Earth system, especially those on the atmosphere and climate (Petersen, Dutton & Lohmann 2016; Veron *et al.* 2009). Modelling work, however, presents the hypothesis that ecological changes and other external forcing functions were at play in different extinctions (Rojas *et al.* 2020), including the role of stratospheric ozone depletion (Marshall *et al.* 2020). Five mass extinctions have long been noted (Table VI.1). In 2020, however, a new mass extinction, was proposed, taking place 233 Mya between the recognized third and fourth mass extinctions. It may have been

Table VI.1 Mass extinctions

Extinctions (* recently advanced; awaits consensus)	Date (Mya)	Description
Ordovician–Silurian	450–440	60–70% of species, perhaps due to climate changes
Late Devonian	375–360	70% of species but cause and timing unclear
Permian–Triassic	252	+90% of species, due to meteor, volcanic eruptions and/or oceanic methane*
*Carian (Late Triassic)	233	Large-scale volcanic eruptions in Canada
Triassic–Jurassic	201.3	70–75% of species, possibly due to volcanic-induced warming and ocean acidification
Cretaceous–Paleogene	66	75% of species, due to meteorite impact and/or volcanic-induced climate change

Source: modified from Lewis & Maslin (2015).

created by rapid warming of the Earth owing to massive volcanic activity in Canada, and perhaps gave rise to the age of dinosaurs (Dal Corso *et al.* 2020). We await consensus of these newer proposed mass extinctions in geological antiquity.

About 60–95 per cent of land or sea species disappeared in these extinction events (Jablonski & Chaloner 1994). The Permian–Triassic extinction 252 Mya, for example, witnessed a loss of up to 95 per cent of known species, apparently owing to asteroid impacts and intensive volcanic activity. Such losses in biota have major consequences for subsequent ecosystem functioning, biogeochemical cycling, including the loss of species with carbonate shells, and the evolution of biota (Hull 2015). In addition, a mass extinction is associated with the Anthropocene because of human changes in the global environment, foremost land-cover change **(Q55)**. Climate change, strongly associated with the accumulation of CO_2, is also projected to link to this extinction (Youngsteadt, López-Uribe & Sorenson 2019).

Major changes in vegetation have been examined in Section III, foremost in the land cover of the lithosphere. As such, fauna, microorganisms and corals are emphasized in this section, although various links to flora are noted, foremost to the human impacts on primary productivity.

References

Dal Corso, J. *et al.* (2020). "Extinction and dawn of the modern world in the Carian (Late Triassic)". *Science Advances* 6(38): eaba0099. doi:10.1126/sciadv.aba0099.

Hull, P. (2015). "Life in the aftermath of mass extinctions". *Current Biology* 25(19): PR941–52. doi.org/10.1016/j.cub.2015.08.053.

Hutchinson, G. (1970). "The biosphere". *Scientific American* 223(3): 44–53. www.jstor.org/stable/24925892.

Jablonski, D. & W. Chaloner (1994). "Extinctions in the fossil record and discussion". *Philosophical Transactions of the Royal Society B: Biological Sciences* 344(1307): 11–17. doi.org/10.1098/rstb.1994.0045.

Lewis, S. & M. Maslin (2015). "Defining the Anthropocene". *Nature* 519(7542): 171–80. doi:10.1038/nature14258.

Marshall, J. *et al.* (2020). "UV-B radiation was the Devonian–Carboniferous boundary terrestrial extinction kill mechanism". *Science Advances* 6(22): eaba0768. doi:10.1126/sciadv.aba0768.

Olson, D. *et al.* (2001). "Terrestrial ecoregions of the world: a new map of life on Earth". *BioScience* 51(11): 933–8. doi.org/10.1641/0006-3568(2001)051[0933:TEOTWA]2.0.CO;2.

Petersen, S., A. Dutton & K. Lohmann (2016). "End-Cretaceous extinction in Antarctica linked to both Deccan volcanism and meteorite impact via climate change". *Nature Communications* 7(1): 1–9. doi.org/10.1038/ncomms12079.

Rojas, A. *et al.* (2020). "A multiscale view of the Phanerozoic fossil record reveals the three major biotic transitions". *Communications Biology* 4(309): 866186. doi.org/10.1038/s42003-021-01805-y.

Veron, J. *et al.* (2009). "The coral reef crisis: the critical importance of < 350ppm CO_2". *Marine Pollution Bulletin* 58(10): 1428–36. doi:10.1016/j.marpolbul.2009.09.009.

Youngsteadt, E., M. López-Uribe & C. Sorenson (2019). "Ecology in the sixth mass extinction: detecting and understanding biotic interactions". *Annals of the Entomological Society of America* 112(3): 119–21. doi.org/10.1093/aesa/saz007.

Q55

IS HUMAN ACTIVITY CREATING A NEW MASS EXTINCTION?

Short answer: yes. Changes in land covers and their fragmentation, climate change and ocean acidification–warming are reducing biota globally, leading to a sixth mass extinction.

The extinction of biota is a continual process, but so is speciation (the evolution of new species). A mass extinction, in contrast, is a rapid and widespread loss in multicellular species (i.e. biodiversity) on Earth at rates far higher than speciation. The number of mass extinctions – commonly described in terms of global fauna – that have taken place differs somewhat by the measures involved in the assessments, although five are commonly recognized across geological time, and an additional one in the ancient past has been proposed (Table VI.1).

In contrast to past mass extinction events or ecological conditions in nature, humankind has long been involved in the extinction of biota, from the large-scale megafauna loss event **(Q10)** to multiple individual species extinctions, especially on small islands throughout the world (Steadman *et al.* 2020). Today, however, human-induced loss of biota is considered by many experts to constitute a sixth mass extinction, assuming the conventional standard of loss occurred in the five previous events (Fig. 55.1 and Fig. 55.2). This Anthropocene extinction has escalated as human activity has increasingly transformed the Earth system (Ceballos & Ehrlich 2018).

Accounting for the >450,000 plant species, 5–11 million animal species (the range created by uncertainty, especially for insects and fungi) and 0.7–2.2 million marine animals lost, an overall but conservative estimate of the rate of the current extinction is ~100 per million species years (or MSY) compared with a background rate of ~1 extinction per MSY (Pimm *et al.* 2014). Recent research indicates that another 515 species worldwide are currently on the brink of extinction because their population is below 1,000 individuals (Ceballos, Ehrlich & Raven 2020). Many of these endangered species are megafauna (carnivores ≥15 kg or 33 lbs and herbivores ≥100 kg or 220 lbs), species for which their losses have especially negative impacts on ecosystem metabolism and functions, with implications for the biosphere (Enquist *et al.* 2020).

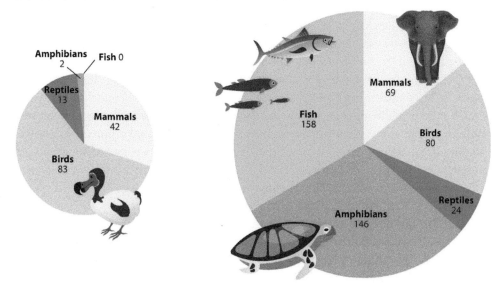

1500–1900
Extinctions: 140

Since 1900
Extinctions: 477

Figure 55.1 Mass extinction in the Anthropocene: the vertebrates
Source: adapted by author based on Ceballos & Ehrlich (2018).

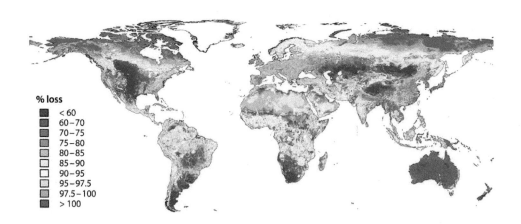

% loss
- < 60
- 60–70
- 70–75
- 75–80
- 80–85
- 85–90
- 90–95
- 95–97.5
- 97.5–100
- > 100

Figure 55.2 Global terrestrial biodiversity losses compared to intact ecosystems
Biodiversity loss in oceans is excluded.
Source: Purvis (2016).

Human activity that transforms and fragments landscapes and ecosystems worldwide (Q16), and that generates global climate (Q46) and ocean warming and ocean acidification (Q39, Q41), triggers losses in biota at various rates of loss. The adaptations by terrestrial biota to these changes tend to be slower than the environmental changes under way, leading to the loss of biodiversity. Many terrestrial biota cannot move as rapidly as their favoured environments under climate change, especially given the impediments of habitat fragmentation in their movement. Underestimation of losses exists for various reasons (Hortal & Santos 2020; Neubauer *et al.* 2021), such as the increasing pace of climate change (Chen *et al.* 2011). In addition, organisms in the highly diverse tropical environments may be more physiologically sensitive to climate change, leading to increased extinctions in the low latitudes (Deutsch *et al.* 2008). Warmer climates also increase fire patterns, frequency and intensity (i.e. fire regime) from natural and human causes, with impacts on biota. In tandem with landscape fragmentation, the new fire regimes are extinguishing biota globally (Kelly *et al.* 2020). Perhaps as much as 32 per cent of animal species is decreasing (Ceballos, Ehrlich & Dirzo 2017), although such assessments have been challenged (Briggs 2017). Likewise, carbon dioxide uptake is warming and acidifying the oceans (Q41), with special threats to the shell-based biota (Turley & Findlay 2016).

Since several mass extinctions have taken place in the past and life has recovered, why the concern over that ongoing in the Anthropocene? A major answer resides in the role of biota in ecosystem functioning and the environmental services (Q7) on which society depends, from food production to clean water and climate regulation. Recognition of this role and the Anthropocene mass extinction has given rise to the Intergovernmental Science-Policy Platform on Biodiversity and Ecosystem Services (IPBES) to combat and mitigate the losses under way (Pascual *et al.* 2017). Various assessments project that up to 64 Mkm^2 or 44 per cent of the terrestrial surface of the planet require various conservation strategies to halt current biodiversity losses (Allan *et al.* 2022).

References

Allan, J. *et al.* (2022). "The minimum land area requiring conservation attention to safeguard biodiversity" *Science*, 376 (6597): 1094-1101. doi/10.1126/science.abl9127

Briggs, J. (2017). "Emergence of a sixth mass extinction?" *Biological Journal of the Linnean Society* 122(2): 243–8. doi.org/10.1093/biolinnean/blx063.

Ceballos, G. & P. Ehrlich (2018). "The misunderstood sixth mass extinction". *Science* 360(6393): 1080–81. doi:10.1126/science.aau0191.

Ceballos, G., P. Ehrlich & R. Dirzo (2017). "Biological annihilation via the ongoing sixth mass extinction signaled by vertebrate population losses and declines". *Proceedings of the National Academy of Sciences* 114(30): E6089–96. doi.org/10.1073/pnas.1704949114.

Ceballos, G., P. Ehrlich & P. Raven (2020). "Vertebrates on the brink as indicators of biological annihilation and the sixth mass extinction". *Proceedings of the National Academy of Sciences* 117: 13596–602. doi.org/10.1073/pnas.1922686117.

Chen, I. *et al.* (2011). "Rapid range shifts of species associated with high levels of climate warming". *Science* 333(6045): 1024–6. doi:10.1126/science.1206432.

Deutsch, C. *et al.* (2008). "Impacts of climate warming on terrestrial ectotherms across latitude". *Proceedings of the National Academy of Sciences* 105(18): 6668–72. doi.org/10.1073/pnas.0709472105.

Enquist, B. *et al.* (2020). "The megabiota are disproportionately important for biosphere functioning". *Nature Communications* 11: 699. doi.org/10.1038/s41467-020-14369-.

Hortal, J. & A. Santos (2020). "Rethinking extinctions that arise from habitat loss". *Nature* 584: 238–43. doi.org/10.1038/d41586-020-02210-x.

Kelly, L. *et. al* (2020). "Fire and biodiversity in the Anthropocene". *Science* 370(96519): eabb0355. doi.org/10.1126/science.abb0355.

Neubauer, T. *et al.* (2021). "Current extinction rate in European freshwater gastropods greatly exceeds that of the late Cretaceous mass extinction". *Communications Earth & Environment* 2(1): 1–7. doi.org/10.1038/s43247-021-00167-x.

Pascual, U. *et al.* (2017). "Valuing nature's contributions to people: the IPBES approach". *Current Opinion in Environmental Sustainability* 26/27: 7–16. doi.org/10.1016/j.cosust.2016.12.006.

Pimm, S. *et al.* (2014). "The biodiversity of species and their rates of extinction, distribution, and protection". *Science* 344(6187): 1246752. doi:10.1126/science.1246752.

Purvis, A. (2016). "Vast biodiversity database now available to all". Natural History Museum. www.nhm.ac.uk/discover/news/2016/december/vast-biodiversity-database-now-available-to-all.html.

Steadman, D. *et al.* (2020). "Bird population and species lost to Late Quaternary environmental change and human impact in the Bahamas". *Proceedings of the National Academy of Sciences* 117(43): 26833–41. doi.org/10.1073/pnas.2013368117.

Turley, C. & H. Findlay (2016). "Ocean acidification". In T. Letcher (ed.), *Climate Change: Observed Impacts on Planet Earth*, 271–93. Amsterdam: Elsevier.

Q56

HAS THE HUMAN MOVEMENT OF DOMESTICATED AND OTHER BIOTA CHANGED THE EARTH SYSTEM?

Short answer: yes. The worldwide movement of biota by humankind has transformed landscapes and primary productivity, decreased the diversity of domesticates and increased the emission of greenhouse gases.

Significant changes in environments and the Earth system at large have resulted from the intentional and unintentional societal redistribution of plants and animals worldwide. This redistribution escalated from the sixteenth to the eighteenth century, focused on domesticated plants and animals (i.e. domesticates), with consequences from the diseases and pests carried by the biota and those who delivered them (**Q12**). This global movement of domesticates and the technologies associated with their use reconfigured landscapes and ecosystems globally and, in the Americas, significantly reduced the indigenous populations (Turner & Butzer 1992). From the twentieth century to the present, the impacts have been strongly linked to the areal extent and intensity of contemporary agriculture and its impacts on primary productivity, possible changes in marine productivity, the reduction in the domesticated plants provisioning society, and Earth system impacts from dietary changes and production systems.

Net primary productivity (NPP) is the organic material remaining after photosynthesis – the transformation of sunlight energy into chemical energy by plants and phytoplankton. The alteration of the world's land cover and human-induced climate change affect the amount of NPP generated on the land surface (Moore *et al.* 2019). That amount, over-whelmingly for agriculture, constitutes the human appropriation of NPP (HANPP). This appropriation was calculated to be 1.8 GtC yr^{-1} by 2005 (Krausmann *et al.* 2013), although local-to-regional, fine-scale HANPP assessments report much higher figures (Khoury *et al.* 2014). Figure 56.1 illustrates the small amount of NPP residing in extant terrestrial ecosystems (the purple lines) relative to harvested land systems, whether agriculture or various disturbed ecosystems (e.g. forests logged of timber).

In addition to the land surface, plankton floating in the oceans from the surface to ~300 m (~900 ft) delivers ~60 per cent of global NPP (~50 x 10^{15} gC yr^{-1}), taking CO_2 from the atmosphere, thus warming and acidifying the waters. These changes and the impacts on nutrient upwelling, which plankton needs for photosynthesis, suggest negative impacts on

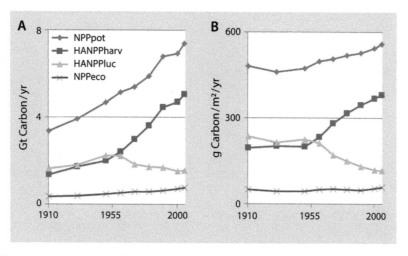

Figure 56.1 Human appropriated net primary productivity (NPP)
Harvested NPP (HANPPharv, red lines), land-change losses in NPP (HANPPlc, green lines) and NPP remaining in ecosystems (NPPeco, purple lines) compared with potential NPP (NPPpot, blue lines), before human interventions. A and B are two different measures of NPP change.
Source: Krausmann *et al.* (2013).

NPP as well. Recognizing the uncertainty in projecting the impacts, it is likely that ocean NPP will moderately decrease as warming takes place (Boyd, Sundby & Pörtner 2014).

Currently, domesticates fulfil an ever-increasing demand for food, fibre and fodder, affecting the Earth system in various ways, three of which draw special attention. First, industrial agriculture – following the green revolution (hybrid crops) to the new, genomic era – has focused on increasing the calories and protein in a few staple foods and fodder, such as wheat, rice, maize, potato (Table 56.1), creating similarities in food and feed stocks worldwide and leading to the loss of many other plant domesticates that were grown historically (Gold & McBurney 2012). Second, the dominant high yielding crops invariably require irrigation and high levels of fertilization. Water withdrawals stress hydrology globally as ~70 per cent of water use worldwide is for cultivation (**Q32–34**). Irrigation, foremost for wet rice production, is a major source of methane (CH_4). Fertilizers, in turn, are a major source of nitrous oxide (N_2O). In short, to produce the food we eat generates significant emissions of greenhouse gases (**Q44**). Third, the global demand for meat consumption continues to rise, especially for beef. Cattle, sheep and goats are major ruminants, animals whose stomach structure produces CH_4, providing ~17 per cent of emissions of this greenhouse gas (Ellis 2007; Knapp 2014). As the consumption of beef, lamb, mutton and dairy products increases globally, so does livestock-sourced methane (Hayek & Miller 2021), although various research seeks to reduce the amount of methane emitted per production unit (Eshel 2021).

In addition, and consistent with the global spread of domesticates, human activity has altered ecosystems by the intentional and unintentional transfer of vegetation globally.

Table 56.1 Major staple crops in 2020, ranked by production

1	Maize (corn)
2	Wheat
3	Rice
4	Potato
5	Soy(bean)
6	Sorghum
7	Sweet potato
8	Yam
9	Cassava
10	Plantain/banana

Source: adapted by author from FAOSTAT 2021 data: www.fao.org/faostat/en/#data/QCL.

Several examples illustrate this. Cattle grazed from southern Mexico to South America historically confronted the problem of inappropriate native grasses for foraging, in part because this vast area had few herd-grazing animals before European contact. Over time, African grasses were introduced to serve as pasture, but these alien or exotic species have had various unintended impacts. The African grass, *Brachiaria decumbens* (palisades grass), used in Amazonian pastures, is a source for especially intensive fires that burn into the forest edge, reducing tropical forests (Silvério *et al.* 2013). Another, *Cenchrus ciliaris* (buffel grass), introduced in northern Mexico and Arizona for ranching, is an aggressive invader that has spread across the landscapes of the southern United States, Central America and various parts of South America, with similar results in Australia (Marshall, Lewis & Ostendorf

Figure 56.2 Buffel grass invasion in southern Arizona
Source: Aaryn Olsson, Arizona-Sonoran Desert Museum.

2012) (Fig. 56.2). It alters wildfire regimes and rates of soil erosion, among other impacts (Brenner & Franklin 2017). Many other examples exist of alien species becoming invasive, such as cacti in southern and eastern Africa, which reduce grassland fodder for livestock and wildlife (Shackleton *et al.* 2017). Invasive grasses now cover about one-fifth (>77,000 km^2) of the Great Basin rangelands in the western United States (Smith *et al.* 2021). Alien species, assisted by land transformations connecting ecosystems, are homogenizing what were once distinctive floras worldwide (Yang *et al.* 2021). Overall, invasive species change ecosystems and generate a large economic burden worldwide, estimated to approach 5 per cent of the global GDP in 2001 (Pimentel *et al.* 2001).

References

Boyd, P., S. Sundby & H.-O. Pörtner (2014). "Cross-chapter box on net primary production in the ocean". In C. Field *et al.* (eds), *Climate Change 2014: Impacts, Adaptation, and Vulnerability. Part A: Global and Sectoral Aspects. Contribution of Working Group II to the Fifth Assessment Report of the Intergovernmental Panel on Climate Change*, 133–6. Cambridge: Cambridge University Press, Cambridge.

Brenner, J. & K. Franklin (2017). "Living on the edge: emerging environmental hazards on the peri-urban fringe". *Environment: Science and Policy for Sustainable Development* 59(6): 16–29. doi:10.1080/00139157.2017.1374793.

Ellis, J. (2007). "Prediction of methane production from dairy and beef cattle". *Journal of Dairy Science* 90(7): 3456–66. doi.org/10.3168/jds.2006-675.

Eshel, G. (2021). "Small-scale integrated farming systems can abate continental-scale nutrient leakage". *PLoS biology* 19(6): e3001264. doi.org/10.1371/journal.pbio.3001264.

Gold, K. & R. McBurney (2012). "Conservation of plant diversity for sustainable diets". In B. Burlingame & S. Dernini (eds), *Sustainable Diets and Biodiversity: Directions and Solutions for Policy, Research and Action*. Rome: FAO.

Hayek, M. & S. Miller (2021). "Underestimates of methane from intensively raised animals could undermine goals of sustainable development". *Environmental Research Letters* 16(6): 063006. doi.org/10.1088/1748-9326/ac02ef.

Knapp, J. (2014). "Enteric methane in dairy cattle production: quantifying the opportunities and impact of reducing emissions". *Journal of Dairy Science* 97(6): 3231–61. doi.org/10.3168/jds.2013-7234.

Khoury, C. *et al.* (2014). "Increasing homogeneity in global food supplies and the implications for food security". *Proceedings of the National Academy of Sciences* 111(1): 4001–06. doi.org/10.1073/pnas.1313490111.

Krausmann, F. *et al.* (2013). "Global human appropriation of net primary production doubled in the 20th century". *Proceedings of the National Academy of Sciences* 110(25): 10324–9. doi.org/10.1073/pnas.1211349110.

Marshall, V., M. Lewis & B. Ostendorf (2012). "Buffel grass (*Cenchrus ciliaris*) as an invader and threat to biodiversity in arid environments: a review". *Journal of Arid Environments* 78: 1–12. doi.org/10.1016/j.jaridenv.2011.11.005.

Moore, C. *et al.* (2019). "Human appropriated net primary productivity of complex mosaic landscapes". *Frontiers in Forests and Global Change* 2: 38. doi.org/10.3389/ ffgc.2019.00038.

Pimentel, D. *et al.* (2001). "Economic and environmental threats of alien plant, animal, and microbe invasions". *Agriculture, Ecosystems & Environment* 84(1):1–20. doi.org/10.1016/ S0167-8809(00)00178-X.

Shackleton, R. *et al.* (2017). "Distribution and socio-ecological impacts of the invasive alien cactus *Opuntia stricta* in eastern Africa". *Biological Invasions* 19(8): 2427–41. doi.org/10.1007/ s10530-017-1453-x.

Silvério, D. *et al.* (2013). "Testing the Amazon savannization hypothesis: fire effects on invasion of a neotropical forest by native Cerrado and exotic pasture grasses". *Philosophical Transactions of the Royal Society B: Biological Sciences* 368(1619): 20120427. doi.org/10.1098/rstb.2012.0427.

Smith, J. *et al.* (2021). "The elevation ascent and spread of exotic annual grass dominance in the Great Basin, USA". *Diversity and Distributions* 28(1): 83–96. doi:10.1111/ddi.13440.

Turner, B. & K. Butzer (1992). "The Columbian encounter and land-use change". *Environment: Science and Policy for Sustainable Development* 34(8): 16–44. doi.org/10.1080/ 00139157.1992.9931469.

Yang, Q. *et al.* (2021). "The global loss of floristic uniqueness". *Nature Communications* 12: 7290. doi.org/10.1038/s41467-021-27603-y.

DO POLLUTANTS FROM HUMAN ACTIVITIES DEGRADE LAND BIOTA WORLDWIDE?

Short answer: yes. Beyond greenhouse gas impacts, atmospheric pollution kills vegetation and retards plant growth, whereas nano- and microplastics accumulate in plant foods, reducing plant growth and food-plant quality.

Beyond greenhouse gas emissions, a myriad of pollutants from human activities endanger biota globally, ranging from toxic spills to tropospheric pollutants. Two of the more important tropospheric impacts derive from acid rain and ozone, generated by industrial-energy production and petrol engines. Acid rain (Q43) has destroyed forests and affected agricultural production especially in the northeastern United States, and southeastern Canada, the Black Triangle in the Czech Republic, Germany and Poland, and northeastern China (Liu *et al.* 2019; Navrátil *et al.* 2016). Various responses to acid rain in the 1990s have significantly reduced the problem, however, and presuming that clean air regulations remain in place and enacted, the larger effects of sulphur dioxide and nitrogen oxides on flora can be largely mitigated.

Ozone (O_3) plays an important role in the stratosphere (Q53) to protect Earth from harmful ultraviolet radiation and serves as a greenhouse gas. In the troposphere, however, it is a pollutant that is harmful to life and a major constituent of photochemical smog (Fig. 57.1). These consequences exist even though approximately only 10 per cent of atmospheric O_3 resides in the troposphere. It damages plant foliage, including crop plants and trees, reducing their productivity (Krupa *et al.* 2001), among other problems, which include delays in phenology (i.e. seasonal or cyclic change in biota). Ozone is assisted in these problems with other pollutants (e.g. NO_2, NO_x, and PM) (Jochner *et al.* 2015). Recent studies in China indicate that O_3 reduces tree biomass by 11–13 per cent, and rice and wheat yield by 8 per cent and 6 per cent, respectively (Feng *et al.* 2019). Ozone also affects the human respiratory system in a number of ways, and may be linked to heart disease and type 2 diabetes. It is estimated to increase mortality in China by 28,000 to 74,000 annually. Such losses alongside those in crop yields decreases China's GDP by about 7 per cent (Feng *et al.* 2019).

Added to the well-known array of pollutants, it now appears that land plants can accumulate nanoplastics (<100 nm) because they can be internalized by vascular plants through

Figure 57.1 Ozone pollution, Mexico City
Source: Gonzales (2010).

their roots, inhibiting the plant's ability to absorb water and reducing plant growth (Sun *et al.* 2020) as well as increasing the plants' toxicity (Yin *et al.* 2021). Microplastics (<10 μm) within water, fruits and vegetables are ingested by humans (Conti *et al.* 2020), as much as 5 g each week. The health ramifications are not yet clear (de Wit & Bigaud 2019). Nevertheless, the huge array of sources of plastics entering the environment, and their decomposition to microplastics and nanoplastics finding their way into all forms of biota, water and soil, pose large concerns for life, and are a focus of review (Kumar *et al.* 2020).

References

Conti, G. *et al.* (2020). "Micro- and nano-plastics in edible fruit and vegetables. The first diet risks assessment for the general population". *Environmental Research* 187: 109677. doi.org/10.1016/j.envres.2020.109677.

de Wit, W. & N. Bigaud (2019). "No plastic in nature: assessing plastic ingestion from nature to people". Gland, Switzerland: World Wildlife Fund. awsassets.panda.org/downloads/plastic_ingestion_press_singles.pdf.

Gonzales, F. (2010). "Aerial view of Mexico City". en.wikipedia.org/wiki/File:AerialViewMexicoCity. jpg.

Jochner, S. *et al.* (2015). "The effects of short-and long-term air pollutants on plant phenology and leaf characteristics". *Environmental Pollution* 206: 382–9. doi.org/10.1016/j.envpol.2015.07.040.

Feng, Z. *et al.* (2019). "Economic losses due to ozone impacts on human health, forest productivity and crop yield across China". *Environment International* 131: 104966. doi.org/10.1016/j.envint.2019.104966.

Kumar, M. *et al.* (2020). "Current research trends on micro-and nano-plastics as an emerging threat to global environment: a review". *Journal of Hazardous Materials* 409: 124967. doi.org/10.1016/j.jhazmat.2020.124967.

Krupa, S. *et al.* (2001). "Ambient ozone and plant health". *Plant Disease* 85(1): 4–12. doi.org/10.1073/pnas.1814880116.

Liu, M. *et al.* (2019). "Ammonia emission control in China would mitigate haze pollution and nitrogen deposition, but worsen acid rain". *Proceedings of the National Academy of Sciences* 116(16): 7760–65.

Navrátil, T. *et al.* (2016). "Soil mercury distribution in adjacent coniferous and deciduous stands highly impacted by acid rain in the Ore Mountains, Czech Republic". *Applied Geochemistry* 75: 63–75. doi.org/10.1016/j.apgeochem.2016.10.005.

Sun, X. *et al.* (2020). "Differentially charged nanoplastics demonstrate distinct accumulation in *Arabidopsis thaliana*". *Nature Nanotechnology* 15: 755–60.

Yin, L. *et al.* (2021). "Interactions between microplastics/nanoplastics and vascular plants". *Environmental Pollution* 290: 117999. doi.org/10.1016/j.envpol.2021.117999.

HAS HUMAN ACTIVITY REDUCED
MARINE FISH STOCKS?

Short answer: yes. Overfishing and ocean warming and acidification have reduced the amount and size of fish stocks, increased their movement poleward and degraded marine ecosystem dynamics.

Human activity affects marine biota both directly and indirectly. About 160 tonnes of marine foods are taken from catch fishing and aquaculture/mariculture each year (Brander 2007). Increases in aquaculture produce, perhaps, about one-half of all seafood consumed today (Campbell & Pauly 2013). The marine stock, however, has been overfished, reducing catches throughout the oceans of the world. About 55–66 per cent of exclusive economic zones – up to 200 nautical miles (~370 km) from the basal coastline – worldwide were overfished by 2004 (Srinivasan *et al.* 2010). Unsustainable catch fishing was in play at least 40 years earlier (Coll *et al.* 2008).

In addition to direct human impacts, human-induced ocean warming and acidification (**Q38, Q41**) are changing marine ecosystems and their dynamics, and driving fish stocks, commercial and otherwise, toward the polar waters (Engelhard, Righton & Pinnegar 2014). One estimate holds that the maximum sustainable yield of different fish stocks fell globally by 4.1 per cent between 1930 and 2010, with some ecoregions decreasing by 15–35 per cent (Free *et al.* 2019). Either estimate is linked to ocean warming (Fig. 58.1). Overfishing and ocean warming are decreasing the body size of commercial fish (Genner *et al.* 2010). Open-ocean aquaculture, in contrast, may have positive and negative impacts from ocean warming, depending on the region in question (Klinger, Levin & Watson 2017).

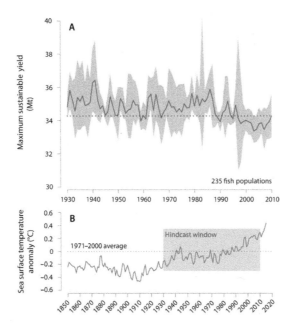

Figure 58.1 Maximum sustainable yield and sea surface temperature
To hindcast or backcast is to use a model to determine if an outcome is robust for the empirically verifiable past. Such modelling demonstrates a declining maximum sustainable yield (A) for 235 fish populations with increasing sea surface temperature anomaly (B) for the 1971–2000 average. Grey area = uncertainty range.
Source: Free *et al.* 2019. Reprinted with permission of the American Association for the Advancement of Science.

References

Brander, K. (2007). "Global fish production and climate change". *Proceedings of the National Academy of Sciences* 104(50): 19709–14. doi.org/10.1073/pnas.0702059104.

Campbell, B. & D. Pauly (2013). "Mariculture: a global analysis of production trends since 1950". *Marine Policy* 39: 94–100. doi.org/10.1016/j.marpol.2012.10.009.

Coll, M. *et al.* (2008). "Ecosystem overfishing in the ocean". *PLoS One* 3(12): e3881. doi.org/10.1371/journal.pone.0003881.

Engelhard, G., D. Righton & J. Pinnegar (2014). "Climate change and fishing: a century of shifting distribution in North Sea cod". *Global Change Biology* 20(8): 2473–83. doi.org/10.1111/gcb.12513.

Free, C. *et al.* (2019). "Impacts of historical warming on marine fisheries production". *Science* 363(6430): 979–83. doi:10.1126/science.aau1758.

Genner, M. *et al.* (2010). "Body size-dependent responses of a marine fish assemblage to climate change and fishing over a century-long scale". *Global Change Biology* 16(2): 517–27. doi.org/10.1111/j.1365-2486.2009.02027.x.

Klinger, D., S. Levin & J. Watson (2017). "The growth of finfish in global open-ocean aquaculture under climate change". *Proceedings of the Royal Society B: Biological Sciences* 284(1864): 20170834. doi.org/10.1098/rspb.2017.0834.

Srinivasan, U. *et al.* (2010). "Food security implications of global marine catch losses due to overfishing". *Journal of Bioeconomics* 12(3): 183–200. doi.org/10.1007/s10818-010-9090-9.

Q59

HAS HUMAN ACTIVITY AFFECTED CORAL REEFS?

Short answer: yes. Near-shore reef degradation from pollution and other direct human impacts has been superseded by ocean warming and acidification that bleaches and kills ecosystems.

Corals are marine invertebrates living in ocean colonies that construct reefs. These reefs occupy no more than 0.1 per cent of the ocean area, but support up to 25 per cent of marine biota, making them hotspots of biodiversity (Stuart-Smith *et al.* 2013). They constitute a significant source of income for people worldwide, ranging from fishing to tourism. The Coral Triangle (Fig. 59.1) in the far southwest Pacific Ocean is considered the epicentre of marine

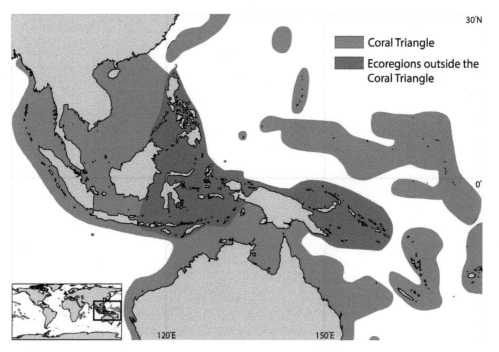

Figure 59.1 The Coral Triangle and nearby major coral areas
Source: Huang *et al.* (2018).

biodiversity and a region in which a significant number of people obtain their livelihoods from the coral (Allen 2008).

Corals are not only threatened worldwide, but by 2011 about 19 per cent of known reefs were dead. This loss follows from local impacts on near-shore reefs, such as that from runoff pollution and coastal development, and from ocean warming and acidification that damage corals regardless of location (Hedley *et al.* 2016) **(Q38, Q41)**. Warming oceans leads to coral bleaching, which inhibits organic carbon development and turns the coral white (Fig. 59.2). Acidification, in turn, reduces the rate of biogenic calcium carbonate production – that which constitutes the skeleton of the coral. Research indicates that the current scale of bleaching events and its severity is unprecedented since observations began in the 1870s (Glynn 1993). With sustained greenhouse gas emissions, some projections envision large-scale losses (Cornwall *et al.* 2021) and even the extinction of coral reefs in this century (Pandolfi *et al.* 2011), a possibility that has set in motion a great number of efforts to protect and conserve them (Hedley *et al.* 2016). Local stressors, such as macroalgae, amplify coral responses to ocean warming, which suggests that management of the stressors may provide a conservation strategy in some cases (Donovan *et al.* 2021). Other research, however, notes that a 2°C increase in ocean warming appears to exceed the limits in which high-latitude corals may live (Glynn 1993).

Figure 59.2 Coral bleaching
Source: Roberts (2014).

References

Allen, G. (2008). "Conservation hotspots of biodiversity and endemism for Indo-Pacific coral reef fishes". *Aquatic Conservation: Marine and Freshwater Ecosystems* 18(5): 541–56. doi.org/10.1002/aqc.880.

Cornwall, C. *et al.* (2021). "Global declines in coral reef calcium carbonate production under ocean acidification and warming". *Proceedings of the National Academy of Sciences* 118(21): e2015265118. doi.org/10.1073/pnas.2015265118.

Donovan, M. *et al.* (2021). "Local conditions magnify coral loss after marine heatwaves". *Science* 372(6545): 977–80. doi:10.1126/science.abd9464.

Glynn, P. (1993). "Coral reef bleaching: ecological perspectives". *Coral Reefs* 12(1): 1–17. doi.org/10.1007/BF00303779.

Hedley, J. *et al.* (2016). "Remote sensing of coral reefs for monitoring and management: a review". *Remote Sensing* 8(2): 118. doi.org/10.3390/rs8020118.

Huang, D. *et al.* (2018). "The origin and evolution of coral species richness in a marine biodiversity hotspot". *Evolution* 72(2): 288–302. doi.org/10.1111/evo.13402.

Pandolfi, J. *et al.* (2011). "Projecting coral reef futures under global warming and ocean acidification". *Science* 333(6041): 418–22. doi:10.1126/science.1204794.

Roberts, K. (2014). "Bent sea rod bleaching". www.flickr.com/photos/usgeologicalsurvey/1501 1207807/.

Stuart-Smith, R. *et al.* (2013). "Integrating abundance and functional traits reveals new global hotspots of fish diversity". *Nature* 501(7468): 539–42. doi.org/10.1038/nature12529.

HAS HUMAN ACTIVITY AFFECTED MICROORGANISMS AND THEIR LINKS TO THE EARTH SYSTEM?

Short answer: yes, with caveats. Human-induced climate change appears to degrade the capacity of the microbial world for photosynthesis, biogeochemical cycling and various other functions that sustain the Earth system, but some impacts enhance this capacity.

Microorganisms or microbes, the 1.2×10^{30} single-celled prokaryotic (unicellular) organisms smaller than 50 μm, are the microworld that supports the entire biosphere (Flemming & Wuertz 2019). Including bacteria, archaea, fungi, protozoa, algae and viruses, they occur in abundance in all parts of the biosphere, about evenly split between the marine and terrestrial worlds. Bacteria and archaea alone may approach $\sim 1 \times 10^{30}$ cells (Flemming & Wuertz 2019).

Microbes play key roles in carbon and nutrient cycling, biota health and all food webs. For example, microbes in soils maintain more carbon than that in the atmosphere and vegetation, and are essential for the capacity of vegetation to pull carbon dioxide from the atmosphere and regulate nutrients for vegetation (Fig. 60.1). Soil microbes are affected by climate change directly through impacts on soil carbon stocks, and indirectly by, for example, vegetation changes with carbon feedbacks (Classen *et al.* 2015). In addition, over multiple hundreds of years, this carbon moves deep into the Earth's crust, ultimately being subducted by ocean plates into the mantle (Section III introduction), where some of it is released back into the atmosphere as CO_2. Recent research, however, indicates that between 2–22 per cent of this soil carbon is maintained by microbial activity near the surface where it can be released back into the atmosphere (Fullerton *et al.* 2021). Significantly, 90 per cent of marine biomass is microbial, fixing carbon and nitrogen, and generating about 50 per cent of global oxygen (Cavicchioli *et al.* 2019) (Fig. 60.2). In addition, marine microbes consume a large majority of CH_4 that percolates from the ocean's floor, preventing the greenhouse gas from reaching the atmosphere (Marlow *et al.* 2021). Climate warming processes, in turn, affect microbes in the oceans via warming and acidification and on land via interactive temperature increases.

These impacts on the microbial world alter the functioning of microbes and the consequences of this alteration affects carbon storage, food webs and productivity. This functioning and the associated consequences are complex. For instance, the majority of

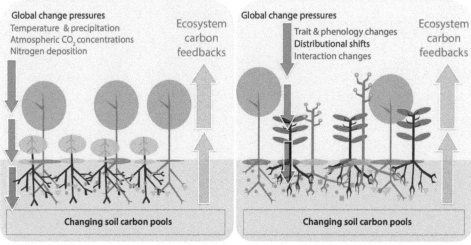

Global change pressures
Temperature & precipitation
Atmospheric CO₂ concentrations
Nitrogen deposition

Ecosystem carbon feedbacks

Global change pressures
Trait & phenology changes
Distributional shifts
Interaction changes

Ecosystem carbon feedbacks

Changing soil carbon pools

Changing soil carbon pools

A. The **direct effects of climate change** act on organism process rates and inorganic resources availability to directly influence carbon pools and feedbacks to the atmosphere.

B. The **indirect effects of climate change** shift a number of above- and belowground properties (e.g. diversity, community composition and functional traits of plants, microbes and other organisms) as well as the interactions among organisms, thus indirectly influencing carbon pools and feedbacks.

Figure 60.1 Direct and indirect climate change interactions on microbes
Source: Classen *et al.* (2015).

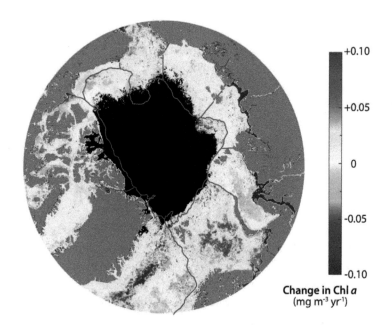

+0.10

+0.05

0

-0.05

-0.10

Change in Chl *a*
(mg m⁻³ yr⁻¹)

Figure 60.2 Increasing phytoplankton productivity in the Arctic, 1998–2018
Increases in primary productivity of marine microbes from 1998–2018 measured as mg m³ yr⁻¹ of chlorophyll a (Chl ***a***).
Yellow to red equal increases to +0.1 mg m³ yr⁻¹. Aqua to blue equal decreases. Black (much of the Arctic Ocean) = no data.
Grey lines are subregions.
Source: Lewis, van Dijken & Arrigo 2020.

research indicates a negative impact on the microbial world owing to the effects of warming and acidification. The primary productivity of the Arctic Ocean, however, increased 75 per cent between 1998 and 2018 due to climate reductions in ice cover and increased phytoplankton biomass (Ardyna & Arrigo 2020; Frey *et al.* 2018). This response, of course, follows from the loss of Arctic sea ice and its high albedo (**Q5**) and the subsequent warming of this ocean, with feedbacks on global warming (**Q38**).

References

Ardyna, M. & K. Arrigo (2020). "Phytoplankton dynamics in a changing Arctic Ocean". *Nature Climate Change* 10: 892–903. doi.org/10.1038/s41558-020-0905-y.

Cavicchioli, R. *et al.* (2019). "Scientists' warning to humanity: microorganisms and climate change". *Nature Reviews Microbiology* 17(9): 569–86. doi.org/10.1038/ s41579-019-0222-5.

Classen, A. *et al.* (2015). "Direct and indirect effects of climate change on soil microbial and soil microbial-plant interactions: what lies ahead?" *Ecosphere* 6(8): 1–21. doi.org/10.1890/ ES15-00217.1.

Flemming, H. & S. Wuertz (2019). "Bacteria and archaea on Earth and their abundance in biofilms". *Nature Reviews Microbiology* 17(4): 247–60. doi.org/10.1038/s41579-019-0158-9.

Frey, K. *et al.* (2018). "Arctic Ocean primary productivity: the response of marine algae to climate warming and sea ice decline". Arctic Report Card. www.arctic.noaa.gov/Report-Card/Report-Card-2018/ArtMID/7878/ArticleID/778/Arctic-Ocean-Primary-Productivity-The-Response-of-Marine-Algae-to-Climate-Warming-and-Sea-Ice-Decline.

Fullerton, K. *et al.* (2021). "Effect of tectonic processes on biosphere–geosphere feedbacks across a convergent margin". *Nature Geoscience* 14(5): 301–06. doi.org/10.1038/s41561-021-00725-0.

Lewis, K., G. van Dijken & K. Arrigo (2020). "Changes in phytoplankton concentration now drive increased Arctic Ocean primary production". *Science* 369(6500): 198–202. doi.org/10.1126/ science.aay8380.

Marlow, J. *et al.* (2021). "Carbonate-hosted microbial communities are prolific and pervasive methane oxidizers at geologically diverse marine methane seep sites". *Proceedings of the National Academy of Sciences* 118(25): e2006857118. doi:10.1073/pnas.2006857118.

Q61

DOES THE BUILT ENVIRONMENT OF CITIES ALTER BIOTA DYNAMICS?

Short answer: yes. Urban areas can reduce biomass, alter species and their dynamics, and drive signatures of phenotypic change across taxa with unknown biodiversity impacts.

Given the sheer amount of constructed buildings and impervious surfaces, and the reconfiguration of flora, such as yards and gardens planted with turf grass and the use of exotic or alien species, cities constitute transformed environments. Cities lost about 40,000 ha of trees globally between 2012 and 2017, largely to impervious surfaces and buildings (Nowak & Greenfield 2020). For arid land cities, however, irrigated residential parcels and public spaces may actually increase tree and turf grass (biomass) cover (Hope *et al.* 2003), typically with large impacts on water withdrawal. The loss or gain in "green land cover" notwithstanding, in most cases cities are characterized by amplified temperatures compared with the ambient rural temperature (**Q54**).

Conventional wisdom holds that cities tend to exhibit reductions in the richness (number) of non-domesticated animal species present in them, but increases in the abundance of the species that remain. A recent meta-analysis of terrestrial animals, however, finds abundances decrease and richness trends are uncertain in the process of urbanization (Saari *et al.* 2016). A study of an arid-land city found that, over a five-year period, the number of bird species declined for both native species on xeric-landscaped areas and for exotic species on mesic-landscaped areas (Warren *et al.* 2019). Such conditions and dynamics of the built (artificial) environment of cities have given rise to a relatively new field of study, urban ecology (McPhearson *et al.* 2016),

Recent work indicates that urban-driven phenotypic (observable properties of an organism) change creates distinctive signatures across taxa, a change that appears to be greater than natural or non-urban ones (Alberti *et al.* 2017). Such conditions may accelerate eco-evolutionary change in biota (Alberti, Marzluff & Hunt 2017). These conditions include habitat modification, social interactions, biotic interactions, heterogeneity and novel disturbances (Hope *et al.* 2013). Among the consequences is that urban land-cover complexity may increase rapid phenotypic change among species with consequences on ecosystem functioning (Alberti, Marzluff & Hunt 2017).

Other aspects of urban-biodiversity appear to be contentious or affected by the species in question. For example, some studies conclude that, accounting for species richness, urban area avian fauna maintain more functional diversity (roles played in the environment) than non-urban avian fauna (Warren *et al.* 2019), whereas other studies report less functional diversity, which in turn reduces environmental services (Oliveira Hagen *et al.* 2017; Sol *et al.* 2020). As urban ecology remains a relatively new field of study, the role that an increasingly urbanizing planet will have on biotic diversity and evolution locally to globally requires much more research (Des Roches *et al.* 2012; McPhearson *et al.* 2016).

References

Alberti, M., *et al.* (2017). "Global urban signatures of phenotypic change in animal and plant populations". *Proceedings of the National Academy of Sciences* 114(34): 8951–56. doi.org/10.1073/pnas.1606034114.

Alberti, M., J. Marzluff & V. Hunt (2017). "Urban driven phenotypic changes: empirical observations and theoretical implications for eco-evolutionary feedback". *Philosophical Transactions of the Royal Society B: Biological Sciences* 372(1712): 20160029. doi.org/10.1098/rstb.2016.0029.

Des Roches, S. *et al.* (2021). "Socio-eco-evolutionary dynamics in cities". *Evolutionary Applications* 14(1): 248–67. doi.org/10.1111/eva.13065.

Hope, D. *et al.* (2003). "Socioeconomics drive urban plant diversity". *Proceedings of the National Academy of Sciences* 100(15): 8788–92. doi.org/10.1073/pnas.1537557100.

McPhearson, T. *et al.* (2016). "Advancing urban ecology toward a science of cities". *BioScience* 66(3): 198–212. doi.org/10.1093/biosci/biw002.

Nowak, D. & E. Greenfield (2020). "The increase of impervious cover and decrease of tree cover within urban areas globally (2012–2017)". *Urban Forestry & Urban Greening* 49: 126638. doi.org/10.1016/j.ufug.2020.126638.

Oliveira Hagen, E. *et al.* (2017). "Impacts of urban areas and their characteristics on avian functional diversity". *Frontiers in Ecology and Evolution* 5: 84. doi.org/10.3389/fevo.2017.00084.

Saari, S. *et al.* (2016). "Urbanization is not associated with increased abundance or decreased richness of terrestrial animals-dissecting the literature through meta-analysis". *Urban Ecosystems* 19(3): 1251–64. doi.org/10.1007/s11252-016- 0549-x.

Sol, D. *et al.* (2020). "The worldwide impact of urbanisation on avian functional diversity". *Ecology Letters* 23(6): 962–72.

Warren, P. *et al.* (2019). "The more things change: species losses detected in Phoenix despite stability in bird–socioeconomic relationships". *Ecosphere* 10(3): e02624. doi.org/10.1002/ecs2.2624.

SECTION VII

THE HUMAN CAUSES OF THE ANTHROPOCENE

The environmental sciences, especially those addressing the operation of the Earth system or ecosystems, refer to phenomena or processes that influence an outcome in the system as *forcings* and *drivers*. A natural forcing on climate, for instance, results from small changes in solar energy arriving at the outermost portions of the atmosphere; climate change, in turn, may be described as a driver of ecosystem change. Forcings and drivers may also be used to infer causes. These two terms are occasionally used in the social sciences addressing environmental concerns, although the use of the term *cause* has been historically more common. This use across the sciences at large implies that the factors (forcings or drivers) responsible for the outcomes are not only empirically demonstrated but are or potentially understood sufficiently to be set within an explanatory structure, otherwise known as theory. Much of the research on anthropic changes in the Earth system uses the term drivers, rather than causes. That dealing with societal factors that create the human activity generating the changes commonly refers to causes.

Different research communities tend to be interested in different ultimate or primary causes as they relate to mediating variables within causal sets of explanatory factors. These sets or causal chains are common in explanations of human–environmental relationships, and various explanatory chains may employ the same variable (or factor) as either primary or mediating in the chain. Depending on their placement in the chain, the independent variables may be labelled proximate or distal in kind. Those most immediate to the outcome are proximate and tend to have strong empirical relationships to the outcome or dependent variable. A recent meta-analysis of global deforestation, for example, found strong empirical associations among (a) agricultural activity, (b) soil suitability, (c) roads, and local (d) urban areas and (e) population (Busch & Ferretti-Gallon 2017). These factors can be considered proximate causes because they are the immediate activities or events replacing forest cover (a), the environmental or infrastructural condition of the land favouring its selection to be deforested (b, c), and the common source of demand for the action taken (d, e). Distal factors, in contrast, tend to maintain less strong relationships with the outcome owing to the complexity and statistical noise (random irregularities) created by the number of factors in the explanatory chain (Lambin *et al.* 2001; Turner *et al.* 2020). A hypothetical deforestation case linked to international policy provides an example as follows: forest lost to (1) orchard cultivation generated by (2) new roads opening those lands to markets in response to (3) expanding markets for orchard crops, assisted by (4) government policy subsidizing and insuring speciality tree crop production owing to (5) an international agreement to maintain tree cover while improving country-level economy. In this explanatory chain, the more distal causes are (5) and (4) and the more proximate are (1) and (2). For a specific international policy–deforestation link, strong empirical associations among the chain of factors – often pursued through a process called progressive contextualization (Vayda 1983) – may be established, even between the primary (most distal, 5) factor and the deforestation. The associations between the international policy agreement and deforestation across multiple cases, however, tend to weaken the distal link to forests owing to variations in the intervening variables or their parameters or the overriding socio-economic and political context in question.

Much, if not most, theory and hypotheses of human-induced causes of environmental change were formulated historically to address natural resource concerns (e.g. land cultivated or freshwater pollution) in somewhat simple cause–consequence formulations, often called middle-range theory in the social sciences and theory in the natural sciences (Meyfroidt *et al.* 2018; Chowdhury & Turner 2019). Increasingly, they have been expanded to address environmental services (e.g. water filtration) (**Q7**) or the conditions of the Earth system (e.g. climate change) (**Q3**). This history, however, is challenged by the current realization that human–environmental problems need to be addressed as social–environmental systems (SESs) (**Q71**) in which the factors at play are interactive and maintain more complex relationships than are advanced in the many chain explanations. Using the illustration above, the consequence of deforestation may feedback to various parts of the explanatory chain, such as tree crops reducing environmental services that lower the quality of the produce in question and requiring increases in subsidies to maintain the cultivation. Such linkages and feedbacks are common in SES models. They are less apparent among social science explanations (Busch & Ferretti-Gallon 2017) but are central to general systems approaches dealing with the broad dynamics of SESs.

Finally, it is noteworthy that the ultimate causes of the anthropogenic changes to the Earth system are contested on various fronts. These challenges involve the focus and approaches of different disciplines or research communities interested in different aspects of societal behaviour that change the starting points (ultimate causes) and mediating factors in question (Hertz & Schlüter 2015; Verburg *et al.* 2019). These disciplines and communities also maintain different means of determining the validity of the explanatory propositions in which they engage (e.g. statistical robustness of relationships or the efficacy of the argument). Finally, interest in different spatiotemporal scales of problems, such as long-term and global versus short-term and local, may influence differences in the relationships found among proximate–distal causes.

These disputes are apparent in the Q&As of this section. The various factors at play have champions and opponents, and prove useful at some spatial and temporal scales and not at others. The aim here is not to promote any causal factor but to present the broader character of their uses. The emphasis is on causes as they have been articulated in science and the social sciences that adhere to normal science-based explanatory perspectives. As such, this section does not delve into the intricacies of structural (e.g. critical theory) and constructivist (e.g. certain social theory) causes or general system theory, although various issues raised in these ways of knowing are identified.

References

Busch, J. & K. Ferretti-Gallon (2017). "What drives deforestation and what stops it? A meta-analysis". *Review of Environmental Economics and Policy* 11(1): 3–23. doi.org/10.1093/reep/rew013.

Chowdhury, R. & B. Turner II (2019). "The parallel trajectories and increasing integration of landscape ecology and land system science". *Journal of Land Use Science* 14(2): 135–54. doi.org/10.1080/1747423X.2019.1597934.

Hertz, T. & M. Schlüter (2015). "The SES-framework as boundary object to address theory orientation in social–ecological system research: the SES-TheOr approach". *Ecological Economics* 116: 12–24. doi.org/10.1016/j.ecolecon.2015.03.022.

Lambin, E. *et al.* (2001). "The causes of land-use and land-cover change: moving beyond the myths". *Global Environmental Change* 11(4): 261–69. doi.org/10.1016/S0959-3780(01)00007-3.

Meyfroidt, P. *et al.* (2018). "Middle-range theories of land system change". *Global Environmental Change* 53: 52–67. doi.org/10.1016/j.gloenvcha.2018.08.006.

Turner II, B. *et al.* (2020). "Framing the search for a theory of land use". *Journal of Land Use Science* 15: 1–20. doi.org/10.1080/1747423X.2020.1811792.

Vayda, A. (1983). "Progressive contextualization: methods for research in human ecology". *Human Ecology* 11(3): 265–81. doi.org/10.1007/BF00891376.

Verburg, P. *et al.* (2019). "Beyond land cover change: towards a new generation of land use models". *Current Opinion in Environmental Sustainability* 38: 77–85. doi.org/10.1016/j.cosust.2019.05.002.

DOES IPAT EXPLAIN ENVIRONMENTAL CHANGE?

Short answer: technically no; inferentially perhaps? IPAT is an identity associating population, affluence and technology to environmental degradation; its correlation with that degradation, in some cases, infers causation, however.

The original IPAT formulation was offered as an identity – an equation that is always true. The equation states that environmental impact (I) is the product of the multiplicative relationship of population (P), affluence (A) and technology (T), or I = P × A × T (Ehrlich & Holdren 1971), where P constitutes the amount of the human population, A is per capita consumption, and T is production and consumption intensity or efficiency, commonly captured in terms of energy use. Whereas P constitutes the multiplier of average per capita consumption, each member of the population has some minimum physiological requirements, and for an overwhelming proportion of the global population, per capita consumption is much larger than this minimum. As such, A may be seen as consumption beyond the P minimum and is related to economic wealth. The intensity or efficiency of the materials and energy used to produce and consume, including waste, is T. Recognizing the interactions among PAT, the equation can be restructured as I = f(PAT). Either formulation of IPAT essentially denotes that two variables – P & A – capture the demand for resources and one – T – represents the impacts of that demand in the provisioning process. IPAT is a less formalized version of the Kaya identity developed to account for CO_2 emissions (Ma & Cai 2018). Reworking IPAT in the Kaya way, that is, $I = P \, x \dfrac{A}{P} x \dfrac{T}{A} x \dfrac{I}{T}$, means all the terms cancel one another, leaving I = I. The original formulation, therefore, was not offered as a causal expression but one of logic in which PAT combined was I and thus a change in PAT generated a change in I.

PAT variables tend to be strongly correlated with I over the long haul and at the regional or global levels, capturing broad-stroke environmental changes under similar PAT conditions. For example, deforestation and agricultural expansion in imperial China and in medieval western Europe were similar (Tuan 1968). Similarly, the scale of environmental changes in antiquity in the Nile Basin, Tigris–Euphrates river valleys, Basin of Mexico and the Andean region were comparable, given the PAT conditions prevailing in each case. Today, global assessments demonstrate strong correlations between more populous and more wealthy urban areas (i.e. high levels of P, A and T) and carbon footprints. Indeed, the 100 urban areas with the highest carbon emissions generate 18 per cent of the global carbon footprint (Moran *et al.* 2018) (Fig. 62.1). The statistical strength of population, affluence and technology with environmental change at macro-levels of assessments lends these variables in many global-level modelling exercises engaging Earth system impacts (Meadows, Randers & Meadows 2004; Weyant 2017) and raises the inference in some communities that they are the underlying causes of the impacts.

Such inferences generate evidential and conceptual challenges, especially among some social science research communities. PAT, for example, may have weak correlations with I at the sub-global scale or over short time spans. In such cases, the PAT variables may not match the spatial resolution (location, area) of the environmental change or involve a considerable lag time between the changes in PAT versus I. Others claim that the formulation is too simplistic for the complexity of human–environmental relationships, involving not only interactions among PAT but the differing causal forces underpinning these three variables, such as governance structures, political economies and cultures, which account for the ultimate causes. These challenges have led to various reformulations of identity (Dasgupta & Ehrlich 2013; Waggoner & Ausubel 2002; York, Rosa & Dietz 2003). Yet others find the

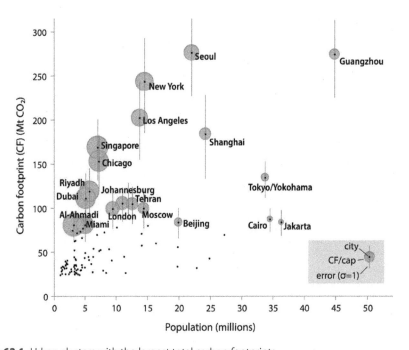

Figure 62.1 Urban clusters with the largest total carbon footprints

The urban clusters with the largest carbon footprint (top 20 cities labelled, typically calculated on an annual basis) are created by a combination of population size and high per capital footprints. Vertical lines = one standard deviation for each labelled city footprint estimate. Different carbon footprint calculators derive different footprints for the same city. *Source*: Moran *et al.* (2018).

formulation problematic, even dangerous, owing to the consequences that may follow from its use. For example, rapidly growing, impoverished, rural populations may be seen as the cause of environmental problems owing to high levels of population growth as opposed to the inequality of opportunities, access to land and capital, or nearby urban population demands on rural resources (DeFries *et al.* 2010; Hosonuma *et al.* 2012). In addition, the strong role of P in the formulation may be used to support decisions to control population, raising various moral and religious concerns (Cripps 2017).

IPAT and the role of population, affluence and technology as identity, driver or cause is not likely to be resolved owing to the varied, context-driven correlations with environmental change, the scales of analysis of interests, and the world views of the analysts.

References

Cripps, E. (2017). "Population, climate change, and global justice: a moral framework for debate". *Journal of Population and Sustainability* **1: 23–36. doi:10.3197/jps.2017.1.2.23.**

Dasgupta, P. & P. Ehrlich (2013). "Pervasive externalities at the population, consumption, and environment nexus". *Science* **340(6130): 324–8. doi:10.1126/science.1224664.**

DeFries, R. *et al.* (2010). "Deforestation driven by urban population growth and agricultural trade in the twenty-first century". *Nature Geoscience* 3(3): 178–81. doi.org/10.1038/ngeo756.

Ehrlich, P. & J. Holdren (1971). "Impact of population growth: complacency concerning this component of man's predicament is unjustified and counterproductive". *Science* **171(3977): 1212–17. doi:10.1126/science.171.3977.1212.**

Hosonuma, N. *et al.* (2012). "An assessment of deforestation and forest degradation drivers in developing countries". *Environmental Research Letters* 7(4): 044009. doi:10.1088/1748-9326/7/4/044009.

Ma, M. & W. Cai (2018). "What drives the carbon mitigation in Chinese commercial building sector? Evidence from decomposing an extended Kaya identity". *Science of the Total Environment* 634: 884–99. doi.org/10.1016/j.scitotenv.2018.04.043.

Meadows, D., J. Randers & D. Meadows (2004). *Limits to Growth: The 30-Year Update.* London: Chelsea Green Publishing.

Moran, D. *et al.* (2018). "Carbon footprints of 13,000 cities". *Environmental Research Letters* 13(6): 064041. doi.org/10.1088/1748-9326/aac72a.

Tuan, Y. (1968). "Discrepancies between environmental attitude and behaviour: examples from Europe and China". *Canadian Geographer/Le Géographe Canadien* 12(3): 176–91. doi.org/10.1111/j.1541-0064.1968.tb00764.x.

Waggoner, P. & J. Ausubel (2002). "A framework for sustainability science: a renovated IPAT identity". *Proceedings of the National Academy of Sciences* 99(12): 7860–65. doi.org/10.1073/pnas.122235999.

Weyant, J. (2017). "Some contributions of integrated assessment models of global climate change". *Review of Environmental Economics and Policy* 11(1): 115–37. doi.org/10.1093/reep/rew018.

York, R., E. Rosa & T. Dietz (2003). "STIRPAT, IPAT and ImPACT: analytic tools for unpacking the driving forces of environmental impacts". *Ecological Economics* 46(3): 351–65.

Q63

DOES AN INCREASING POPULATION ALTER ENVIRONMENTS?

Short answer: yes. Increasing population increases demands on environmental services owing to the base subsistence that each person requires, although the means of accounting for this demand is contested.

Recognition of the role of population on resource (natural capital) demand extends to antiquity but was formalized in the western world by Thomas Malthus (**Q74**) in the eighteenth century, and subsequently advanced to address environmental services at large in the late twentieth century. The long-term evidence of global population growth is not questioned. The interpretation of this growth on human–environmental dynamics is, however, shaped by different lenses through which those relationships are viewed (**Q73**), commonly associated with different presentations (graphs) of long-term global population.

The most common expression of the global human population is captured in a linear graph (Fig. 63.1), revealing the slow growth in the population until that growth exploded exponentially from the twentieth century onwards. This graph signifies the unprecedented rates of growth in population and its demand for natural capital and environmental services (**Q7**) over the past century plus, with inferences about the environmental repercussions of the Anthropocene (**Q1**) (Desvaux 2007; Harte 2007). In contrast, a logarithmic graph of global population growth produces three major steps in growth, each of which has been associated with the great technological eras of human history: mastery of tools/fire, domestication of biota and fossil fuel technology (Fig. 63.2 and Fig II.1). The logarithmic scale itself, however, generates the steps, or the appearance of a levelling of the population when it is actually increasing (Schulze & Mealy 2001). This issue is overcome through the use of a logarithmic-linear graph demonstrating three periods of associations between increasing *rates* of population growth and technological change (Fig. 63.3). Similar to the logarithmic plot, three steps emerge associated with technological eras. Each era increased the resource base to support increasing population, but invariably led to various changes to environments and the Earth system, many of which have been degrading.

These impacts follow because population constitutes a demand variable on natural capital initially and environmental services ultimately. Each of us requires a biological base of

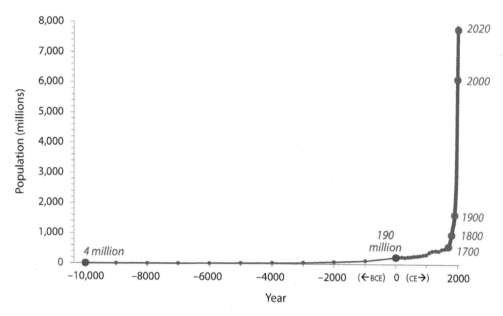

Figure 63.1 Linear plot of global population growth

Different assessments have somewhat different global population estimates for the distant past. This depiction and that for Figures 63.2 and 63.3 demonstrate these distinctions. The telling issue in the three figures is the different slopes of the growth in global population.

Source: Our World Data (2021).

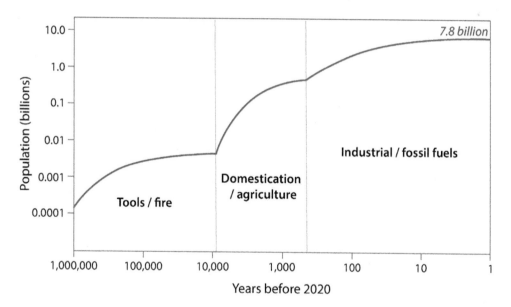

Figure 63.2 Logarithmic plot of global population growth

Three steps in global population growth associated with advances in technology eras. See Figure 63.1 regarding the population estimates.

Source: adapted from Desvaux (2007).

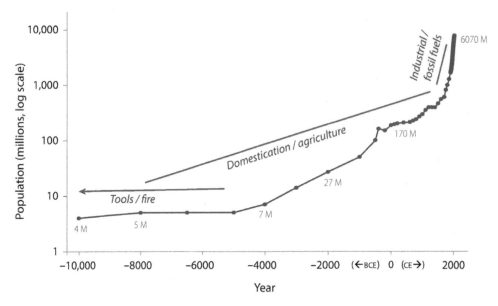

Figure 63.3 Log-linear plot of global population growth
BCE dates to the left of 0; CE dates to the right. Note the increased growth rate associated with the advent of agriculture (−4000 to 1800s) and the exponential rate with the industrial-fossil fuel phases afterwards. The arrow for the tool/fire era indicates its beginning long before 10,000 BCE. See Figure 63.1 regarding the population estimates.
Source: log population from Waldir (2010).

food, fibre and fuel; the more of our species, the larger this demand becomes, especially illustrated by holding affluence and technology constant. As an example, CO_2 emissions across countries appear to be proportional to population size (Dietz & Rosa 1997). The role of population is so self-evident that one may ask, why is it challenged as a cause of environmental change at the Earth system level or otherwise? The answer resides partly in the role of the other factors (i.e. affluence, technology, institutions, political economy) that also influence demand, including non-linear interactions that can exist among them (Harte 2007). Another part of the answer resides in calls for population control – interpreted as involuntary reductions in fertility or birth rates – which raise various societal concerns. These concerns include control of minority populations or on impoverished peoples whose livelihoods are, in many cases, assisted by large families. In addition, large inequalities exist in consumption globally. That 690 to 868 million people worldwide suffer from calorific malnutrition and two billion from insufficient micronutrients masks the reality that global food production is sufficient to feed everyone well if food were distributed equitably and food wastes were reduced, even without major changes to diets (FAO/WFP/IFAD 2012).

Importantly, growth rates in population need not always remain high, as articulated in the demographic transition theory that deals with the changes between the birth and death rates for different stages of socio-economic conditions. Prior to the industrial-science era (Fig. 63.4, Stage 1), both birth and death rates were high, creating slow rates of population growth.

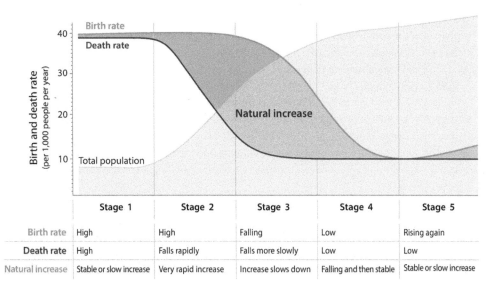

	Stage 1	Stage 2	Stage 3	Stage 4	Stage 5
Birth rate	High	High	Falling	Low	Rising again
Death rate	High	Falls rapidly	Falls more slowly	Low	Low
Natural increase	Stable or slow increase	Very rapid increase	Increase slows down	Falling and then stable	Stable or slow increase

Figure 63.4 Demographic transition theory
Source: Donoso, Cordero & Cordova (2017).

Transitioning to the industrial era (Stages 2 and 3), death rates fell owing to advances in and access to health care, whereas the birth rates lagged behind, generating high rates of population growth. As development continues towards service economies (Stages 3 and 4), the birth rates fall for various reasons (e.g. lower levels of infant and child mortality and increased costs of raising children), especially as the educational level of women rises (Lutz 2019), generating slower rates in population growth (Lesthaeghe 2014), although some arguments contradict this decline (Maier 2015; Nielsen 2016). A transition to Stage 5 is uncertain as no country has entered this proposed stage. Some projections have population modestly increasing, others not.

No part of the world remains in Stage 1, however, and most countries have begun or are well into the lower fertility conditions (Stages 3 and 4), save parts of Africa. The current world average fertility rate is low, 2.4 per cent; that to replace the current global population is 2.1 per cent. As such, the global growth rate is slowing but produces a surge in population because of a large global population base approaching 8 billion people. Such trends indicate that population as a demand on resources and the Earth system will continue to grow, if at slower rates than in the immediate past at least, until about 2050, reaching 9 billion or more people globally.

References

Desvaux, M. (2007). "The sustainability of human populations: how many people can live on Earth?" *Significance* 4(3): 102–07. doi.org/10.1111/j.1740-9713.2007.00241.x.

Dietz, T. & E. Rosa (1997). "Effects of population and affluence on CO_2 emissions". *Proceedings of the National Academy of Sciences* 94(1): 175–9. doi.org/10.1073/pnas.94.1.175.

Donoso, M., P. Cordero & N. Cordova (2017). "Economic consequences in the gross domestic product of the population decrease in the countries that are going through the final phase of their demographic transition". *Maskana* 8(2): 31–50. doi.org/10.18537/mskn.08.02.03.

FAO (Food and Agricultural Organization), WFP (World Food Programme) & IFAD (International Fund for Agricultural Development) (2012). *The State of Food Insecurity in the World.* Rome: FAO. www.fao.org/3/i3027e/i3027e00.htm.

Harte, J. (2007). "Human population as a dynamic factor in environmental degradation". *Population and Environment* 28(4/5): 223–36. doi.org/10.1007/s11111-007-0048-3.

Lesthaeghe, R. (2014). "The second demographic transition: a concise overview of its development". *Proceedings of the National Academy of Sciences* 111(51): 18112–15. doi.org/10.1073/pnas.1420441111.

Lutz, W. (2019). "Education rather than age structure brings demographic dividend". *Proceedings of the National Academy of Sciences* 116(26): 12798–803. doi.org/10.1073/pnas.1820362116.

Maier, D. (2015). "The great myth: why population growth does not necessarily cause environmental degradation and poverty". *The Public Sphere*: 150–58. psj.lse.ac.uk/articles/abstract/37/.

Nielsen, R. (2016). "Demographic transition theory and its link to the historical economic growth". *Journal of Economics and Political Economy* 3(1): 32–49.

Our World in Data (2021). "Population 10,000 BCE to 2021". ourworldindata.org/grapher/population.

Schulze, P. & J. Mealy (2001). "Population growth, technology and tricky graphs". *American Scientist Magazine* 89: 209–11. link.gale.com/apps/doc/A74455151/AONE?u=googlescholar&sid=googleScholar&xid=a2bd2234.

Waldir (2010). "World population growth (lin-log scale)". commons.wikimedia.org/wiki/File:World_population_growth_(lin-log_scale).png.

DOES INCREASING AFFLUENCE ALTER ENVIRONMENTS?

Short answer: yes. Consumption above biological need accounts for the majority of environmental and Earth system change.

In reference to IPAT (**Q63**) and its variations, affluence refers to levels of per capita consumption, commonly expressed as the average of some population unit. Simplistically, that average times the total population derives the demand of our species for natural capital and environmental services. Alternatively, affluence can be seen as the level of consumption above (or below) the minimum physiological requirements to sustain an individual. Affluence has increased dramatically with advances in technology and societal organization, such that the global average material consumption has never been higher in humankind's history, continues to increase, and probably constitutes the largest driver of resource consumption and surely greenhouse gas emissions globally (Wiedmann *et al.* 2020). The relative change in the global material footprint from 1970 to 2017 virtually matches global gross domestic product, having increased about 180 per cent between 1970 and 2017 (Fig. 64.1) (Bleischwitz *et al.* 2018).

Variations from the global average (mean) and median consumption among individuals, countries and regions exist. These variations tend to be associated with higher to lower wealth (individuals), and levels of economic development and the safety nets protecting society (countries and regions). For individuals, affluence is linked to income and economic assets, with various indicators of their impact. For example, the energy footprints of the top 10 per cent of individuals in wealth indicate a consumption of energy equal to the bottom 80 per cent (Oswald, Owen & Steinberger 2020), and that same 10 per cent generated 52 per cent of $GtCO_2$ emitted between 1990–2015 (Oxfam 2020). High income individuals have energy footprints >200 gigajoules per year (up to 300 GJ yr^{-1}), whereas 77 per cent of people globally consume <30 GJ yr^{-1}. Essentially, the Gini coefficient (metric denoting in/equality) (**Q95**) for energy footprints and CO_2 emission is large.

The ecological (or environmental) footprint – measured by the land area or global hectares (gha) needed to sustain populations with different levels of affluence (Fig. 64.2) – is another measure addressing affluence. The ecological footprint by country ranged from <1 to 16 gha

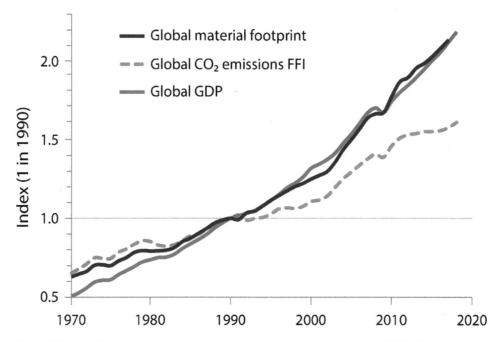

Figure 64.1 Relative change in global economic and environmental indicators, 1970–2017
Global GDP (gross domestic product in constant 2010 US dollars); Global MF (global material footprint = raw material extraction); Global CO₂ FFI (emissions from fossil fuel combustion and industrial processes); *y* axis = change relative to 1990 as 1.
Source: Wiedmann *et al.* (2020).

per capita across the world in 2020, with a global average of 2.75 gha per capita (GFN 2022). Interestingly, however, country-level footprints correspond less precisely with country GDP owing to, for example, different levels of country-specific engagements in non-fossil fuel energy use and other environmentally friendly pursuits. Nonetheless, countries with lower levels of GDP tend to have much less of an ecological footprint than those with higher levels. In addition, countries that have long been developed have produced the most greenhouse gases, a critical part of the calculation of the ecological footprint.

While economic development increases societies' affluence, its impact on the environment may not continue upwards and onwards, owing to the hypothesized environmental Kuznets curve (Dinda 2004). This hypothesis asserts that as locations or populations develop economically, their environments initially degrade (e.g. deforestation and air pollution), but once well developed, advanced technologies permit policies that alleviate much of the degradation or at least the rate of that degradation (e.g. reforestation, reduced air pollution, efficient energy production and consumption) (Fig. 64.3). The evidence to date supporting the hypothesis is equivocal (Apergis & Ozturk 2015); various works support and others do not support the hypothesis, depending on the country analysed and the environmental issues at play (Özokcu & Özdemir 2017). The curve is apparent owing to improvements in the efficiency of energy and materials used in production and waste treatment, such as

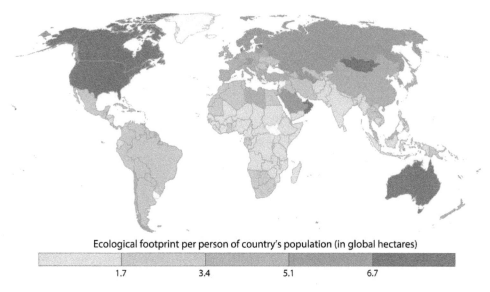

Ecological footprint per person of country's population (in global hectares)

| 1.7 | 3.4 | 5.1 | 6.7 |

Figure 64.2 Ecological footprint per capita
The ecological footprint (EF) per person is a country's total EF divided by the total population of the country.
Source: GFN (2022); basemap: Esri.

greenhouse gas emissions, especially if governments engage in appropriate environmental regulations. One study of country-scale CO_2 emissions finds that emissions level off proportionally at high levels of GDP ($10,000 GDP per capita) (Dietz & Rosa 1997). On the other hand, environmental leakage is common, in which improved environmental performance

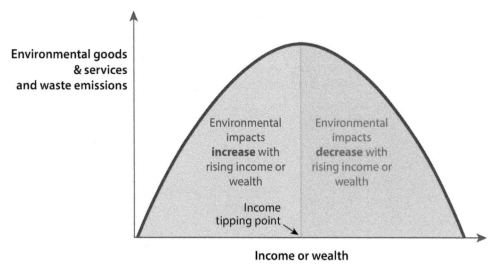

Environmental goods & services and waste emissions

Environmental impacts **increase** with rising income or wealth

Environmental impacts **decrease** with rising income or wealth

Income tipping point

Income or wealth

Figure 64.3 Environmental Kuznets curve
Source: author.

in one country, local or economic sector, such as reforestation or reduced greenhouse gas emissions, leads to deforestation or increased emissions elsewhere **(Q23)** (Kauppi *et al.* 2006; Meyfroidt & Lambin 2009).

References

Apergis, N. & I. Ozturk (2015). "Testing environmental Kuznets curve hypothesis in Asian countries". *Ecological Indicators* 52: 16–22. doi.org/10.1016/j.ecolind.2014.11.026.

Bleischwitz, R. *et al.* (2018). "Extrapolation or saturation: revisiting growth patterns, development stages and decoupling". *Global Environmental Change* 48: 86–96. doi.org/10.1016/j.gloenvcha.2017.11.008\.

Dietz, T. & E. Rosa (1997). "Effects of population and affluence on CO_2 emissions". *Proceedings of the National Academy of Sciences* 94(1): 175–9. doi.org/10.1073/pnas.94.1.175.

Dinda, S. (2004). "Environmental Kuznets curve hypothesis: a survey". *Ecological Economics* 49(4): 431–55. doi.org/10.1016/j.ecolecon.2004.02.011.

GFN (Global Footprint Network) (2022). "Ecological deficit/reserve". www.footprintnetwork.org/.

Kauppi P. *et al.* (2006). "Returning forests analyzed with the forest identity". *Proceedings of the National Academy of Sciences* 103(26): 17574–9. doi.org/10.1073/pnas.0608343103.

Meyfroidt, P. & E. Lambin (2009). "Forest transition in Vietnam and displacement of deforestation abroad". *Proceedings of the National Academy of Sciences* 106(38): 16139–44. doi.org/10.1073/pnas.0904942106.

Oswald, Y., A. Owen & J. Steinberger (2020). "Large inequality in international and intranational energy footprint between income groups and across consumption categories". *Nature Energy* 5: 231–39. doi.org/10.1038/s41560-020-0606-9.

Oxfam (2020). "Confronting carbon inequity". Oxfam Media Briefing, 21 September. oxfamilibrary.openrepository.com/bitstream/handle/10546/621052/mb-confronting-carbon-inequality-210920-en.pdf.

Özokcu, S. & Ö. Özdemir (2017). "Economic growth, energy, and environmental Kuznets curve". *Renewable and Sustainable Energy Reviews* 72: 639–47. doi.org/10.1016/j.rser.2017.01.059.

Wiedmann, T. *et al.* (2020). "Scientists' warning on affluence". *Nature Communications* 11(1): 3107. doi.org/10.1038/s41467-020-16941-y.

DOES TECHNOLOGY CREATE ENVIRONMENTAL CHANGE?

Short answer: yes. It influences the types of materials and energy used, the efficiency of production and environmental impacts of provisioning humankind, and can generate demand itself.

Technology determines the kind and amount of energy, labour and materials used in production and consumption, including waste, and the environmental impacts of these activities. The energy used or the amount of carbon emitted per unit of activity are common gross measures of technological impacts. For much of the developed world, the process of dematerialization – a reduction in the material intensity of the economy – may be taking place (Steinberger *et al.* 2013). Dematerialization means less material (e.g. wood, steel, plastic), energy and water are used to produce the products and services consumed, consistent with the concept of a green economy (Loiseau *et al.* 2016). Dematerialization within some developed economies follows from improved production processes, the substitution of new materials that are more efficient than those replaced, recycling and better use of waste. For example, total annual water withdrawal in the United States has been relatively stable since the mid-1970s owing to technological efficiencies, yet has led to no major reductions in agricultural and industrial production, and residential uses (Fig. 65.1) (Gleick & Palaniappan 2010).

Despite efficiencies in many production processes (e.g. solar energy), the evidence for dematerialization is contested, owing to the array of factors considered, long- or short-term assessment periods, and the parts of the world examined (Fix 2019; Hanley *et al.* 2009; Magee & Devezas 2017; Shao *et al.* 2017). While some economically developed countries show evidence of dematerialization (Steinberger *et al.* 2013), decarbonization is not taking place globally (Fix 2019). Various arguments maintain that economic growth and use of materials are so strongly associated in a capitalist world that they cannot be fully decoupled (Kemp-Benedict 2018). The implication is that economic development (affluence) entails increased consumption which translates to increased material consumption of resources (Kallis 2017), even though efficiency of production and consumption may take place.

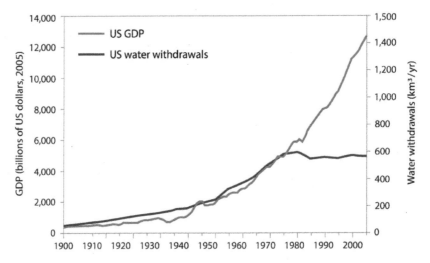

Figure 65.1 GDP and water withdrawals in the United States, 1900–2005
Source: Gleick & Palaniappan (2010).

Technology can also affect demand itself. New technologies can enlarge the global reach of items, creating new sources of products and demand for them. Historically, for example, fifteenth-century advances in ship construction permitted increased capacity for open ocean transport, rescaling the movement of resources around the world (**Q12**). Technology can

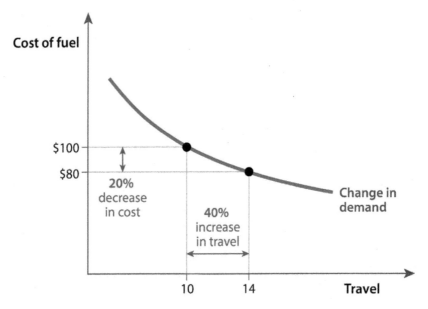

Figure 65.2 Jevons paradox
A smaller percentage decrease in the fuel creates a much larger percentage increase in travel (i.e. distance travelled or number of trips taken).
Source: en.wikipedia.org/wiki/Jevons_paradox (accessed 3 March 2022)BIB_ch65_0002.

also affect demand through the rebound phenomenon or effect as captured in the Jevons paradox. In this paradox, technological advances improving efficiency tend to lower the price of the item. This new cost, in turn, increases consumption of the item or its associated impacts, such as the travel distances associated with hybrid engine cars (Fig. 65.2), presumably not reducing gas consumption (Magee & Devezas 2017; Small & van Dender 2005). The Khazzoom–Brookes postulate elaborates this paradox whereby consumption of energy efficient items at the individual or micro-level (e.g. more individuals purchasing hybrid cars) actually increases energy use at the aggregate or macro-level (more cars used to drive further per day), leading to a new supply and demand at large that requires more energy than existed previous to the efficiency increases (Saunders 2000).

The magnitude of the technological world has given rise to the concept of the technosphere – the physical infrastructure and technological artefacts that support our complex social structures. These entities are estimated to have a mass of ~30 trillion tonnes worldwide or >50 kg/m^2 of the surfaces of the Earth (Zalasiewicz *et al.* 2017), a mass about five times greater than that of the human population (Smil 2011). Recognizing that such estimates involve a large degree of uncertainty, the sheer scale of the physical technosphere drives home the enormity of Earth's resources to support humankind, and the concern that such consumption cannot be sustained (Hoekstra & Wiedmann 2014; Schramski, Gattie & Brown 2015).

References

Fix, B. (2019). "Dematerialization through services: evaluating the evidence". *BioPhysical Economics and Resource Quality* 4(2): 6. doi.org/10.1007/s41247-019-0054-y.

Gleick, P. & M. Palaniappan (2010). "Peak water limits to freshwater withdrawal and use". *Proceedings of the National Academy of Sciences* 107(25): 11155–62. doi.org/10.1016/j.jclepro.2016.08.024.

Hanley N. *et al.* (2009). "Do increases in energy efficiency improve environmental quality and sustainability?" *Ecological Economics* 68(3): 692–709. doi.org/10.1016/j.ecolecon.2008.06.004.

Hoekstra, A. & T. Wiedmann (2014). "Humanity's unsustainable environmental footprint". *Science* 344(6188): 1114–17. doi:10.1126/science.1248365.

Kallis, G. (2017). "Radical dematerialization and degrowth". *Philosophical Transactions of the Royal Society A: Mathematical, Physical and Engineering Sciences* 375(2095): 20160383. doi.org/10.1098/rsta.2016.0383.

Kemp-Benedict, E. (2018). "Dematerialization, decoupling, and productivity change". *Ecological Economics* 150: 204–16. doi.org/10.1016/j.ecolecon.2018.04.020.

Loiseau, E. *et al.* (2016). "Green economy and related concepts: an overview". *Journal of Cleaner Production* 139: 361–71. doi.org/10.1016/j.jclepro.2016.08.024.

Magee, C. & T. Devezas (2017). "A simple extension of dematerialization theory: incorporation of technical progress and the rebound effect". *Technological Forecasting and Social Change* 117: 196–205. doi.org/10.1016/j.techfore.2016.12.001.

Saunders, H. (2000). "A view from the macro side: rebound, backfire, and Khazzoom–Brookes". *Energy Policy* 28(6/7): 439–229. doi.org/10.1016/S0301-4215(00)00024-0.

Schramski, J., D. Gattie & J. Brown (2015). "Human domination of the biosphere: rapid discharge of the earth-space battery foretells the future of humankind". *Proceedings of the National Academy of Sciences* 112(31): 9511–17. doi.org/10.1073/pnas.1508353112.

Shao, Q. *et al.* (2017). "The high 'price' of dematerialization: a dynamic panel data analysis of material use and economic recession". *Journal of Cleaner Production* 167: 120–32. doi.org/10.1016/j.techfore.2016.12.001.

Small, K. & K. van Dender (2005). "The effect of improved fuel economy on vehicle miles travels: estimating the rebound effect using US State Data 1966–2001". University of California Energy Institute. escholarship.org/uc/item/1h6141nj.

Smil, V. (2011). "Harvesting the biosphere: the human impact". *Population and Development Review* 37: 613–36. doi.org/10.1111/j.1728-4457.2011.00450.x.

Steinberger, J. *et al.* (2013). "Development and dematerialization: an international study". *PLoS One* 8(10): e70385. doi.org/10.1371/journal.pone.0070385.

Zalasiewicz, J. *et al.* (2017). "Scale and diversity of the physical technosphere: a geological perspective". *The Anthropocene Review* 4(1): 9–22. doi.org/10.1177/2053019616677743.

DO INSTITUTIONS CREATE ENVIRONMENTAL CHANGE?

Short answer: yes. The rules of access to and use of natural capital and environmental services can degrade or enhance environmental outcomes.

Institutions are the structure of governance, constituting rules, norms and values that constrain or enable societal (any collection of individuals) behaviour. They may be formal, codified rules that, for example, determine how much groundwater state authorities permit to be pumped by a farmer, or informal, social rules associated with traditions, customs, religious beliefs or moral values, such as a community's understanding of the distributions of canal water to individual farm fields. A large range of institutions govern activities involving natural capital that hold environmental consequences. The associated rules address resource access, use, and products from that use, such as rights to sow land, fish or deforest. They also address the impacts of various uses, such as air pollution emissions or limits on resource extraction to maintain resource sustainability.

Institutions contribute to the "means" by which the demand for human activity affects various parts of the environment up to and including the Earth system. For example, given similar levels of demand, private, communal or state property rights can directly influence the amount of timber harvested from the landscape. In contrast, zoning regulations indirectly impact the urban heat island effect and energy use **(Q54)** by influencing the density of buildings and impervious surfaces relative to green spaces. Variations in institutions focused on similar resources or environmental concerns can lead to substantially different impacts at fine-grain spatial and temporal resolutions. For instance, different water institutions applied across the states straddling the Ogallala aquifer of the Southern Great Plains of the United States have led to major distinctions in the aquifer's water losses regionally **(Q35)** (Winter & Foster 2014).

The tragedy of the commons is, perhaps, the most familiar concept linking institutions and the environment (Araral 2014). It proposes that, in the absence of governing institutions, resources become "open access", in which the demands for the resources leads to their depletion (Gopalakrishnan 2009). Historic recognition of such a fate by communities worldwide has led to the formation of institutions specifically to adjudicate access to the resources, such as grazing restrictions on grasslands and water access for irrigation. These restrictions and access involves different kinds of institutions, such as private property and common property regimes. While general characteristics that enhance the performance of these institutions have been identified (e.g. Ostrom 2008), such as enforcement of the rules of resource use, no institutional panaceas exist that serve all resources or environmental services (Ostrom, Janssen & Anderies 2007), owing to complexities and variances of economies, cultures and places (Cole, Epstein & McGinnis 2014).

The well-being of the Earth system experiences at least two major problems in terms of open access resources or environments. These involve the oceans and atmosphere, which historically have been viewed as almost infinite in their capacities to absorb emissions, material waste and effluents from human activity with minimal consequences for the biosphere. As noted in Sections IV and V, limits to these capacities exist, especially given the scale of contemporary human impacts (Vogler 2012), leading to international institutions to address them. International agreements and laws applicable to marine resource use – fishing, whaling and mining – have emerged that operate reasonably well (Appendix), whereas those dealing with ocean pollution have not. Local to country level regulation of air quality, such as emission reductions to counter acid rain and tropospheric pollution (Q43) have also proven effective in many cases. International regulations addressing the atmosphere, such as reducing greenhouse gas emissions, have proven difficult (Q98). In contrast, the Montreal Protocol (Q53, Q98) appears to have saved the ozone layer, despite the recent increases in the emissions of ozone-depleting CFC-11 (Montzka *et al.* 2018). International accords dealing with atmosphere and marine pollution have proven difficult owing to a number of factors that make their design complex (Vogler 2012; Young 2011), such as how past and future emissions and effluents affect the equity and economic competitiveness among countries (Brown, Adger & Cinner 2019).

References

Araral, E. (2014). "Ostrom, Hardin and the commons: a critical appreciation and a revisionist view". *Environmental Science & Policy* 36: 11–23. doi.org/10.1016/j.envsci.2013.07.011.

Brown, K., W. Adger & J. Cinner (2019). "Moving climate change beyond the tragedy of the commons". *Global Environmental Change* 54: 61–3. doi.org/10.1016/j.gloenvcha.2018.11.009.

Cole, D., G. Epstein & M. McGinnis (2014). "Digging deeper into Hardin's pasture: the complex institutional structure of 'the tragedy of the commons'". *Journal of Institutional Economics* 10(3): 353–69. doi.org/10.1017/S1744137414000101.

Gopalakrishnan, C. (2009). "Classic papers revised: symposium on Garrett J. Hardin". *Journal of Natural Resources Policy Research* 1(3): 243–53. doi.org/10.1080/19390450903038219.

Montzka, S. *et al.* (2018). "An unexpected and persistent increase in global emissions of ozone-depleting CFC-11". *Nature* 557(7705): 413–17. doi.org/10.1038/s41586-018-0106-2.

Ostrom, E. (2008). "The challenge of common-pool resources". *Environment: Science and Policy for Sustainable Development* 50(4): 8–21. doi.org/10.3200/ENVT.50.4.8-21.

Ostrom, E., M. Janssen & J. Anderies (2007). "Going beyond panaceas". *Proceedings of the National Academy of Sciences* 104(39): 15176–8. doi.org/10.1073/pnas.0701886104.

Vogler, J. (2012). "Global commons revisited". *Global Policy* 3(1): 61–71. doi.org/10.1111/j.1758-5899.2011.00156.x.

Winter, M. & C. Foster (2014). "Ogallala Aquifer: lifeblood of the High Plains". CoBank ACB. aquadoc.typepad.com/files/ke_ogallalaaquifer_reportpt1-oct2014.pdf.

Young, O. (2011). "Effectiveness of international environmental regimes: existing knowledge, cutting-edge themes, and research strategies". *Proceedings of the National Academy of Sciences* 108(50): 19853–60. doi.org/10.1073/pnas.1111690108.

ARE SOME ECONOMIES AND POLITICAL ECONOMIES MORE ENVIRONMENTALLY DEGRADING THAN OTHERS?

Short answer: contested answer. Claims about the environmental superiority of different economic/political economic systems are difficult to substantiate given their histories.

The production and consumption of goods and services – the economy – are integrated with law, custom and government to form the political economy. Idealized free-market and command structures represent the polar extremes of political economies, although in practice many political economies mix dimensions of the two. In free-market or capitalist systems, production and consumption are largely determined by the market through supply and demand, driven in part by decisions of individuals and businesses. Command or centrally planned structures rely on different levels of government control of production and consumption. Mixed systems have elements of capitalist and command systems, especially if the command element also considers various regulated safety nets and redistribution of wealth. For example, the capitalist system of the United States employs such safety nets as agricultural subsidies and insurance, health insurance for the elderly, and interventions in regional and national emergencies such as warfare, recessions and depressions (Moon & Pino 2018). The countries of the European Union, in contrast, have an even larger array of state interventions and redistributions. Similarly, if from a different starting point, the former command economy of China increasingly privatizes its production activities – agriculture and various business sectors – to its state-owned enterprises and state-controlled planning (Hammoudeh *et al.* 2014).

Various arguments propose that the ideal types of political economies lead to differential impacts on the environment and Earth system. The free-market ideal of capitalist systems, for example, are critiqued for their need for constant economic growth, which elicits increasing demands on natural capital. One view holds that this growth and its impacts are most efficiently acquired through the privatization of natural capital, but another argues that infinite growth is not possible (Bell 2015), especially at the Earth system level. Without regulation, various environmental concerns, often labelled externalities, prove vulnerable in this system. Externalities are damages not incurred to the producer or paid by the user,

Figure 67.1 Abandoned mine, Leadville, Colorado
Fortunes made in silver mining 140 years ago left toxic landscapes of mine tailings for which no mining unit had to remediate and rendered the land unusable save for tourism.
Source: Railfan & Railfan (2020).

such as atmospheric or water pollution from industrial production and consumption. The case of Leadville, Colorado (US) in the nineteenth century demonstrates the damages that may result from unregulated activities in free markets. Once a global capital of silver, its wealthy mine owners left a scarred landscape of toxic mine pilings that render much of the landscape useless today, save for tourism to view the results, given the costly detoxification required for other uses (Muller 2005) (Fig. 67.1). Accounting for environmental costs in capitalist systems has been the aim of ecological and environmental/resource economics (Dasgupta 2014; Gowdy & Erickson 2005).

Another view holds that command economies, especially those historically dominated by state-controlled production and professed social equity, can internalize externalities in their accounting, reducing environmental impacts. However, few true command economies remain, such as the former Soviet Union, and evidence of environmental damages from these economies has been and remains significant. The destruction of the Aral Sea is illustrative (Fig. 67.2) **(Q33)**. The state command structure redirected the waters of the two rivers feeding the sea to supply massive irrigation systems, largely for export crops to assist the finances of the state. The result was the desiccation of the world's fourth largest lake and the loss of its productive fishing industry to a salt-laden desert (Micklin 2007).

Evidence at the Earth system level is equivocal as well, especially in the era of a globalizing economy. The United States and the European Union, with capitalist-based, mixed systems, have contributed almost 50 per cent of the cumulative atmospheric carbon dioxide, registering their long history of industrial development. In 2016, however, China's mixed

Figure 67.2 Desiccation of the Aral Sea, Central Asian Republics
Former Soviet Union decisions to use the water sources that fed the sea for irrigated agriculture instead desiccated the sea, killed its fishing industry and created various negative feedbacks on the regional environment.
Source: Kulenov (2006).

command system became the largest emitter, contributing about a quarter of all annual emissions, 10 per cent more than the United States (Ritchie & Roser 2017). Similar results follow global environmental footprints. China has by far the largest footprint, over 5 billion ha, with the United States just under 3 billion ha (Ritchie & Roser 2017). On a per capital basis, however, the footprint largely aligns with per capita wealth.

Overall, it appears that the political economy has less to do with stresses on the global environment than does the level of economic development and technology on a per capita and national basis and the size of the population in question (Mavragani, Nikolaou & Tsagarakis 2016). Indeed, the Environmental Performance Index ranking 180 countries on 11 environmental concerns indicates that well-developed economies with strong environmental regulations outperform less-developed economies and command-oriented political economies (EPI 2020). Various models and tests indicate that political economies delivering high levels of development and good institutions (Scruggs 1999), apparently those emanating from voluntary cooperation or consensus of the state, labourers and business, tend to perform well in terms of environmental performance. In addition, effective governance, such as strong national safety nets to mitigate or recover from disaster events, may reduce vulnerabilities to such events, more so than strong economic well-being (Tennant & Gilmore 2020).

Regardless of these outcomes, political economy directly influences the means by which environmental performance is achieved and indirectly affects this performance through the impacts on affluence and equity. Highly distributive decision-making systems can create variance in environmental performance, illustrated by the Leadville case, whereas less distributive systems may have less variance per issue, but can result in significant negative outcomes as illustrated above for the Aral Sea.

References

Bell, K. (2015). "Can the capitalist economic system deliver environmental justice?" *Environmental Research Letters* 10(12): 125017. doi:10.1088/1748-9326/10/12/125017.

Dasgupta, P. (2014). "Measuring the wealth of nations". *Annual Review of Resource Economics* 6(1): 17–31. doi.org./10.1146/annurev-resource-100913-012358.

EPI (Environmental Performance Index). Yale Center for Environmental Law and Policy & the Earth Institute, Columbia University. epi.yale.edu/epi-results/2020/component/epi.

Gowdy, J. & J. Erickson (2005). "The approach of ecological economics". *Cambridge Journal of Economics* 29(2): 207–22. doi.org/10.1093/cje/bei033.

Hammoudeh, S. *et al.* (2014). "Dependence of stock and commodity futures markets in China: implications for portfolio investment". *Emerging Markets Review* 21: 183–200. doi.org/10.1016/j.ememar.2014.09.002.

Kulenov, Z. (2006). "The Aral Sea is drying up. Bay of Zhalanash, Ship Cemetery, Aralsk, Kazakhstan". commons.wikimedia.org/wiki/File:The_Aral_sea_is_drying_up._Bay_of_Zhalanash,_Ship_Cemetery,_Aralsk,_Kazakhstan.jpg.

Mavragani, A., I. Nikolaou & K. Tsagarakis (2016). "Open economy, institutional quality, and environmental performance: a macroeconomic approach". *Sustainability* 8(7): 601. doi.org/10.3390/su8070601.

Micklin, P. (2007). "The Aral Sea disaster". *Annual Review of Earth and Planetary Sciences* 35: 47–72. doi.org/10.1146/annurev.earth.35.031306.140120.

Moon, W. & G. Pino (2018). "Do US citizens support government intervention in agriculture? Implications for the political economy of agricultural protection". *Agricultural Economics* 49(1): 119–29. doi.org/10.1111/agec.12400.

Muller, B. (2005). "Leadville: the struggle to revive an American town". *Journal of the American Planning Association* 71(3): 346–7.

Railfan, D. & R. Railfan (2020). "Abandoned mine tailings near Leadville, Colorado". commons.wikimedia.org/wiki/File:Abandoned_Mine_Tailings_near_Leadville,_Colorado.jpg.

Ritchie, H. & M. Roser (2017). "CO2 and greenhouse gas emissions". Our World in Data. ourworldindata.org/co2-and-other-greenhouse-gas-emissions.

Scruggs, L. (1999). "Institutions and environmental performance in seventeen western democracies". *British Journal of Political Science* 29(1): 1–31. doi.org/10.1017/S0007123499000010.

Tennant, E. & E. Gilmore (2020). "Government effectiveness and institutions as determinants of tropical cyclone mortality". *Proceedings of the National Academy of Sciences* 117: 28692–9. doi.org/10.1073/pnas.2006213117.

Q68

DO CULTURAL VALUES AND NORMS SHAPE ENVIRONMENTAL BEHAVIOUR?

Short answer: no at the cultural level; perhaps at the individual level. Different categories of individual values tend to exhibit consistent environmental impacts; those aggregated at regional or country levels, however, are difficult to discern.

Values refer to relatively stable principles that guide decisions and actions, shaping worldviews. Shared collectively, these values constitute cultural worldview-values, although different values may be held individually within cultures. Various interpretations hold that these views and values shape the treatment of nature (**Q72**). For instance, one argument advances that environmental degradation and unsustainable practices are embedded in the worldview-values of Western culture's adherence to Christianity's belief that nature is to be controlled (White 1967). In contrast, values based on alternative prescriptions, such as Eastern religions professing harmony or living with nature, maintain more positive environmental outcomes (Bourdeau 2004). Only a few efforts have attempted to explore this worldview-value distinction toward nature, and the results are unclear or inconsistent.

Environmental degradation in antiquity was worldwide in its scope, especially where land pressures were high (Butzer & Endfield 2012; Haldon 2018; Tuan 1970). The view of living with nature professed in the three religions of imperial China (i.e. Confucianism, Daoism, Buddhism) led to levels of deforestation and otherwise transformed environments as much as that in medieval Europe with its Christian faiths (Tuan 1968). Both Eastern and Western worldviews in the past and present have long "built" environments enhancing and degrading environmental services. Cultural distinctions relative to environmental consequences are difficult to detect today, exemplified in greenhouse gas emissions or marine plastic waste, especially if population, affluence and technological capacity are controlled.

In contrast, substantial work has been undertaken on individual and cohort values, and environmental worldviews (i.e. the relationship between people and nature) as they affect environmental perceptions and actions. This work indicates that values underpin environmental worldviews, norms and behaviour (Dietz, Fitzgerald & Shwom 2005). Altruistic, biocentric and biospheric values, reflecting concern for, respectively, the welfare of others, nonhuman species and the nature–Earth system, are more accepting of environmental

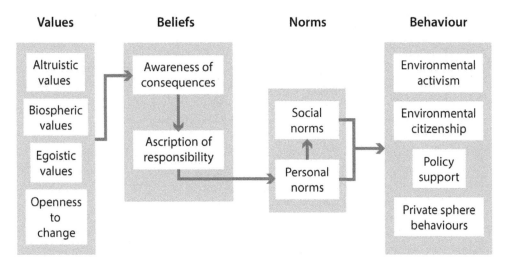

Figure 68.1 Modified value-belief-norms model/theory
Values about the environment shape different beliefs – awareness and responsibility – which translate to personal and social norms that lead to different environmental behaviours.
Source: author after Dietz, Fitzgerald & Shwom (2005).

concerns and behaviour, compared with egoistic values or those based on maximizing outcomes (Steg 2011). This understanding gives rise to the value-belief-norm (VBN) model/theory related to the environment in which pro-environmental behaviour is closely tied to personal norms (Fig. 68.1).

Different individual values shape the awareness of problems and their consequences, translated to personal and social norms that lead to different behaviours affecting the environment (Fig. 68. 1) (Ghazali 2019; Stern 1999). Such relationships have been documented across diverse countries (Chen 2015; Schultz 2005), and may also hold across ethnic groups as well (Pearson 2018). Assuming that certain values lead to human–environment relationships that are, for a given technology, less stressful for the environment and Earth system, the question remains about how to move mass culture toward those values, as well as understanding the ancillary consequences of adopting them (Dietz, Fitzgerald & Shwom 2005; Stern 1999).

References

Bourdeau, P. (2004). "The man–nature relationship and environmental ethics". *Journal of Environmental Radioactivity* 72(1/2): 9–15. doi.org/10.1016/S0265-931X(03)00180-2.

Butzer, K. & G. Endfield (2012). "Critical perspectives on historical collapse". *Proceedings of the National Academy of Sciences* 109(10): 3628–31. doi.org/10.1073/pnas.1114772109.

Chen, M. (2015). "An examination of the value-belief-norm theory model in predicting pro-environmental behaviour in Taiwan". *Asian Journal of Social Psychology* 18(2): 145–51. doi.org/10.1111/ajsp.12096.

Dietz, T., A. Fitzgerald & R. Shwom (2005). "Environmental values". *Annual Review of Environment Resources* 30: 335–72. doi.org/10.1146/annurev.energy.30.050504.14444.

Ghazali, E. (2019). "Pro-environmental behaviours and value-belief-norm theory: assessing unobserved heterogeneity of two ethnic groups". *Sustainability* 11(12): 3237. /doi.org/10.3390/su11123237.

Haldon, J. (2018). "History meets palaeoscience: consilience and collaboration in studying past societal responses to environmental change". *Proceedings of the National Academy of Sciences* 115(13): 3210–18. doi.org/10.1073/pnas.1716912115.

Pearson, A. (2018). "Diverse segments of the US public underestimate the environmental concerns of minority and low-income Americans". *Proceedings of the National Academy of Sciences* 115(49): 12429–34. doi.org/10.1073/pnas.1804698115.

Schultz, P. (2005). "Values and their relationship to environmental concern and conservation behavior". *Journal of Cross-Cultural Psychology* 36(4): 457–75. doi.org/10.1177/0022022105275962.

Steg, L. (2011). "General antecedents of personal norms, policy acceptability, and intentions: the role of values, worldviews, and environmental concern". *Society and Natural Resources* 24(4): 349–67. doi.org/10.1080/08941920903214116.

Stern, P. (1999). "A value-belief-norm theory of support for social movements: the case of environmentalism". *Human Ecology Review* 6(2): 81–97. cedar.wwu.edu/hcop_facpubs/1.

Tuan, Y.-F. (1968). "Discrepancies between environmental attitudes and behavior: examples from Europe and China". *Canadian Geographer* 12: 176–91. doi.org/10.1111/j.1541-0064.1968.tb00764.x.

Tuan, Y.-F. (1970). "Views: our treatment of the environment in ideal and actuality: a geographer observes man's effect on nature in China and in the pagan and Christian West". *American Scientist* 58(3): 244–9.

White, L. (1967). "The historical roots of our ecologic crisis". *Science* 155(3767): 1203–07. doi:10.1126/science.155.3767.1203.

SECTION VIII

UNDERSTANDING OUR RELATIONSHIP WITH NATURE

The survival of our species is a product of the interactions with other biota and the abiotic environment of the biosphere, and the Earth system more broadly. These interactions have changed as humankind has moved through its major technological phases (Section I), which have enhanced our access to and depleted the stock of natural capital, altered the flows of environmental services and, over time, increased our impacts on the Earth system. Throughout these phases, diverse worldviews about humanity, nature and the relationships between the two have emerged that profess to separate us from nature or unify us with it. Embedded in culture, religion or political economy, these worldviews are linked to values and norms about, and institutions (rules) directing, societal behaviour (**Q68**) and proclaim different outcomes with nature.

A common argument holds that pre-science and alternative, especially indigenous, and Eastern-world understanding of nature created or could create different outcomes of human–environment interactions (e.g. Chen & Wu 2009; van der Velden 2018). Others trace the separation of people and nature to Greek philosophy undergirding science, Judeo–Christian views on controlling nature (**Q68**) and, subsequently, evolutionary theory secularizing science from spirituality (Smith 1972). Such arguments about the pros and cons of moral philosophies, spirituality and ontology, however, are difficult to reconcile with the actual evidence. Environmental impacts appear to be rather consistent over the long term, holding population size and technological capacity constant (**Q62, Q68**). Aggregated at the global level, of course, the outcomes display a significant, long-term trajectory of environmental change at the Earth system level.

Such histories notwithstanding, distinctive worldviews continue to shape our understanding of human–environment relations and influence our behaviour toward the environment, its resources, services and functions. They underlie both the positive and negative interpretations of our current conditions and prognostications of those we will confront in the future, such as the significance of global climate change on future generations. This section identifies various facets of these worldviews as they have been manifested in science-based approaches to understanding and the role that they play in our interpretations of human–environmental and sustainability themes.

References

Chen, X. & J. Wu (2009). "Sustainable landscape architecture: implications of the Chinese philosophy of 'unity of man with nature' and beyond". *Landscape Ecology* 24(8): 1015–26. doi.org/10.1007/s10980-009-9350-z.

Smith, R. (1972). "Alfred Russel Wallace: philosophy of nature and man". *British Journal for the History of Science* 6(2): 177–99. doi.org/10.1017/S0007087400012279.

van der Velden, M. (2018). "Indigenous philosophies". *Philosophy Now* 127: 32–5.

WHAT IS – AND OUGHT TO BE – OUR RELATIONSHIP WITH NATURE?

Short answer: "is" involves the domain of empiricism and science, while "ought to be" is in the realm of individual and societal worldviews-values. Science practitioners can blur the two questions when they enter the environmental policy arena.

What constitutes our relationship with nature – the human–environment condition – is a foundational question of the science concerned with climate change, global environmental change, the Anthropocene and sustainability. Antiquity is replete with observations about the status of that relationship in Eastern and Western thought (Carone 1998; Fang & Mu 2005). A global stocktaking of sorts was undertaken in the nineteenth century by George Perkins Marsh (Koelsch 2012), subsequently systematized in various twentieth-century works, reaching assessments of the Earth system and its biogeochemical flows in the 1990s (Turner *et al.* 1990). Taken as a whole, human-induced changes in environments and the Earth system are the product of increasing demands on that system and advances in technology across space, time, cultures and political systems to improve the vagaries of nature and the provisioning, security and aesthetic demands of society. In this sense, the global-level history of our relationships with nature is anthropocentric, elevating our species over all others and the Earth system (Cocks & Simpson 2015; Kopnina *et al.* 2018), complementing the coinage of our current condition as the Anthropocene (Q7). Recognition of the environmental and Earth system consequences of the Anthropocene, while we seek to improve societal well-being, has gained worldwide concern, spawning a multitude of international efforts to address the problems in question (Appendix). For the most part, these efforts are anthropocentric in kind.

The "ought to be" question takes us beyond the science of "is" into the moral and often spiritual realms about human–environment relationships, largely embedded within individual and societal values and worldviews. Are Western worldviews, embedded in Judeo–Christian spirituality, anthropocentric in that they elevate humanity above or beyond nature, justifying control of the environment for the benefit of society? Are Middle Eastern and Eastern worldviews and various indigenous spiritualities ecocentric or biocentric, valuing the environment or attributing intrinsic value to all biota and supporting the harmony of people with nature (Palmer, McShane & Sandler 2014)? Does the evidence demonstrate distinctions among these worldviews and the human–environment conditions they render over the long term and as mediated by human pressures and technology?

These "centric" categories are simplifications of an array of worldviews that the field of environmental ethics has addressed since its emergence in middle of the twentieth century (Palmer, McShane & Sandler 2014). These ethics are grounded in the instrumental and intrinsic values given to our species versus other species and the environment at large. The instrumental or extrinsic case is the value given to entities owing to their services for other entities or ends. In contrast, the value given to an entity itself is the intrinsic value. The anthropocentric view, which has dominated much of Western discourse, promotes an intrinsic value on humankind and the remainder of the biophysical world is treated largely in terms of its instrumental value to people. Ecocentric and biocentric views, in contrast, tend to give the biophysical world intrinsic value; for example, biota have value independent of their services for people. That these views alter human impacts on the environment or Earth system over the long term, given comparable pressures on the landscape and similar technological capacities, is questionable (Q62). Alternatively, evidence exists that individual human–environment actions, such as conserving resource uses, are linked to values consistent with worldviews, and may translate to interpretations consistent with the Cornucopian or the Cassandra positions about population and the environment (Q71).

The distinction between what is and what should be can be blurred among science practitioners, especially when science seeks to influence practice. Aldo Leopold (1887–1948), Rachel Carson (1907–64) and Vandana Shiva (1952–), a conservation biologist, marine biologist and philosopher, respectively, have each made handsome contributions to environmental and agricultural science. In the public eye, however, they are largely recognized as activists for environmental and biodiversity movements, with impacts on government programmes. As such, their science connected or continues to connect to the "ought to be" question (Manfredo et al. 2017; Noss 2007). Today, many scientists are engaged in the formalization of frameworks, platforms and conventions aimed at environmental policy with direct human links to human activities, ranging from climate change to biodiversity (Appendix). In many cases, this involvement transcends science-to-inform decision-making (e.g. the Intergovernmental Panel on Climate Change) and engages with policy objectives, such as those reached in the Paris Accords. Such efforts, of course, underscore worldviews about human–environment relationships.

References

Carone, G. (1998). "Plato and the environment". *Environmental Ethics* 20(2): 115–33. doi.org/10.5840/enviroethics199820227.

Cocks, S. & S. Simpson (2015). "Anthropocentric and ecocentric: an application of environmental philosophy to outdoor recreation and environmental education". *Journal of Experiential Education* 38(3): 216–27. doi.org/10.1177/1053825915571750.

Fang, X. & S. Mu (2005). "The ancient Chinese ideas on the relationship between man and natural environment". *Human Geography* 20(4): 110–13. doi:10.13959/j.issn.1003-2398.2005.04.026.

Koelsch, W. (2012). "The legendary 'rediscovery' of George Perkins Marsh". *Geographical Review* 102(4): 510–24. doi.org/10.1111/j.1931-0846.2012.00172.x.

Kopnina, H. *et al.* (2018). "Anthropocentrism: more than just a misunderstood problem". *Journal of Agricultural and Environmental Ethics* 31: 109–27. doi.org/10.1007/s10806-018-9711-1.

Manfredo, M. *et al.* (2017). "Why social values cannot be changed for the sake of conservation". *Conservation Biology* 31(4): 772–80. doi.org/10.1111/cobi.12855.

Noss, R. (2007). "Values are a good thing in conservation biology". *Conservation Biology* 21(1): 18–20. doi:10111/j1523-1739.2006.00637.x.

Palmer, C., K. McShane & R. Sandler (2014). "Environmental ethics". *Annual Review of Environment and Resources* 39: 419–42. doi.org/10.1146/annurev-environ-121112-094434.

Turner II, B. *et al.* (1990). *The Earth as Transformed by Human Action: Global Change and Regional Changes in the Biosphere Over the Past 300 Years*. Cambridge: Cambridge University Press.

Q70

HOW HAS SCIENCE TREATED HUMAN–ENVIRONMENT RELATIONSHIPS CONCEPTUALLY AND ANALYTICALLY?

Short answer: relationships once considered unidirectional are now treated as interactive, although different explanatory perspectives treat the interactions differently.

The emergence of modern science from the mid-sixteenth century through to the eighteenth century did not necessarily bifurcate the natural (or biophysical) and social worlds. Human–environment integration was examined in various ways, from Alexander von Humboldt's "unity of nature" – foremost integrating the sciences of the physical world but with links to the human world, such as human-induced climate change (Fränzle 2001; NEE 2019) – through the subsequent emergence of human–environment geography in Germany, France and the United States (Turner 2002). Such integration or synthesis, however, took a back seat to the subsequent specialization of science, which ultimately led to the separation of the natural and social sciences and the subsequent splintering and reconfiguration of the two. The depth of understanding of phenomena and processes gained with this specialization has transformed knowledge.

For human–environmental (or HE) science, however, this division led to various unidirectional approaches to the HE relationship (Fig. 70.1). The social sciences largely focused on the H→E relationship as it relates to human needs for resources and the transformation and maintenance of landscapes. E→H relationships involved the "geographic factor" in which the abiotic realm helped to explain the biotic one (Turner 2002), ultimately moving to the doctrine of environmental determinism in which the environment itself explained societal development, characteristics and behaviour. The Anthropocene, however, constitutes a reawakening of the integrative sciences, recognizing that many of the problems confronting the world are, at their base, a product of human–environmental interactions, such as those focusing on sustainability (Kates *et al.* 2001). This recognition has given rise to the concept of the social–environmental (or ecological) systems (SESs) (Moran 2010) (**Q71**).

The basic relationships noted in Figure 70.1 are consistent across a large range of research fields and topics of study, including those in anthropology and geography historically (Moran 2010; Vayda & McCay 1975), ranging from the abiotic influences on the biotic world (E→H)

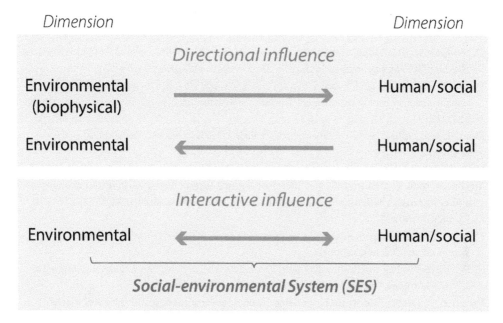

Figure 70.1 Basic directional approaches applied to human–environmental relationships
Source: author.

to societal adaptations reconfiguring the environment (E←H) (Turner 2002). In a more contemporary example, the people-and-nature orientation of conservation science (Mace 2014) is founded on the interactive SES (E←→H).

While various system approaches are used to grapple with SESs (**Q69**), components of either subsystem are not necessarily treated analytically in the same way. Biophysical phenomena and processes may involve non-linear dynamics in which parts of the subsystem or the entire subsystem transform to a new state (i.e. conditions and functions). Many of the processes within any one state, however, operate mechanistically, involving regular cause-and-effect relationships (i.e. A causes B). In such cases, probabilities account for the uncertainty that the mechanistic conditions will be met. Such relationships are not so common in the social realm because humans are reflexive actors – those who think, anticipate and change behaviour; they possess agency (i.e. independent or free choice) (Cohen-Cole 2005). In addressing these actors, probabilities refer to the uncertainty that a specific cause (A) renders a specific outcome (B). The two subsystems operationally involve decidedly different phenomena and processes, although both they and their interactions can be treated through the lens of science (McGinnis & Ostrom 2014). **Question 71** elaborates the SES approach.

References

Cohen-Cole, J. (2005). "The reflexivity of cognitive science: the scientist as model of human nature". *History of the Human Sciences* 18(4): 107–39. doi.org/10.1177/0952695105058473.

Fränzle, O. (2001). "Alexander von Humboldt's holistic world view and modern inter- and transdisciplinary ecological research". *Northeastern Naturalist* 8(1): 57–90. www.jstor.org/stable/4130727.

Kates, R. *et al.* (2001). "Sustainability science". *Science* 292(5517): 641–42. doi:10.1126/science.1059386.

Mace, G. (2014). "Whose conservation?". *Science* 345(6204): 1558–60. doi:10.1126/science.1254704.

McGinnis, M. & E. Ostrom (2014). "Social-ecological system framework: initial changes and continuing challenges". *Ecology and Society* 19(2): 30. dx.doi.org/10.5751/ES-06387-190230.

Moran, E. (2010). *Environmental Social Science: Human–Environment Interactions and Sustainability*. Chichester: Wiley.

NEE (Nature, Ecology and Evolution) (2019). "Humboldt's legacy". *Nature, Ecology and Evolution* 3: 1265–6. doi.org/10.1038/s41559-019-0980-5.

Turner II, B. (2002). "Contested identities: human–environment geography and disciplinary implications in a restructuring academy". *Annals of the Association of American Geographers* 92(1): 52–74. doi.org/10.1111/1467-8306.00279.

Vayda, A. & B. McCay (1975). "New directions in ecology and ecological anthropology". *Annual Review of Anthropology* 4(1): 293–306. doi.org/10.1146/annurev.an.04.100175.001453.

WHAT IS THE SOCIAL–ENVIRONMENT SYSTEM IN HUMAN–ENVIRONMENTAL SCIENCE?

Short answer: it is either an aggregate phenomenon to be explained or an approach requiring attention to feedbacks among components in the two subsystems (i.e. social and environmental).

Human–environmental relationships tend to be treated as a system, variously labelled: (coupled) human–environment, nature–society, social–ecological and social–environmental systems (Clark & Harley 2020). Of these labels, the most commonly used is the social–ecological system (McGinnis & Ostrom 2014) in which ecologists recognize ecosystems as composed of biotic and abiotic phenomena and processes (Currie 2011). Throughout this text, however, social–*environmental* system (SES) is used to signal firmly to the non-specialist that the biophysical subsystem involves abiotic phenomena and processes, not just biotic ones that may be associated with the ecological.

The SES constitutes an integrated framing of the relationships between the social world and the environment (Fig. 71.1). Explicitly or implicitly, SESs constitute the fabric of various models, analyses and explanations cross-cutting all facets of science addressing human–environmental relationships. They are used in at least two ways: (1) the analysis and explanation involve interactions of the components and processes within and between both subsystems as positive or negative feedbacks (Colding & Barthel 2019). In this use, various phenomena and processes of the system and theories related to them are addressed independently of the existence of the system itself as phenomenon; the system serves as a framing for the interactions of its components; (2) The system itself constitutes a real, if aggregate, phenomenon to be explained and for which theory can be developed. In this case, the SES or various types of them, such as land systems (Turner *et al.* 2020), are treated no differently than are society or ecosystems as phenomena.

SESs are treated as complex adaptive systems (**Q86**) that learn and change, involve non-linear relationships, and maintain tipping points (**Q89**) that, if exceeded, shift the SES into a new configuration or state, among other characteristics. Owing to such complexity, general explanations or theories of specific types of SESs are lacking, although general system concepts of the structure of SESs and their disruption and renewal have been proposed

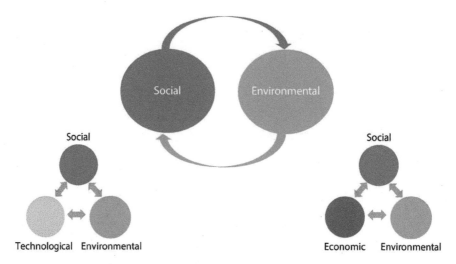

Figure 71.1 Variations of social–environmental systems
The dominant two-subsystem variant involves feedbacks between all social and environmental dimensions, whereas variants may separate the economic or technological from the social.
Source: author

(Allen *et al.* 2014). Broad frameworks explore potential theory development of SESs in general and others focus on specific SES types (Ostrom 2009; Turner *et al.* 2020) or elements within the SES (Cole, Epstein & McGinnis 2014). To date, however, specific theory or theory sets have yet to emerge.

References

Allen, C. *et al.* (2014). "Panarchy: theory and application". *Ecosystems* 17(4): 578–89. doi.org/10.1007/s10021-013-9744-2.

Clark, W. & A. Harley (2020). "Sustainability science: toward a synthesis". *Annual Review of Environment and Society* 45: 331–86. doi.org/10.1146/annurev-environ-012420-043621.

Colding, J. & S. Barthel (2019). "Exploring the social-ecological systems discourse 20 years later". *Ecology and Society* 24(1): 2. doi.org/10.5751/ES-10598-240102.

Cole, D., G. Epstein & M. McGinnis (2014). "Toward a new institutional analysis of social-ecological systems (NIASES): combining Elinor Ostrom's IAD and SES frameworks". *Indiana Legal Studies Research Paper 299*. papers.ssrn.com/sol3/papers.cfm?abstract_id=2490999.

Currie, W. (2011). "Units of nature or processes across scales? The ecosystem concept at age 75". *New Phytologist* 190(1): 21–34. doi.org/10.1111/j.1469-81372011.03646.x.

McGinnis, M. & E. Ostrom (2014). "Social-ecological system framework: initial changes and continuing challenges". *Ecology and Society* 19(2): 30. dx.doi.org/10.5751/ES-06387-190230.

Ostrom, E. (2009). "A general framework for analyzing sustainability of social-ecological systems". *Science* 325(5939): 419–22. doi:10.1126/science.1172133.

Turner II, B. *et al.* (2020). "Framing the search for a theory of land use". *Journal of Land Use Science* 15(4): 489–508. doi.org/10.1080/1747423X.2020.1811792.

DO VIEWS ABOUT HUMAN–ENVIRONMENT RELATIONSHIPS CROSS-CUT CULTURES AND SOCIAL GROUPS?

Short answer: four cultural views of human–environmental relationships have been postulated but their cross-culture/social group applicability is not proven.

Our views of nature are social constructs, the product of our socio-cultural context as modified by our individuality. Culture theory of risk (e.g. to hurricanes or air pollution) proposes that community cultures create "myths" about the environment that influence individual and cohort perceptions and behaviours applied to the environment (Drake 1992). These myths are associated with broader worldviews categorized as egalitarianism, hierarchy, individualism and fatalism (Chuang, Manley & Petersen 2020; Xue *et al.* 2014). Egalitarians favour social inclusiveness and participatory democracy, and envision the social inequalities associated with environmental risk. Individualists seek personal freedom from regulation, maximization of personal gain and champion human invention to resolve environmental problems. Hierarchists seek to protect their interests through existing power structures and view change with suspicion; environmental risk tends to be dismissed if it conveys negative implications about the power structure. Fatalists are largely disengaged from social bonding and view environmental risk as random.

These four worldviews are proposed to lead to distinctive interpretations about human impacts on the environment and the Earth system (Drake 1992): nature is robust, nature is ephemeral, nature is perverse or tolerant, and nature is capricious or random (Fig. 72.1). In the robust – or stable – view, the environment (or Earth system) is resistant to and recovers from disturbances, fostering less societal concern about changes in the biophysical world and its social implications. This view is associated with individualists and hierarchists, the former championing human invention to overcome problems and the latter as a means to maintain the extant power structure (Xue *et al.* 2014). That nature is ephemeral envisions disturbances that easily degrade nature and is associated with egalitarians. The impacts of these disturbances on society are seen to be unequal and prove difficult to overcome. Between the robust and ephemeral views resides the view that nature is perverse and tolerant; nature has the capacity to absorb substantial disturbances, akin to the robust view, but if critical limits are passed, nature becomes ephemeral (**Q88, Q89**). This more complex view

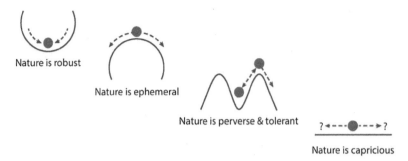

Nature is robust

Nature is ephemeral

Nature is perverse & tolerant

Nature is capricious

Figure 72.1 Views of risk in human–environment relationships
The ball represents disturbances on nature, biophysical or human-induced in kind; the line represents the capacity of nature to respond or recover. Description is provided in the text.
Source: author.

is associated with variants of egalitarians, individualists and hierarchists. Finally, nature is capricious (or random), in the sense that nature's response to disturbances, even some disturbances themselves, are beyond human control, is associated with fatalists.

Some analyses provide modest support that these worldviews and their associations cross-cut cultures (Xue *et al.* 2014). Other analyses indicate that education levels and even risk exposures fail to track as well to views about the environment as does forming beliefs consistent with personal relationships (Kahan *et al.* 2012; Mayer *et al.* 2017), commonly associated with political ideology, suggesting the role of culturally reinforced views and cultural cohorts holding them.

The value-belief-norm (VBN) model/theory **(Q68)** suggests that different environmental beliefs and views about the environment and behaviour toward it can be detected among individuals, not cultures per se, which in turn generates differences in behaviour about the environment (Chen 2015; Wolske, Stern & Dietz 2017). These differences reside in the mix of altruistic, egotistic and biospheric values, denoting concerns for self, others and the environment, respectively. Inasmuch as different mixes of these values are held by cultural cohorts worldwide, VBN theory may be interpreted to support the role of cultural worldviews and the environment.

References

Chen, M. (2015). "An examination of the value-belief-norm theory model in predicting pro-environmental behaviour in Taiwan". *Asian Journal of Social Psychology* 18(2): 145–51. doi.org/10.1111/ajsp.12096.

Chuang, F., E. Manley & A. Petersen (2020). "The role of worldviews in the governance of sustainable mobility". *Proceedings of the National Academy of Sciences (PNAS)* 117(8): 4034–42. doi.org/10.1073/pnas.1916936117.

Drake, K. (1992). "Myths of nature: culture and the social construction of risk". *Social Issues* 48(4): 21–37. doi.org/10.1111/j.1540-4560.1992.tb01943.x.

Kahan, D. *et al.* (2012). "The polarizing impact of science literacy and numeracy on perceived climate change risks". *Nature Climate Change* 2(10): 732–5. doi.org/10.1038/nclimate1547.

Mayer, A. *et al.* (2017). "Environmental risk exposure, risk perception, political ideology and support for climate policy". *Sociological Focus* 50(4): 309–28. doi.org/10.1080/00380237.2017.1312855.

Wolske, K., P. Stern & T. Dietz (2017). "Explaining interest in adopting residential solar photovoltaic systems in the United States: toward an integration of behavioral theories". *Energy Research and Social Science* 25: 134–51. doi.org/10.1016/j.erss.2016.12.023.

Xue, W. *et al.* (2014). "Cultural worldviews and environmental risk perceptions: a meta-analysis". *Journal of Environmental Psychology* 40: 249–58. doi.org/10.1016/j.jenvp.2014.07.002.

HOW DO THE CASSANDRA AND CORNUCOPIAN PERSPECTIVES SHAPE VIEWS ON HUMAN–ENVIRONMENT RELATIONSHIPS?

Short answer: Cassandras are concerned about human demands overtaxing nature; Cornucopians envision human ingenuity as overcoming environmental problems while sustaining those demands.

In Greek mythology, Cassandra had the gift of prophecy but carried the curse of being unbelievable. Those who worry about human impacts degrading environments and the Earth system share concerns with a community self-labelled as Cassandras. The label is appropriate, according to this community, because humanity does not appreciate sufficiently the potential dangers in which it engages nature, especially viewed over the long run and at the Earth-system level. The Cassandra position is consistent with the "nature is ephemeral" or "nature is perverse" views discussed in **Question 72**. The formal Cassandra position was initially aimed at global population growth and its implications about the negative consequences on the environment from the pressures that this growth entails (**Q78**). That position has been recalculated to encompass the full range of human drivers of environmental change, with concerns about the planetary boundaries that these changes may surpass (Rockström *et al.* 2009; Steffen *et al.* 2015). Often linked to practitioners in environmental science and conservation biology, elements of the Cassandra perspective actually cross-cuts a wide spectrum of researchers concerned that traditional economic and other accountings devalue nature and that directly accounting for the services of the Earth system raises concerns about planetary sustainability (Arrow *et al.* 2004).

The Cornucopian (a mythical horn providing abundant food), technological fix or Pollyanna perspective are labels applied by Cassandras to their intellectual opponents. Cornucopians interpret human–environment history as one in which human ingenuity has prevailed and will continue to do so, foremost in the capacity to innovate technologically and substitute for scarce resources (Chenoweth & Feitelson 2005; Locher 2020) and, presumably, for environmental services at large. Cornucopians can be linked to the "nature is robust" view, inasmuch as society is seen as overcoming the stresses it places on the environment through innovations. This view, they argue, is consistent with the past performance of humankind and the environment, one in which the environment is altered but the biosphere maintains its base functions and, with technological innovation, increases the material resources of society.

These worldviews – the Cassandras of the "glass half empty" and the Cornucopians of the "glass half full" – prove powerful in interpretations of societal relationships with nature. Minimal, if any, debate exists over the evidence of the history of the increases in global population or the loss of natural capital and environmental services, save perhaps that about human-induced climate change. How these increases and declines have affected nature and humankind, however, are invariably interpreted at the level at which the glass is filled.

References

Arrow, K. *et al.* (2004). "Are we consuming too much?" *Journal of Economic Perspectives* 18(3): 147–72. doi:10.1257/0895330042162377.

Chenoweth, J. & E. Feitelson (2005). "Neo-Malthusians and Cornucopians put to the test: Global 2000 and The Resourceful Earth revisited". *Futures* 37(1): 51–72. doi.org/ 10.1016/j.futures.2004.03.019.

Locher, F. (2020). "Neo-Malthusian environmentalism, world fisheries crisis, and the global commons, 1950s–1970s". *Historical Journal* 63(1): 187–207. doi.org/10.1017/S0018246X19000116.

Rockström, J. *et al.* (2009). "A safe operating space for humanity". *Nature* 461(7263): 472–5. doi.org/ 10.1038/461472a.

Steffen, W. *et al.* (2015). "Planetary boundaries: guiding human development on a changing planet". *Science* 347(6223): 736. doi:10.1126/science.1259855.

HOW DOES THE MALTHUSIAN THESIS AND ITS IMPLICATIONS APPLY TO HUMAN–ENVIRONMENT RELATIONSHIPS?

Short answer: that food crises – famines – follow from human population growth exceeding the capacity of food produced is difficult to prove.

The Malthusian thesis is one of the best known and most researched propositions about human–environment relationships in the Western world. It originated in eighteenth-century Britain, based on interpretations of population pressures resulting from the migration of rural labour to cities (Barrows 2010). Thomas Malthus proposed that population grows geometrically (change by a constant ratio; 2, 8, 32, 128), whereas the resource base or food does so arithmetically (change by constant number; 3, 5, 7, 9), creating a food ceiling or limit on population (Fig. 74.1).

Given this relationship, an unchecked growth in population ultimately surpasses the capacity to produce food (i.e. technological advances in food production are exogenous to the relationship), creating a Malthusian crisis or famine, commonly accompanied by warfare or migration as a solution to alleviate the predicament. A decreasing population, in turn, may fall below the food ceiling but eventually recovers to conditions of rapid growth, ultimately surpassing the food ceiling and generating another crisis. This growth, however, might be checked by socio-economic and political factors that decrease fertility (e.g. contraceptive use; delayed marriage) or increase mortality (e.g. diseases, warfare), keeping population below the food ceiling.

The mid-nineteenth-century Irish potato famine and the explosive migration of the Irish off their island has been equated to a Malthusian crisis, an interpretation failing to recognize the huge export of food from Irish estates to England during the famine period. Models and tests based on historical economic and demographic data, however, variously support or refute different parts of the applications of the Malthusian thesis (Madsen, Robertson & Ye 2019; Møller & Sharp 2014). Contemporary exponential (as opposed to geometric) population growth has not yielded a true Malthusian crisis (Trewavas 2002; Urdal 2005), despite malnutrition persisting throughout the world. In many – if not most – cases, supplies of

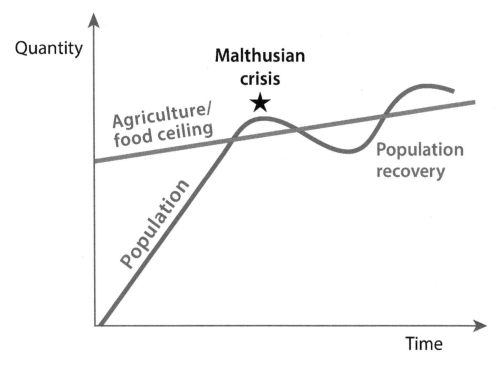

Figure 74.1 Simplified Malthusian thesis
Population (red line); food production (blue line); until a Malthusian crisis (*); see text for explanation. The population and food lines are drawn for impact and are not necessarily arithmetic and geometric in kind.
Source: author.

food retained or produced in areas experiencing famine have been sufficient to support the population if that population had access to it. This evidence has led to an alternative famine proposition, the entitlement thesis, in which famine is caused by inequalities in access to food stocks (Sen 1980). Entitlement appears to be more robust than the Malthusian thesis to explain famine, be it the Irish potato famine or those in rural, postwar China (Lin & Yang 2000).

It is generally recognized that contemporary famine and food shortages are not simple population-surpassing-food ceiling events. Variants on the Malthusian thesis, or neo-Malthusian theses, continue to this day, however, for at least two reasons. First, they are hypothetically plausible as the demands of global population press upon the Earth system at large. Second, they also undergird the critical "ought to" questions of human–environment relationships **(Q69)** and the concerns about provisioning the global population equitably within the capacity of the Earth system. In addition, the modern twist on the original thesis (i.e. population–food) is to cast the question in regard to all environmental services and economic externalities (Dasgupta & Ehrlich 2013) at the Earth-system level, focusing on the planetary boundaries **(Q85)**.

References

Barrows, S. (2010). "The law of population and the Austrian School: how Austrian economists interacted with Thomas Robert Malthus". *American Journal of Economics and Sociology* 69(4): 1178–205. doi.org/10.1111/j.1536-7150.2010.00740.x.

Dasgupta, P. & P. Ehrlich (2013). "Pervasive externalities at the population, consumption, and environment nexus". *Science* 340(6130): 324–8. doi:10.1126/science.1224664.

Lin, J. & D. Yang (2000). "Food availability, entitlements and the Chinese famine of 1959–61". *Economic Journal* 110(460): 136–58. doi.org/10.1111/1468-0297.00494.

Madsen, J., P. Robertson & L. Ye (2019). "Malthus was right: explaining a millennium of stagnation". *European Economic Review* 118: 51–68. doi.org/10.1016/ j.euroecorev.2019.05.004.

Møller, N. & P. Sharp (2014). "Malthus in cointegration space: evidence of a post-Malthusian pre-industrial England". *Journal of Economic Growth* 19(1): 105–40. doi.org/10.1007/ s10887-013-9094-0.

Sen, A. (1980). "Famines". *World Development* 8(9): 613–21. doi.org/10.1016/ 0305-750X(80)90053-4.

Trewavas, A. (2002). "Malthus foiled again and again". *Nature* 418(6898): 668–70. doi.org/10.1038/ nature01013.

Urdal, H. (2005). "People vs. Malthus: population pressure, environmental degradation, and armed conflict revisited". *Journal of Peace Research* 42(4): 417–34. doi.org/10.1177/0022343305054089.

DOES CARRYING CAPACITY APPLY TO CONDITIONS IN THE ANTHROPOCENE?

Short answer: carrying capacity – the size of human populations that can be sustained indefinitely by a defined area of the physical environment – is largely a heuristic concept.

Highly complementary to the Malthusian thesis **(Q74)**, but having various origins and applications, carrying capacity refers to the maximum number of the population of a species that can be supported by the environment (i.e. ecosystem or landscape) without degrading it for future support (Chapman & Byron 2018). The concept has been used robustly in ecology, perhaps introduced as a way of managing game, in which unchecked species growth surpasses the food ceiling, leading to a decline in that species, as occurred in St Matthew Island (Text box 75.1) (Klein 1968).

Applied to humankind, the concept has a chequered history and is challenged on various grounds. It was applied by British and French colonial offices in land-use policies for indigenous Africans, used in cultural ecological research addressing population–agricultural dynamics, especially among subsistence cultivators, and, ultimately, linked to various assessments of the global limits of population and sustainable ecological footprints **(Q96)** (Sayre 2008). In this last application, environmental limits on humankind reflect in a number of ways arguments commonly labelled neo-Malthusian, owing to the similarities between Malthus's original proposition **(Q74)** and the use of carrying capacity (Sayre 2008). Carrying capacity is problematic as applied to human populations, however, for a variety of reasons (Cohen 1995). The provisioning of a population, no matter where it is located, is not merely based on the physical environment, but on the technology and management strategies used, as well as innovations that permit substitutions for diminishing resources, such as the invention of synthetic nitrogen

Text box 75.1 Carrying capacity and St Matthew Island

In 1944, 29 reindeer were introduced to St Matthew Island in the Bering Sea in Alaska. Having no predators on the island, the reindeer increased to 6,000 head by 1963. Grazing pressures on the island were so enormous that the vegetation degraded and an exceptional snowfall in 1964 reduced food stocks for the reindeer. With the island's carrying capacity so degraded, 99 per cent of the reindeer population died (Klein 1968).

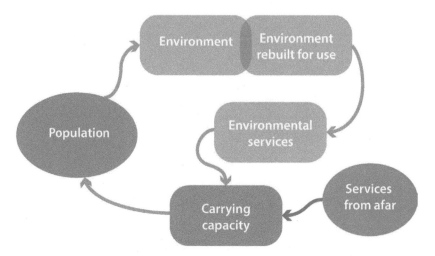

Figure 75.1 Carrying capacity applied to humans
The human carrying capacity of some portion of the biosphere is a product of the interactions of human activities, built environment and ambient environmental conditions on environmental services, accounting for imported resources and export of emissions and waste.
Source: author.

for fertilizers and, of course, the influx of resources obtained from elsewhere. Carrying capacity is, at best, a fluid, not fixed measure (Cohen 1995; Sayre 2008) (Fig. 75.1).

Considerations of the Anthropocene have turned attention away from carrying capacity in its simple or crude sense to a more inclusive assessment of the human drivers of demand, accounting for affluence, technology and social equity (Dasgupta & Ehrlich, 2013), and to the direct impacts of this demand on environmental services and the functioning of the Earth system (**Q88**). Importantly, however, the broader implications of the carrying capacity concept are not likely to dissipate given that humankind confronts a closed system of finite resources at the planetary level.

References

Chapman, E. & C. Byron (2018). "The flexible application of carrying capacity in ecology". *Global Ecology and Conservation* 13: e00365. doi.org/10.1016/j.gecco.2017.e00365.

Cohen, J. (1995). "Population growth and Earth's human carrying capacity". *Science* 269(5222): 341–6. doi:10.1126/science.7618100.

Dasgupta, P. & P. Ehrlich (2013). "Pervasive externalities at the population, consumption, and environment nexus". *Science* 340(6130): 324–8. doi:10.1126/science.1224664.

Klein, D. (1968). "The introduction, increase, and crash of reindeer on St. Matthew Island". *Journal of Wildlife Management*: 350–67. doi.org/10.2307/3798981.

Sayre, N. (2008). "The genesis, history, and limits of carrying capacity". *Annals of the Association of American Geographers* 98(1): 120–34. doi.org/10.1080/00045600701734356.

HOW DOES THE BOSERUPIAN THESIS AND ITS IMPLICATIONS APPLY TO HUMAN–ENVIRONMENT RELATIONSHIPS?

Short answer: the Boserupian thesis that population pressures increase agricultural intensity, enlarging the food ceiling, has substantial support.

Ester Boserup developed an anti-Malthusian thesis based on her observation of agriculture in the developing world. The intensity of food production and hence food supply (i.e. the food ceiling in the Malthus case; **Q74**) is a response to land pressures, largely from provisioning population. As these pressures increase, so does the food ceiling or supply

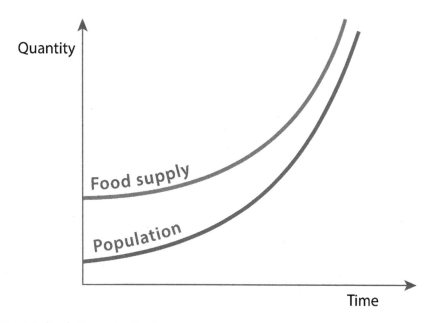

Figure 76.1 Simple Boserupian thesis

Holding place constant, changes in population, creating food demand and food supply are associated with the rise (or fall) in land pressures and agricultural intensity.

Source: author.

(Fig. 76.1). This dynamic follows because the techno-managerial applications for production (e.g. irrigation, crop rotation, no-till cultivation) are endogenous to (built into or part of) the food system. As such, agricultural intensity is fluid, given the land pressures in question. Various examinations demonstrate that population pressures and agricultural intensity have been and tend to remain correlated, especially among subsistence and mixed subsistence–market cultivators (Turner & Shajaat Ali 1996), but with variations (Allen 2001; Börjeson 2007). Boserup, however, extended her thesis to cover the shift into commercial cultivation, whereby the market determines land pressures, driving changes in the techno-managerial strategies applied, a position subsequently known as induced intensification (Meyfroidt *et al.* 2018). Support for this thesis varies across different economies, in which rebound effects (i.e. improved technology changes cultivators' behaviour; Q70) are apparent, but appears to hold across low-income countries (Garcia *et al.* 2020). The "Boserupian" thesis is consistent with the Cornucopian position (Q73) that techno-managerial improvements follow land pressures or demand. Complementary to this thesis is that of induced innovation (Liu & Shumway 2006), which associates land and labour scarcity to investments in agricultural innovations.

Interestingly, and in contrast to the neo-Malthusian position, the Boserupian case has not been directly expanded to address questions of environmental services and the capacity of the Earth system to sustain the biosphere. A Boserupian-like thesis that techno-managerial innovations are linked to increasing population pressures, however, undergirds some Cornucopian views.

References

Allen, B. (2001). "Boserup and Brookfield and the association between population density and agricultural intensity in Papua New Guinea". *Asia Pacific Viewpoint* 42(2/3): 236–54. doi.org/10.1111/1467-8373.00147.

Börjeson, L. (2007). "Boserup backwards? Agricultural intensification as 'its own driving force' in the Mbulu highlands, Tanzania". *Geografiska Annaler: Series B, Human Geography* 89(3): 249–67. doi.org/10.1111/j.1468-0467.2007.00252.x.

Garcia, V. *et al.* (2020). "Agricultural intensification and land use change: assessing country-level induced intensification, land sparing and rebound effect". *Environmental Research Letters* 15(8): 08507. doi.org/10.1088/1748-9326/ab8b14.

Liu, Q. & C. Shumway (2006). "Geographic aggregation and induced innovation in American agriculture". *Applied Economics* 38(6): 671–82. doi.org/10.1080/00036840500397457.

Meyfroidt, P. *et al.* (2018). "Middle-range theories of land system change". *Global Environmental Change* 53: 52–67. doi.org/10.1016/j.gloenvcha.2018.08.006.

Turner II, B. & A. Shajaat Ali (1996). "Induced intensification: agricultural change in Bangladesh with implications for Malthus and Boserup". *Proceedings of the National Academy of Sciences* 93(26): 14984–91. doi.org/10.1073/pnas.93.25.14984.

CAN THE MALTHUSIAN AND BOSERUPIAN THESES BE RECONCILED AND WHAT ARE THE IMPLICATIONS FOR UNDERSTANDING HUMAN–ENVIRONMENT RELATIONSHIPS?

Short answer: yes. Combined with the concept of involution, the Malthus and Boserup theses provide a framework of the trajectories of population–resource relationships.

Malthus envisioned population's capacity to surpass food production (i.e. the existence of a carrying capacity; **Q74**), whereas Boserup proposed that population pressures were a driver of agricultural growth with the potential to provision food (**Q76**). Either case implies the economic concept of marginal product or the rate of return (output) per a unit of input. Malthus infers that these returns can be negative, triggering the crisis period, whereas Boserup foresaw sustained positive returns, although the margins might narrow in high pressure-intensity conditions. Linking these two positions is the concept of involution proposed by Clifford Geertz (Odijie 2016), in which farmers may approach and sustain conditions in which one unit of input returns only one unit of output, one that is viewed as stagnation.

The long history of population–food relationships can be conceived as a fusion of the three views (Fig. 77.1) (Ellis *et al.* 2013; Turner & Shajaat Ali 1996). For most of human history, Boserup conditions have reigned, in which land pressures, initially by immediate population growth and subsequently by the addition of the growth in affluence, increase innovations within a techno-managerial regime of procurement, generating growth in provisions. Ultimately, the marginal returns (output gained from an additional unit of input) of these techno-managerial strategies, given increasing demands, may approach the Geertz conditions of involution (one added unit of input gains only one unit of output). At this juncture, increasing demands may lead to two broad possible outcomes. In the first, migrations, warfare, colonialism and other means to expand the global arena for procuring provisions have taken place, akin to the Malthus case; these activities, however, create new spaces for the Boserupian conditions to exist. In the second, technological regime shifts occur, in some cases accompanied by changes in socio-economic regimes, that move humankind into Boserupian conditions, such as that of the green revolution (**Q22**) and perhaps the current revolution in genomics (DeSchutter & Vanloqueren 2011), alleviating potential food stresses. Globally, the expansion of agricultural lands may be approaching its limits, focusing attention on techno-managerial regimes to create the Boserupian space.

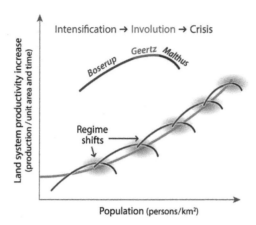

Figure 77.1 Population–resources (food) relationships: historical and global

Population growth (green); Boserup conditions of advances in food production (blue); Geertz conditions of low marginal returns (red); critical conditions (orange ovals) either moving to Malthusian crisis (black) or shifting to a new Boserup space. See text for explanation.

Source: Ellis *et al.* (2013), after author.

Mathematical modelling, however, suggests that highly intensive agriculture approaching Earth system limits could render a Malthusian outcome of insufficient production leading to malnutrition or famine (Lee 1988).

In principle, the same "fusion" schema of Boserup–Geertz–Malthus (Fig. 77.1) reflects the role of increasing global demand on the Earth system beyond food provisioning in which innovation shifts the global capacity of techno-managerial strategies to maintain and increase natural capital. Sustaining this capital, however, has degraded the Earth system and increasingly amplified the total amount of net primary productivity that humankind takes from that system, illustrating potential stress on provisioning society in the longer term (Haberl *et al.* 2007).

References

DeSchutter, O. & G. Vanloqueren (2011). "The new green revolution: how twenty-first-century science can feed the world". *Solutions* 2(4): 33–44. hdl.handle.net/10535/7482.

Ellis, E. *et al.* (2013). "Used planet: a global history". *Proceedings of the National Academy of Sciences* 110(20): 7978–85. doi.org/10.1073/pnas.1217241110.

Haberl, H. *et al.* (2007). "Quantifying and mapping the human appropriation of net primary production in earth's terrestrial ecosystems". *Proceedings of the National Academy of Sciences* 104(31): 12942–7. doi.org/10.1073/pnas.0704243104.

Lee, R. (1988). "Induced population growth and induced technological progress: their interaction in the accelerating stage". *Mathematical Population Studies* 1(3): 265–88. doi.org/10.1080/08898488809525278

Odijie, E. (2016). "Diminishing returns and agricultural involution in Côte d'Ivoire's cocoa sector". *Review of African Political Economy* 43(149): 504–17. doi.org/10.1080/03056244.2015.1085381.

Turner II, B. & A. Shajaat Ali (1996). "Induced intensification: agricultural change in Bangladesh with implications for Malthus and Boserup". *Proceedings of the National Academy of Sciences* 93(25): 14984–91. doi.org/10.1073/pnas.93.25.14984.

WHAT WAS "THE BET" AND WHAT ARE ITS IMPLICATIONS FOR UNDERSTANDING HUMAN–ENVIRONMENT RELATIONSHIPS?

Short answer: a Cornucopian and Cassandra bet about global population stresses on resources, determined by the market value of minerals, illustrates the dichotomous interpretations applied to the relationships.

"The Bet" is a longstanding conceptual icon in the Cassandra–Cornucopian debate (Turner 2014). During the late 1970s, the Cassandra – Paul Ehrlich – and the Cornucopian – Julian Simon – had been debating publicly about the dangers or not of global population growth. Ehrlich, an ecologist, drew on the carrying capacity concept (Q75), leading to a neo-Malthusian position: more people induced more demand for resources, damages to the environment, and the potential for resource stresses. Simon, an economist examining the history of population–resource relationships, derived a Boserupian-like view (Q76): more people lead to more innovations, which in turn expand and substitute for resources. In 1980, Simon challenged Ehrlich to a bet in *The New York Times*. Ehrlich would pick five minerals and the value of them after ten years of rapid population growth would determine which of their two positions was correct. If the minerals increased in costs, the resource stresses championed by Ehrlich would be supported; a decrease in costs would confirm Simon's innovation position. The results are captured in Table 78.1. All five minerals had declined in price (inflation adjusted), as trumpeted by *The New York Times*. The Cornucopians were right, at least for the moment!

In reality, The Bet was little more than a stunt, divorced from the reality that the two positions entertained. Why? Because the value of commodities in the market is contingent on many factors that need not be immediately tied to the population–resource relationship (Lawn 2010). Price is affected by imperfect markets and information and externalities, among other factors. In addition, the price results, and thus the winner, was determined by the decade employed in the assessment. Subsequent decadal analysis from 1990 to 2008 found Ehrlich, not Simon, to have been the winner: 61.6 per cent in ten-year intervals (Kiel, Mathewson & Golembiewski 2010) (also for 1981–2007; Table 78.2).

Table 78.1 Results of the original Bet

Mineral	Percentage change in value ($), 1980 to 1990
Copper	−18.5
Chrome	−40
Nickel	−3.5
Tin	−72
Tungsten	−57

Source: based on Tierney (1990) and Kedrosky (2010).

Simon recognized the luck (timing) in his win but believed that on average his position would hold over different time periods (Emmett & Grabowski 2022). He offered Ehrlich a new bet which was wisely declined for the reason noted above, although a similar bet engaged by Simon with another individual resulted in Simon's loss. Ehrlich and his colleagues realized that the Cassandra position was not one of market value for resources but of the degradation of the environment and Earth system linked to the demands of increasing population (and subsequently other demand factors). Ehrlich proposed a new bet based on the conditions of different components on the Earth system, such as carbon dioxide in the atmosphere, given another decade of population growth. Simon wisely declined.

The critical lesson of The Bet is how the differences in the worldviews of the participants lent themselves to different interpretations and forecasts of our relationships with nature – the world as cup half full or half empty (**Q73**). The trajectory of population and technology has placed us in the Cornucopian world for much of our history, one in which technological advancement has led to resource substitutions and increased efficiency in resource use. This space, however, has come at the cost of changes in environment that have scaled to Earth system impacts. These impacts, such as major climate changes, loss of the ozone layer and insufficient potable water, raise serious concerns about the Cornucopian vision but do not confirm that a Cassandra position will take place (**Q85**).

Table 78.2 Results of The Bet based on different start years

1981: Simon	1990: Simon	1999: Ehrlich
1982: Simon	1991: Simon	2000: Ehrlich
1983: Simon	1992: Simon	2001: Ehrlich
1984: Simon	1993: Simon	2002: Ehrlich
1985: Ehrlich	1994: Ehrlich	2003: Ehrlich
1986: Ehrlich	1995: Ehrlich	2004: Ehrlich
1987: Simon	1996: Ehrlich	2005: Ehrlich
1988: Simon	1997: Ehrlich	2006: Ehrlich
1989: Simon	1998: Ehrlich	2007: Ehrlich

Source: Kedrosky (2010).

References

Emmett, R. & J. Grabowski (2022). "Better lucky than good: the Simon–Ehrlich bet through the lens of financial economics". *Ecological Economics* 193: 107322. doi.org/10.1016/j.ecolecon.2021.107322.

Kedrosky, P. (2010). "Taking another look at Simon vs. Ehrlich on commodity prices". seekingalpha.com/article/189539-taking-another-look-at-simon-vs-ehrlich-on-commodity-prices.

Kiel, K., V. Matheson & K. Golembiewski (2010). "Luck or skill? An examination of the Ehrlich–Simon bet". *Ecological Economics* 69(7): 1365–7. doi.org/10.1016/j.ecolecon.2010.03.007.

Lawn, P. (2010). "On the Ehrlich–Simon bet: both were unskilled and Simon was lucky". *Ecological Economics* 69(11): 2045–6. doi.org/10.1016/j.ecolecon.2010.07.009.

Tierney, J. (1990). "Betting the planet". *New York Times Magazine*, 2 Dec.: 81.

Turner II, B. (2014). "The Bet: Paul Ehrlich, Julian Simon, and our gamble with the Earth's future". *The AAG Review of Books* 2(1): 1–3. Doi.10.1080/2325548X.2014.894410.

WHY IS HUMAN-INDUCED CLIMATE CHANGE SO SERIOUSLY CHALLENGED DESPITE THE SCIENCE SUPPORTING IT?

Short answer: a vocal naysayer cohort is concerned about the commercial and regulatory impacts associated with actions to combat climate warming.

Science overwhelmingly supports the case for human-induced climate change – overall the Earth is warming and is doing so at unusually rapid rates owing to the accumulation of human changes in the Earth system, foremost the emission of greenhouse gases (Cook *et al.* 2016) (**Q44**). The processes at play in this warming are well established, and the Earth system responses are consistent with them (**Q46**). Few atmospheric scientists and only a small number of scientists at large challenge claims about Earth system warming due to human activity. Their various challenges and alternative explanations to examine have included the role of natural forcings, such as warming due to sun spots, and inconsistencies in the expected Earth system responses to warming, such as differences in ice loss on either side of Antarctica or cooling episodes during the warming trend. The science of climate change, however, increasingly strengthens support for the Earth system trend of warming and the associated role of human activity (**Q46**) (NAS/RS, 2020). The impacts of the pro and con assessments of human-induced climate change have created substantial variation in worldwide views of the public at large (Fig. 79.1).

Most of the naysayers' positions seem to be grounded in socio-political worldviews which are antithetical to the mitigation and adaptation strategies proposed to combat global warming. The most vocal and forceful naysayers are a few conservative and libertarian organizations and various corporations (Boussalis & Coan 2016; Supran & Oreskes 2017) who are concerned about the use of governmental regulation to combat the problem and the potential negative impacts on commerce, especially in the fossil fuel industry (Nasiritousi 2017; Supran & Oreskes 2017), and, in the United States, the general growth of federal control in everyday life (Collomb 2014). In essence, worldviews linked to the political economy colour the responses to human-induced climate change. Various conservative and libertarian think-tank organizations leading the challenges to climate change use the uncertainties required in science studies (Aven & Renn 2015; Drouet, Bosetti & Tavoni 2015)

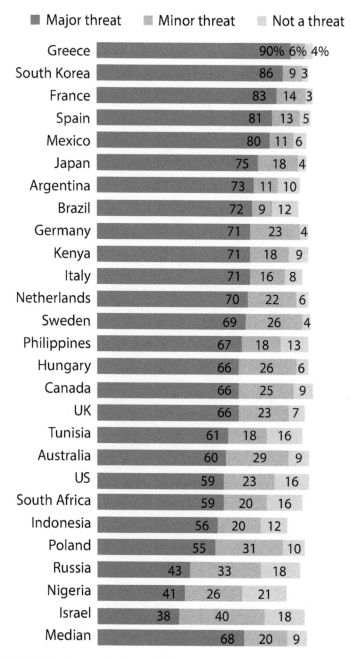

Figure 79.1 Worldwide views on the threat of climate change
Source: Pew Research Center (2019).

to deflect the overall strength of research conclusions (Cann & Raymond 2018; Smith & Leiserowitz 2012), and employ the economic and political leverages of the organizations to influence information sources, decision-makers and the public. These strategies have been effective, especially in "misrepresent[ing] doubt about anything [e.g., in models] to insinuate doubt about everything" (Palmer & Stevens 2019: 24390), with public perception impacts.

These impacts vary by country (Fig. 79.1), however, perhaps associated with the attention that naysayers expend to the subject, such as in the United States and Australia. Examinations of multiple media sources in the US, for example, indicate that climate change contrarians have been featured in 49 per cent more articles than are climate scientists (Petersen, Vincent & Westerling 2019) and that press releases in opposition to climate change actions are cited in major US presses about two times more than press releases calling for action (Wetts 2020). In addition, naysayer use of social media has been effective in shaping public perceptions about the reality of climate change (Smith & Leiserowitz 2012). These strategies were also employed by the same organizations against a series of past environmental concerns, including acid rain damages (Q43) and ozone layer thinning (Q53), all of which proved the science concerns to be correct. Moreover, the ultimate actions (regulations) taken to address these issues have been shown to rectify the problems.

Those organizations challenging the science of climate change as well as many individuals holding minimal concerns about the consequences of that change tend to subscribe to technological-fix worldviews of Cornucopians (Q71). Such views elevate the capacity of humankind to solve potential problems through innovation. Yet others clearly understand the consequences of changing climate, but believe that attention directed to mitigating climate change is largely ineffective. Instead, those holding this view argue that attention should be given to improving socio-economic development to assist in the adaptation to climate warming and its environmental impacts (Bortscheller 2008; Smith & Leiserowitz 2012). The United Nations' Sustainable Development Goals apply this attention but also include concerns about climate change mitigation (Q94).

References

Aven, T. & O. Renn (2015). "An evaluation of the treatment of risk and uncertainties in the IPCC reports on climate change". *Risk Analysis* 35(4): 701–12. doi.org/10.1111/risa.12298.

Bortscheller, M. (2008). "Cool it. the skeptical environmentalist's guide to global warming by Bjørn Lomborg". *Sustainable Development Law & Policy* 8(2): 77–8. doi.org/10.1002/ieam.5630040428.

Boussalis, C. & T. Coan (2016). "Text-mining the signals of climate change doubt". *Global Environmental Change* 36: 9–100. doi.org/10.1016/j.gloenvcha.2015.12.001.

Cann, H. & L. Raymond (2018). "Does climate denialism still matter? The prevalence of alternative frames in opposition to climate policy". *Environmental Politics* 27(3): 433–54. doi.org/10.1080/09644016.2018.1439353\.

Collomb, J. (2014). "The ideology of climate change denial in the United States". *European Journal of American Studies* 9(9-1). doi.org/10.4000/ejas.10305.

Cook, J. *et al.* (2016). "Consensus on consensus: a synthesis of consensus estimates on human-caused global warming". *Environmental Research Letters* 11(4): 048002. doi:10.1088/1748-9326/11/4/048002.

Drouet, L., V. Bosetti & M. Tavoni (2015). "Selection of climate policies under the uncertainties in the Fifth Assessment Report of the IPCC". *Nature Climate Change* 5(10): 937–40. doi.org/10.1038/nclimate2721.

Nasiritousi, N. (2017). "Fossil fuel emitters and climate change: unpacking the governance activities of large oil and gas companies". *Environmental Politics* 26(4): 621–47. doi.org/10.1080/09644016.2017.1320832.

NAS/RS (National Academy of Sciences/Royal Society) (2020). *Climate Change: Evidence and Causes: Update 2020.* Washington, DC: National Academies Press. doi.org/10.17226/25733.

Palmer, T. & B. Stevens (2019). "The scientific challenge of understanding and estimating climate change". *Proceedings of the National Academy of Sciences* 116(49): 24390–95. doi.org/10.1073/pnas.1906691116.

Petersen, A., E. Vincent & A. Westerling (2019). "Discrepancy in scientific authority and media visibility of climate change scientists and contrarians". *Nature Communications* 10(1): 1–14. doi.org/10.1038/s41467-019-09959-4.

PRC (Pew Research Center) (2019). "A look at how people around the world view climate change". www.pewresearch.org/fact-tank/2019/04/18/a-look-at-how-people-around-the-world-view-climate-change/ft_19-04-18_climatechangeglobal_inmostsurveyedcountries_edited_2/.

Smith, N. & A. Leiserowitz (2012). "The rise of global warming skepticism: exploring affective image associations in the United States over time". *Risk Analysis: An International Journal* 32(6): 1021–32. doi.org/10.1111/j.1539-6924.2012.01801.x.

Supran, G. & N. Oreskes (2017). "Assessing ExxonMobil's climate change communications (1977–2014)". *Environmental Research Letters* 12(8): 084019. doi.org/10.1088/1748-9326/ab89d5.

Wetts, R. (2020). "In climate news, statements from large businesses and opponents of climate action receive heightened visibility". *Proceedings of the National Academy of Sciences* 117(32): 19054–60. doi.org/10.1073/pnas.1921526117.

Q80

IN WHAT WAYS DO HUMANISTS INFLUENCE OUR UNDERSTANDING OF HUMAN–ENVIRONMENT RELATIONSHIPS?

Short answer: humanists provide a moral and philosophical foundation for the many worldviews of human–environment relationships.

Over the past two centuries, humanists – including writers, poets, natural historians, philosophers and scientists invoking moral expressions – have been instrumental in moving humankind beyond pure anthropocentrism. Variously expressed, nature (i.e. the environment and Earth system) in their view is a functioning and interconnected world in its own right and, for many, with its own agency. It deserves our respect and admiration, and not simply because of its environmental services (**Q7, Q69**). Such sentiment has been handed down in Western and Eastern philosophies and among Indigenous cultures, calling for environmental and land ethics in which humankind interprets its relationship to the Earth system not merely in terms of economics and privilege, but in terms of stability, integrity, respect and obligation (e.g. following Leopold 1949: 224–5).

Early Western humanist attempts to protect functioning ecosystems – those that were least influenced by human activities – understandably turned their backs on built environments. Nature was at its best when wild, which meant untainted by humans. To appreciate the natural world, you had to be absent from the artificial, human-made one. This ecocentric or biocentric worldview, however, has an anthropocentric undertone. Appreciation of nature, in this view, is required for the well-being of humanity, captured iconically by the title, "Can one love a plastic tree?" (Iltis 1973). In the past few decades, however, scientists, naturalists and environmentalists have tended to reject what they see as a romanticized view of "wilderness" in favour of humanity and the built environment, complete with humanist appreciation for this "wrong nature" (Cronon 1996). If real nature is always somewhere else, and is anti-modern and anti-city, then people will ignore the nature in their own backyards and shirk responsibility for the environment in which they actually live (think social justice, environmental hazards, toxic waste). This view also ignores the long human history and land stewardship that existed, both past and present, in the very places otherwise romanticized as wild nature (**Q10–12**).

Recent emphasis calls for connections to the environment in our everyday often urban lives. An exemplar is work exploring the natural, political and socio-economic history and revitalization of the 50-mile, degraded, storm-sewer that is the Los Angeles River, noting the issue is not nature management, but fair negotiations to achieve sustainability (Price 2008). Others lament putting humans front and centre. For example, McKibben (2005) points to the human role in global climate change as unalterably changing human–environment relationships, in which nature becomes a by-product of human activities in his provocatively titled *The End of Nature*. Others, however, take comfort in the same conclusion, finding an anthropocentric arrogance in the view that nature should be unchanging, recognizing the huge changes in the Earth system over the long run, independent of human impact (Kingsnorth 2017).

Humanists, therefore, do not speak from a worldview silo, but express the range of views found within the sciences and social sciences at large (**Q72, Q73**), helping to shape them in various ways. Critically, they lay a moral and philosophical foundation for these views rather than resolve them per se.

References

Cronon, W. (1996). "The trouble with wilderness: or, getting back to the wrong nature".
Environmental History 1(1): 7–28. doi.org/10.2307/3985059.

Iltis, H. (1973). "Can one love a plastic tree?" *Bulletin of the Ecological Society of America* 54(4): 5–7,
19. www.jstor.org/stable/20165966.

Kingsnorth, P. (2017). *Confessions of a Recovering Environmentalist and Other Essays*. Minneapolis,
MN: Graywolf Press.

Leopold, A. (1949). *A Sand County Almanac: And Sketches Here and There*. Oxford: Oxford
University Press.

McKibben, B. (2005). "The emotional core of the end of nature". *Organization & Environment*
18(2): 182–5. doi:10.1177/1086026605276009.

Price, J. (2008). "Remaking American environmentalism: On the banks of the LA River".
Environmental History 13(3): 536–55. www.jstor.org/stable/25473266.

SECTION IX

SUSTAINABILITY IN THE ANTHROPOCENE

Conceptually, the Anthropocene and sustainability fit hand-in-glove. The proposed epoch or event (**Q2**) constitutes the recognition that human activity is not only a force equivalent to nature in many instances in terms of driving changes in the Earth system, but these changes threaten feedbacks on the maintenance of the biosphere and human well-being. Sustainability, in turn, is concerned with these threats as they affect the maintenance and improvement of human well-being now and for future generations. To be concise, the Anthropocene challenges the sustainability of human–environment relationships.

Concern about the human impacts on the environment can be traced back to antiquity (**Q69**), with escalating scholarship on the subject from the nineteenth century onwards. The 1987 report of the United Nations World Commission on Environment and Development (WCED 1987) is commonly noted as the starting point for contemporary attention to sustainability themes (Clark & Harley 2019). It called for making human development sustainable in the face of stresses placed on the environment, one in which provisioning humankind now would not endanger provisioning for future generations. This call, complemented by the recognition of climate warming through the efforts of the United Nations Intergovernmental Panel on Climate Change, triggered such programmes as the International Geosphere–Biosphere Programme, the International Human Dimensions Programme and the United Nations Sustainable Development Goals and Future Earth (see Appendix) which have provided the science foundation that distinguishes the Anthropocene and creates the foundation for sustainability science.

The various definitions of sustainability coalesce around two connected themes: (1) provisioning humankind more equitably, now and for future generations and (2) without threatening the base functions of the Earth system to sustain life (Kates *et al.* 2001). It is envisioned as a useful science, akin, for instance, to agricultural or medical science, in which the problems to understand and resolve are those identified by global society that are critical to human well-being (Clark 2007). It also recognizes that improvements in the human–environment relationship requires a deep appreciation of the Earth system and the need for innovations in all dimensions of human endeavour – social, economic, political and technological – to provision humankind while sustaining nature (Kates, Travis & Wilbanks 2012; Scoones *et al.* 2020). Such endeavours are daunting, requiring major improvements to address the relationships in question; one of the most significant is the reconciliation of the extremes of the Cassandra and Cornucopian worldviews about those relationships (**Q73**). This section elaborates themes, issues and concerns of sustainability science.

References

Clark, W. (2007). "Sustainability science: a room of its own". *Proceedings of the National Academy of Sciences* 104: 1737–8. doi.org/10.1073/pnas.0611291104

Clark, W. & A. Harley (2019). "Sustainability science: towards a synthesis". *Annual Reviews in Environment and Resources* 45: 331–86. doi.org/10.1146/annurev-environ-012420-043621.

Kates, R. *et al.* (2001). "Sustainability science". *Science* 292(5517): 641–2. doi10.1126/science.1059386.

Kates, R., W. Travis & T. Wilbanks (2012). "Transformational adaptation when incremental adaptations to climate change are insufficient". *Proceedings of the National Academy of Sciences* 109(19): 7156–61. doi.org/10.1073/pnas.1115521109.

Scoones, I. *et al.* (2020). "Transformations to sustainability: combining structural, systemic and enabling approaches". *Current Opinion in Environmental Sustainability* 42: 65–75. doi.org/10.1016/j.cosust.2019.12.004.

WCED (World Commission on Environment and Development) (1987). *Our Common Future.* Oxford: Oxford University Press.

DO DISTINCTIONS EXIST AMONG APPROACHES TO SUSTAINABILITY?

Short answer: with shared interests in reconciling human–environment relationships, distinctions exist between problem-solving to inform decision-making versus hands-on application and management of proposed solutions.

Various definitions of sustainability more or less coalesce on themes about reconciling society's development goals within the constraints of the Earth system (Clark & Dickson 2003; Kates *et al.* 2001; Kates, Travis & Wilbanks 2012) (Text box 81.1). Much sustainability research focuses explicitly on sustainable development **(Q84)**, in which accounting for the health of the environment and Earth system is critical for that development (Clark & Harley 2020). The social subsystem orientation is illustrated by the 17 sustainable development goals identified by the United Nations **(Q99)** (Griggs *et al.* 2013) of which 14 are specific socio-economic issues and only three are aggregated environmental issues (Constanza *et al.* 2016). The number and specification details of the human subsystem goals illuminate the "development" part of sustainability, interestingly counterposed to the numerous former and current international science programmes directed at the Earth system (e.g. Appendix). We might suppose that, having made enormous progress in understanding the Earth system **(Q3)**, attention can now turn to the societal outcomes of human–environment relationships

---

Text box 81.1 Definitions of sustainability

Synthesis definitions interpreted from multiple sources.
- Meeting fundamental human needs while preserving the life-support systems of the planet Earth.
- Meeting the needs of present and future generations while substantially reducing poverty and conserving the planet's life-support systems.
- Provisioning humankind more equitably without threatening the functioning of the Earth system to deliver environmental services.
- Reducing poverty and hunger, improving health and well-being, and creating sustainable production and consumption patterns.

and responses that might be taken toward making those relationships sustainable. This type of integration underscores the international programme, Future Earth (van der Hel 2016).

All research directed toward sustainability questions share at least three broad attributes: (1) the base problems are normative in that they underpin real-world goals, such as that to improve the equity in material well-being among the global population; (2) knowledge is ultimately linked in various ways to action (below); and (3) transdisciplinary research is required, including co-design and implementation with stakeholders at large because their buy-in is essential for solutions to be achieved. Affiliated with these attributes are critical questions addressing the measurement of progress toward the goals (**Q95, Q96**), adaptation of SESs as complex adaptive systems (**Q86**), governance and equality within this complexity, and what constitutes transformative change (Kates 2011; Kates, Travis & Wilbanks 2012; Lubchenco *et al.* 2019; Shi & Moser 2021).

Approaches to sustainability may differ among practitioners, however, somewhat following the classification of research developed by Stokes (Fig. 81.1) (Doran, Golden & Turner 2017; Spangenberg 2011). This classification distinguishes research based on individual inquisitiveness per se, or what is commonly referred to as pure basic research (the Bohr quadrant, Fig. 81.1, named after the physicist making seminal contributions to atomic structure and quantum theory), from use-inspired basic research based on problems confronting civil society (the Pasteur quadrant, named for the microbiologist's research to solve human health problems). These two "basic" research clusters are distinguished from applied research, which draws on basic research for innovation (the Edison quadrant for his many inventions). The early proponents of sustainability science advanced a role within the Pasteur quadrant: to undertake research on sustainability problems confronting society and providing state-of-the-art understanding for decision-makers and application efforts (Clark & Harley 2020). Others carrying out sustainability research actively engage in application or innovation, such as the

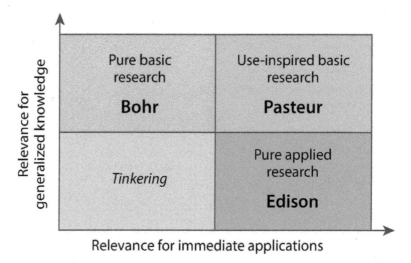

Figure 81.1 Quadrant model of science and practice
Source: Stokes (2011).

practice of no-till cultivation management to reduce soil carbon emissions or innovating technologies to capture carbon from the atmosphere. Yet others identify their work as crossing or fusing the Pasteur–Edison quadrants (Doran, Golden & Turner 2017). For instance, those engaged in basic research on carbon dynamics may apply their knowledge to developing carbon capture technology. In addition, many environmental scientists engaged in basic research have also been involved in developing conventions and protocols about the governance of greenhouse gas emissions, ocean health and biodiversity preservation (e.g. Ruckelshaus *et al.* 2020).

References

Clark, W. & N. Dickson (2003). "Sustainability science: the emerging research program". *Proceedings of the National Academy of Sciences* 100(14): 8059–61. doi.org/10.1073/pnas.1231333100.

Clark, W. & A. Harley (2020). "Sustainability science: toward a synthesis". *Annual Review of Environment and Society* 45: 331–86. doi.org/10.1146/annurev-environ-012420-043621.

Costanza, R. *et al.* (2016). "Modelling and measuring sustainable wellbeing in connection with the UN Sustainable Development Goals". *Ecological Economics* 130: 350–55. doi.org/10.1016/j.ecolecon.2016.07.009.

Doran, E., J. Golden & B. Turner II (2017). "From basic research to applied solutions: are two approaches to sustainability science emerging?" *Current Opinion in Environmental Sustainability* 29: 138–44. doi.org/10.1016/j.cosust.2018.01.013.

Griggs, D. *et al.* (2013). "Sustainable development goals for people and planet". *Nature* 495(7441): 305–07. doi:10.1038/495305a.

Kates, R. (2011). "What kind of a science is sustainability science?" *Proceedings of the National Academy of Sciences* 108(49): 19449–50. doi.org/10.1073/pnas.1116097108.

Kates, R. *et al.* (2001). "Sustainability science". *Science* 292(5517): 641–2. doi:10.1126/science.1059386.

Kates, R., W. Travis & T. Wilbanks (2012). "Transformational adaptation when incremental adaptations to climate change are insufficient". *Proceedings of the National Academy of Sciences* 109(19): 7156–61. doi.org/10.1073/pnas.1115521109.

Lubchenco, J. *et al.* (2019). "Connecting science to policymakers, managers, and citizens". *Oceanography* 32(3): 106–15. doi.org/10.5670/oceanog.2019.317.

Ruckelshaus, M. *et al.* (2020). "The IPBES global assessment: pathways to action". *Trends in Ecology & Evolution* 35(5): 407–14. doi.org/10.1016/j.tree.2020.01.009.

Shi, L. & S. Moser (2021). "Transformative climate adaptation in the United States: trend and prospects". *Proceedings of the National Academy of Sciences* 372(6549): eabc8054. doi10.1126/science.abc8054.

Spangenberg, J. (2011). "Sustainability science: a review, an analysis and some empirical lessons". *Environmental Conservation* 38(3): 275–87. doi:10.1017/S0376892911000270.

Stokes, D. (2011). *Pasteur's Quadrant: Basic Science and Technological Innovation.* Washington, DC: Brookings Institution Press.

van der Hel, S. (2016). "New science for global sustainability? The institutionalisation of knowledge co-production in Future Earth". *Environmental Science & Policy* 61: 165–75. doi.org/10.1016/j.envsci.2016.03.012.

IN WHAT WAYS DOES SUSTAINABILITY DIFFER FROM PREVIOUS ENVIRONMENTAL CONCERNS?

Short answer: it differs in at least two ways – attention to Earth system interactions and complex framings of human–environment interactions.

Recognition of human-induced degradation of environments did not await the scholarly recognition of the Anthropocene at the turn into the twenty-first century. Plato bemoaned land degradation in Greece as early as the fifth century BCE, and George Perkins Marsh provided a broad-ranging assessment of the pros and cons of human impacts on the land in the nineteenth century. The initiation of Earth Day in 1970 (Rome 2010), recognizing the value of nature and human impacts on the environment, preceded attention to the Anthropocene by 30 years.

Sustainability, however, has advanced and amplified past concerns in at least two ways. First, the environment of the Anthropocene (**Q2**) ultimately considers the structure and function of the Earth system – the interacting physical, chemical and biological processes that maintain the biosphere (**Q3**) (O'Neill *et al.* 2018). The Earth system components, linkages and functions have been framed via a wiring diagram (Text box 82.1; Fig. 82.1 A) and are increasingly documented and integrated throughout the environmental sciences. This planetary system generates variations in outcomes by place (**Q83**), owing to the complexity of factors in play, and the cumulative changes in places feed back on the Earth system (**Q8**), such that sustainability concerns link sub-global problems to global ones.

Second, sustainability addresses problems through the lens of the social–environmental system (SES) (**Q71**), in which the interactions between the two subsystems are central to the

---

Text box 82.1 Wiring diagram

The term wiring diagram is borrowed from electrical circuit depictions of the physical layout and connections of an electrical system. The science community that created the Earth system depiction referred to the product as a wiring diagram, a visual representation of the physical elements and the connections among them that maintain the Earth system. It is often called the Bretherton diagram in reference to Francis Bretherton who led in its development.

A. Earth system

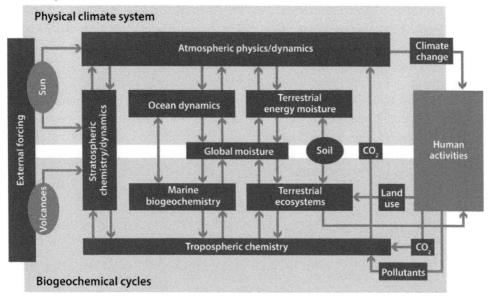

B. Social system

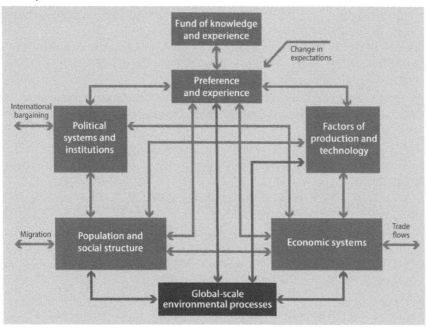

Figure 82.1 Earth system and social system of the social–environmental system

The Earth system (A) and social system (B) depictions. In principle, they connect as subsystems to create a global social–environment system. Note that these two subsystems connect but the fit is highly simplified. Human activities on the Earth system (A, lower right) and global-scale environmental processes on the social system (B, bottom centre) are aggregate and incomplete connections. Note also that the Earth system (A) identifies "forcings" on the system (in blue) whereas the social system (B) does not, although many of the components in B could serve as drivers, such as changes in political systems and institutions.

Source: A: IGBP (2015); B: CIESIN (1992).

answers posed. As such, a social subsystem wiring diagram is required (Text box 82.1), complementary to – and potentially attached to – the Earth system diagram. Such a framework exists (Fig. 82.1 B) but has yet to be embraced by the social sciences at large. Significantly, explicit links between the two subsystems have yet to be rendered in a complete framework, perhaps because the linkages are too complex to illustrate fully.

These issues notwithstanding, the SES approach is employed in a multitude of models that join various parts of the subsystems in a variety of ways (Patt *et al.* 2010). Interestingly, such models applied at the global scale can be traced back at least to early modelling projections of the 1970s, such as that in *The Limits to Growth* (Turner 2008). Today's efforts, however, draw on much more detailed understanding of the operation of the environmental and social subsystems. The analytics and range of models employed permit fine-grain and nuanced assessments of the interactions within and between the two basic subsystems (Gillingham *et al.* 2018; Manning *et al.* 2020; Schlüter, Müller & Frank 2019).

References

CIESIN (Consortium for International Earth Science Information Network) (1992). "Social process diagram". In *Pathways of Understanding: The Interactions of Humanity and Global Environmental Change*, 32–3. CIESIN, Columbia University. www.ciesin.columbia.edu/docume nts/CIESIN1992PathwaysofUnderstanding_sm.pdf.

Gillingham, K. *et al.* (2018). "Modeling uncertainty in integrated assessment of climate change: a multimodel comparison". *Journal of the Association of Environmental and Resource Economists* 5(4): 791–826. doi.org/10.1086/698910.

IGBP (International Geosphere-Biosphere Programme) (2015). "Reflections on Earth-system science". www.igbp.net/news/features/features/reflectionsonearthsystemscience.5.950c2fa149 5db7081ecdc.html.

Manning, D. *et al.* (2020). "Non-market valuation in integrated assessment modeling: the benefits of water right retirement". *Journal of Environmental Economics and Management*: 102341. doi.org/ 10.1016/j.jeem.2020.102341.

O'Neill, D. *et al.* (2018). "A good life for all within planetary boundaries". *Nature Sustainability* 1(2): 88–95. doi.org/10.1038/s41893-018-0021-4.

Patt, A. *et al.* (2010). "Adaptation in integrated assessment modeling: where do we stand?" *Climatic Change* 99(3/4): 383–402. doi.org/10.1007/s10584-009-9687-y.

Rome, A. (2010). "The genius of Earth Day". *Environmental History* 15(2): 194–205. doi.org/10.1093/ envhis/emq036.

Schlüter, M., B. Müller & K. Frank (2019). "The potential of models and modeling for social-ecological systems research". *Ecology and Society* 24(1): 31. doi.org/10.5751/ ES-10716-240131.

Turner, G. (2008). "A comparison of *The Limits to Growth* with 30 years of reality". *Global Environmental Change* 18(3): 397–411. doi.org/10.1038/s41893-018-0021-4.

WHAT DOES "PLACE-BASED" RESEARCH MEAN IN SUSTAINABILITY SCIENCE?

Short answer: the complexity of human–environment relationships and differences among peoples and places create variations in outcomes that must be addressed for successful sustainability solutions.

Place-based research is championed throughout the sustainability sciences, drawing on the need to understand global processes interacting with the social and environmental characteristics of particular peoples, places and sectors (Kates *et al.* 2001). This consensus does not deny the usefulness of the search for general processes operating within or upon social–environmental systems (SESs) (Turner *et al.* 2020) or that sustainability research is restricted to a descriptive or idiographic (individual fact-based) status, failing to yield insights about generic human–environment relationships. Rather, place-based research in sustainability science references the complex adaptive characteristics of SESs (**Q86**), and the variances these qualities create for application or practice.

Spatiotemporal scale mismatches (e.g. environmental processes operating differently from social processes in terms of area and time), interconnected stakeholders (i.e. international organizations to local individuals) with different perceptions and objectives, emergent properties (i.e. entities have properties that their components do not) and hysteresis (the entity is influenced by its history) are just a few of the qualities at play that generate complexity (Balvanera *et al.* 2017; Potschin & Haines-Young 2013; Shrivastava & Kennelly 2013). As such, and despite general processes operating across locations or sectors (e.g. climate change and economic recession), the outcomes for specific places or sectors vary, as do the solutions to confront unwanted outcomes. Sustainability panaceas, therefore, prove difficult and attention is given to the specificity of each case (Ostrom, Janssen & Anderies 2007). Place-based solutions, however, may trigger transformative operations useful for similar sustainability issues in other locales (Shrivastava & Kennelly 2013). Meta-analyses of such studies may provide the foundation for scaling up to broader sets of problems (Balvanera *et al.* 2017), such as the identification of the general rules that make common property regimes successful (Forsyth & Johnson 2014).

References

Balvanera, P. *et al.* (2017). "Interconnected place-based social–ecological research can inform global sustainability". *Current Opinion in Environmental Sustainability* 29: 1-7. doi.org/10.1016/j.cosust.2017.09.005.

Forsyth, T. & C. Johnson (2014). "Elinor Ostrom's legacy: governing the commons, and the rational choice controversy". *Development and Change* 45(5): 1093-110. eprints.lse.ac.uk/56198/.

Kates, R. *et al.* (2001). "Sustainability science". *Science* 292(5517): 641-2. doi:10.1126/science.1059386.

Ostrom, E., M. Janssen & J. Anderies (2007). "Going beyond panaceas". *Proceedings of the National Academy of Sciences* 104(39): 15176-8. doi.org/10.1073/pnas.0701886104.

Potschin, M. & R. Haines-Young (2013). "Landscapes, sustainability and the place-based analysis of ecosystem services". *Landscape Ecology* **28(6): 1053-65. doi.org/10.1007/s10980-012-9756-x.**

Shrivastava, P. & J. Kennelly (2013). "Sustainability and place-based enterprise". *Organization & Environment* 26(1): 83-101. doi.org/10.1177/1086026612475068.

Turner II, B. *et al.* (2020). "Framing the search for a theory of land use". *Journal of Land Use Science* 15(4): 489-508. doi.org/10.1080/1747423X.2020.1811792.

HOW DOES SUSTAINABLE DEVELOPMENT DIFFER FROM SUSTAINABILITY?

Short answer: inspired by the 1987 report of the World Commission on Environment and Development, sustainable development and sustainability research are linked conceptually, but sustainability has significantly enlarged the dimensions and level of research integration applied to their shared problem of a sustainable human–environmental relationship.

The antecedents to the concept of sustainable development have a long intellectual history (Du Pisani 2006). The contemporary emergence of the concept, however, one that crosses several research fields (Clark & Harley 2020; Dasgupta 2007), is related to the 1987 World Commission on Environment and Development report, *Our Common Future*, widely referred to as the Brundtland Report (WCED 1987). At least two derivations of research subsequently evolved that are highly linked but different in their emphasis, especially in regard to the environment.

By the 1990s, sustainable development was emerging as a subfield in some disciplines, such as economics (Dasgupta 2007; Chichilnisky 1997; Lélé 1991; Mitlin 1992) and interdisciplinary research fields, foremost those dealing with development in the Global South (Du Pisani 2006; Waas *et al.* 2011). For the most part, this attention was led by the social sciences and focused on improving material well-being through improved resource management and access to resources. It involved both mainstream economic approaches, supported by various international organizations and agencies involved in economic development (Lélé 1991) and political economic approaches critical of the mainstream. In either case, attention to the biophysical world centred on provisioning services (**Q7**), such food and water, at the local level (Cobbinah, Erdiaw-Kwasie & Amoateng 2015). This work provided a foundation for subsequent consideration of the environment in the development process.

The architects and proponents of what has become sustainability research began to coalesce about a decade beyond that of sustainable development. They too drew on the World Commission's concept, subsequently amplified by way of the United Nations' Sustainable Development Goals (**Q99**) and with explicit calls for a more globally equitable level of development (Costanza *et al.* 2016; Wackernagel, Hanscom & Lin 2017). Indeed, the problem of sustainability science has been identified as "sustainable development" (Clark & Harley 2020). Significantly, however, this science, and sustainability research in general (**Q81**), enlarges the environmental dimension of concern (**Q88, Q89**), seeks understanding through integrated human–environment approaches – the social–environmental system (**Q71, Q86**) – and expands the dimensions of research and applications compared with the earlier sustainable development efforts. These dimensions include a full economic accounting of environmental services via inclusive wealth in resource/environmental economics (**Q95**) or in ecological economics approaches (**Q85**) and linking knowledge to action, with attention to stakeholder co-design of research and application (Cash *et al.* 2003; Kates 2011; Turner *et al.* 2016). Foremost, the role of environment has been strongly influenced by environmental science, especially ecology (e.g. Lade *et al.* 2020; Wu 2013), with its emphasis on vulnerability and resilience of environmental services (**Q7, Q93**), especially at the Earth system level (**Q3**). In addition, inter- and transdisciplinary interactions have advanced attention to complexity for understanding social–environmental systems and their management (**Q71**). An argument can be made that sustainability or sustainability science has now become the moniker for sustainable development (Clark & Harley 2020).

References

Cash, D. *et al.* (2003). "Knowledge systems for sustainable development". *Proceedings of the National Academy of Sciences* 100(14): 8086–91. doi.org/10.1073/pnas.1231332100.

Chichilnisky, G. (1997). "What is sustainable development?" *Land Economics* 73(4): 467–91. www.jstor.org/stable/pdfplus/3147240.

Clark, W. & A. Harley (2020). "Sustainability science: toward a synthesis". *Annual Review of Environment and Resources* 45: 331–86. doi.org/10.1146/annurev-environ-012420-043621.

Cobbinah, P., M. Erdiaw-Kwasie & P. Amoateng (2015). "Rethinking sustainable development within the framework of poverty and urbanisation in developing countries". *Environmental Development* 13: 18–32. doi.org/10.1016/j.envdev.2014.11.001.

Costanza, R. *et al.* (2016). "Modelling and measuring sustainable wellbeing in connection with the UN Sustainable Development Goals". *Ecological Economics* 130: 350–55. doi.org/10.1016/j.ecolecon.2016.07.009.

Dasgupta, P. (2007). "The idea of sustainable development". *Sustainability Science* 2(1): 5–11. doi10.1007/s11625-007-0024-y.

Du Pisani, J. (2006). "Sustainable development: historical roots of the concept". *Environmental Sciences* 3(2): 83–96. doi.org/10.1080/15693430600688831.

Kates, R. (2011). "What kind of a science is sustainability science?" *Proceedings of the National Academy of Sciences* 108(49): 19449–50. doi.org/10.1073/pnas.1116097108.

Lade, S. *et al.* (2020). "Human impacts on planetary boundaries amplified by Earth system interactions". *Nature Sustainability* 3(2): 119–28. doi.org/10.1038/s41893-019-0454-4.

Lélé, S. (1991). "Sustainable development: a critical review". *World Development* 19(6): 607–21. doi.org/10.1016/0305-750X(91)90197-P.

Mitlin, D. (1992). "Sustainable development: a guide to the literature". *Environment and Urbanization* 4(1): 111–24.

Turner II, B. *et al.* (2016). "Socio-environmental systems (SES) research: what have we learned and how can we use this information in future research programs". *Current Opinion in Environmental Sustainability* 19: 160–68. doi.org/10.1016/j.cosust.2016.04.001.

Waas, T. *et al.* (2011). "Sustainable development: a bird's eye view". *Sustainability* 3(10): 1637–61. doi.org/10.3390/su3101637.

Wackernagel, M., L. Hanscom & D. Lin (2017). "Making the sustainable development goals consistent with sustainability". *Frontiers in Energy Research* 5: 18. doi.org/10.3389/ fenrg.2017.00018.

WCED (World Commission on Environment and Development) (1987). *Our Common Future.* Oxford: Oxford University Press.

Wu, J. (2013). "Landscape sustainability science: ecosystem services and human well-being in changing landscapes". *Landscape Ecology* 28(6): 999–1023. doi.org/10.1007/s10980-013-9894-9.

Q85

WHAT IS THE DISTINCTION BETWEEN WEAK AND STRONG SUSTAINABILITY AND ITS IMPLICATIONS FOR SUSTAINABILITY GOALS?

Short answer: weak and strong sustainability are distinguished by the degree to which the substitutability for environmental goods and services is viewed as supporting the goals of sustainability.

Sustainability can be viewed as non-decreasing human welfare or wealth over generations in regard to environmental goods and services (**Q95**). Strong sustainability refers to the actual conservation of those goods and services, whereas weak sustainability refers to substitutions for them (Dietz & Neumayer 2007; Cohen, Hepburn & Teytelboym 2019). As such, weak sustainability entertains the concept of substitutability.

Substitutability is the replacement of one good or service with another. This replacement or substitution commonly takes place owing to hazardous or efficiency characteristics (e.g. CFC replacement for explosive ammonia refrigerant; **Q87**), income limitations on consumer choice (e.g. less expensive, inefficient heating options over more expensive but efficient ones) or the loss or degradation of environmental services, foremost natural capital. While substitutability for sustainability has applied all of these rationales, a critical issue over the long term is the role of technological innovations as substitutes for environmental services at large (Fenichel & Zhao 2015).

Various positions exist about weak sustainability, one of which advances that environmental degradation or losses are acceptable if monetary compensation is maximized for the future. Setting this position aside, weak sustainability at large focuses on the capacity of human and social capital to innovate for the loss or decline of environmental services through the substitutes of manufactured capital, especially for the services provided by natural capital. The invention of synthetic nitrogen to overcome nature's shortages for fertilizer is an iconic example (**Q44**). Weak sustainability fits well within neoclassical approaches for economic growth in human welfare and intergenerational equity (Ang & van Passel 2012). As an example, inclusive wealth (**Q95**), a measure of intergenerational sustainability proposed by resource and environmental economists, accounts for substitutability of environmental services (Polasky *et al.* 2015).

Strong sustainability holds that many environmental services cannot be replaced (i.e. they are irreplaceable) or will prove to be excessively costly to do so, especially for natural capital critical to the functioning of ecosystems and the Earth system (Brand 2009). Maintenance of stratospheric ozone – the protective ozone layer (Q53) – is an exemplar of an environmental service for which technological innovations are difficult to conceive. In addition, much manufactured capital is not a complete replacement of natural capital because it requires that capital for its production. Intergenerationally, therefore, these services must be conserved (Chaminade 2020). Ecological economics tend to align with strong sustainability (van den Bergh 2001).

These distinctions notwithstanding, the difference between weak and strong sustainability should not be viewed in their polar extremes. Many proponents of weak sustainability recognize that not all environmental goods and services are replaceable and must be conserved. Likewise, many proponents of strong sustainability recognize that some goods and services may be replaceable, providing improvements for human–environmental welfare. It is also understood that the level of substitutability varies across different kinds of capitals. For this and other reasons, substitutability involves considerable caveats and cautions for which advances in understanding are warranted (Cohen, Hepburn & Teytelboym 2019).

References

Ang, F. & S. van Passel (2012). "Beyond the environmentalist's paradox and the debate on weak versus strong sustainability". *BioScience* 62(3): 251–9. doi.org/10.1525/bio.2012.62.3.6.

Brand, F. (2009). "Critical natural capital revisited: ecological resilience and sustainable development". *Ecological Economics* 68: 605–12. doi.org/10.1016/j.ecolecon.2008.09.013.

Chaminade, C. (2020). "Innovation for what? Unpacking the role of innovation for weak and strong sustainability". *Journal of Sustainability Research* 2(1): e200007. doi.org/10.20900/jsr20200007.

Cohen, F., C. Hepburn & A. Teytelboym (2019). "Is natural capital really substitutable?" *Annual Review of Environment and Resources* 44: 425–48. doi.org/10.1146/annurev-environment-101718-033055.

Dietz, S. & E. Neumayer (2007). "Weak and strong sustainability in the SEEA: concepts and measurement". *Ecological Economics* 61: 617–26. doi.org/10.1016/j.ecolecon.2006.09.007.

Fenichel, E. & J. Zhao (2015). "Sustainability and substitutability". *Bulletin of Mathematical Biology* 77: 348–67. doi.org/10.1007/s11538-014-9963-5.

Polasky, S. *et al.* (2015). "Inclusive wealth as a metric of sustainable development". *Annual Review of Environment and Resources* 40: 445–66. doi.org/10.1146/annual-environ-101813-013253.

van den Bergh, J. (2001). "Ecological economics: themes, approaches, and differences with environmental economics". *Regional Environmental Change* 2(1): 13–23. doi:10.1007/s101130000020.

Q86

WHAT ARE COMPLEX ADAPTIVE SYSTEMS AND THEIR IMPLICATIONS FOR SUSTAINABILITY?

Short answer: they are networks of interacting components in which the outcomes of the system at large are not consistent with the behaviours of the individual components and which warrant adaptive management for sustainability.

Social–environmental systems (SESs) – the phenomena generating the Anthropocene and the framing of sustainability science (**Q71**) – are complex adaptive systems (Clark & Harley 2020; Preiser *et al.* 2018). They constitute interacting networks in which the behaviour of the individual components comprising them gives rise to divergent outcomes that are not consistent with the behaviour of the individual components and that are difficult to predict. The interactions in these systems involve a number of important characteristics that make them complex (Hagstrom & Levin 2017; Turner & Baker 2020). These include: "path dependencies"(i.e. history matters, aka hysteresis, in some cases considered one type of path dependence) in which the systems' responses are sensitive to their initial conditions and the subsequent history of the systems affect their future behaviour; "non-linearity" in which the systems react disproportionately to and lag behind the influencing operating factors; "emergence" in which the interactions of the systems components give rise to properties or behaviours that the components do not have; and "self-organization" and "adaptation" in which systems move from chaotic to organizational arrangements and change their states to maintain critical functions as conditions beyond the system change.

These characteristics are illustrated by the Dust Bowl of the 1930s in the central-southern Great Plains of the United States (**Q21**). This region was populated in the late 1800s by immigrant farmers during a prosperous economic period, experiencing relatively humid conditions for this otherwise decadal oscillating precipitation regime. Novices to the environment and spurred by the economy, huge areas of deeply ploughed grasslands and the absence of soil protection strategies were used – creating path-dependent conditions. By the 1930s the entire SES changed with the extreme vagaries (i.e. non-linearity) of the Great Depression and climatic drought. The drought, however, was amplified by land surface–atmosphere feedbacks (i.e. emergence from past farming practices), increasing the wind-blown loss of the exposed top soil. Millions of farmers, mostly smallholders, and their service communities went bankrupt (hysteresis), and responded by leaving the chaos on the plains, migrating for work elsewhere, sometimes as caravans (self-organizing). With the passing of the Great Depression and a return of more humid conditions, agriculture was reinstated in the region, based on larger farmsteads in which soil-protecting practices were adopted and irrigation from the Ogallala aquifer expanded. Interestingly, this response foretells yet another complex adaptive SES regarding the future human–environment conditions as the aquifer is depleted (**Q33**).

As reflexive actors or intelligent agents with high levels of contemplation, reflection, anticipation and capacity to learn, humankind's role in SESs invariably increases complexity and the adaptive capacities of the systems (Levin *et al.* 2013). This complexity tends to increase as divergences in stakeholder preferences enlarge (Kallis, Kiparsky & Norgaard 2009). For these and other reasons, the trends and trajectories of SESs are replete with uncertainties. As such, adaptive management is proposed to address the sustainability of SESs (Allen *et al.* 2011). This management requires iterative, systematic monitoring in which strategies change as the system and knowledge of the system changes (Barnard & Elliott 2015; Williams & Brown 2014). One lesson gained from studies of common property regimes such as SESs is the absence of panaceas for their sustained, successful governance (Ostrom, Janssen & Anderies 2007). Rather, successful institutions (**Q66**) or rules of governance may vary by case, reinforcing the relevance of place-based approaches in sustainability science (**Q83**).

References

Allen, C. *et al.* (2011). "Adaptive management for a turbulent future". *Journal of Environmental Management* 92(5): 1339–45. doi.org/10.1016/j.jenvman.2010.11.019.

Barnard, S. & M. Elliott (2015). "The 10-tenets of adaptive management and sustainability: an holistic framework for understanding and managing the socio-ecological system". *Environmental Science & Policy* 51: 181–91. doi.org/10.1016/j.envsci.2015.04.008.

Clark, W. & A. Harley (2020). "Sustainability science: toward a synthesis". *Annual Review of Environment and Society* 45: 331–86. doi.org/10.1146/annurev-environ-012420-043621.

Hagstrom, G. & S. Levin (2017). "Marine ecosystems as complex adaptive systems: emergent patterns, critical transitions, and public goods". *Ecosystems* 20(3): 458–76. doi.org/10.1101/056838.

Kallis, G., M. Kiparsky & R. Norgaard (2009). "Collaborative governance and adaptive management: lessons from California's CALFED Water Program". *Environmental Science & Policy* 12(6): 631–43. doi.org/10.1016/j.envsci.2009.07.002.

Levin, S. *et al.* (2013). "Social-ecological systems as complex adaptive systems: modeling and policy implications". *Environment and Development Economics* 18(2): 111–32. doi.org/10.1017/S1355770X12000460.

Ostrom, E., M. Janssen & J. Anderies (2007). "Going beyond panaceas". *Proceedings of the National Academy of Sciences* 104(39): 15176–8. doi.org/10.1073/pnas.0701886104.

Preiser, R. *et al.* (2018). "Social-ecological systems as complex adaptive systems". *Ecology and Society* 23(4): 46. doi.org/10.5751/ES-10558-230446.

Turner, J. & R. Baker (2020). "Just doing the do: a case study testing creativity and innovative processes as complex adaptive systems". *New Horizons in Adult Education and Human Resource Development* 32(2): 40–61. doi.org/10.1002/nha3.20283.

Williams, B. & E. Brown (2014). "Adaptive management: from more talk to real action". *Environmental Management* 53(2): 465–79. doi10.1007/s00267-013-0205-7.

WHY ARE UNCERTAINTY, SURPRISE AND THE PRECAUTIONARY PRINCIPLE APPLIED THROUGHOUT SUSTAINABILITY PROBLEMS?

Short answer: complexity generates uncertainties and surprises in the outcomes of human–environmental relationships that warrant careful observation and monitoring to avoid unwanted outcomes.

The sustainability of social–environmental systems constitutes what may be labelled a "wicked problem" (Hsiang *et al.* 2017), in part because such systems are complex and adaptive (**Q86**), complete with characteristics that give rise to uncertainties and surprises. Uncertainty arises not only from the value-laden dimensions of the sustainability and sustainable development concepts (e.g. equity and social justice; **Q84**) (Ciuffo *et al.* 2012), but from various facets of their assessments, such as variables defined by ranges and stochastic (randomly determined) processes requiring probabilistic approaches to account for them (Fawcett *et al.* 2012). Co-designed or co-produced research and application developed and applied among multiple stakeholders to produce actionable knowledge add to these uncertainties (Soltani *et al.* 2017; Quaas *et al.* 2007). These characteristics of social–environmental systems (**Q86**), in turn, give rise to surprises, or unanticipated outcomes.

Considerable attention has been given to environmental surprises resulting from human actions, commonly involving innovations otherwise thought to be environmentally benign but proved not to be. Perhaps the archetypal example is the invention and large-scale use of "environmentally non-harmful" CFCs and the industrial "surprise" that these compounds thinned the ozone layer, despite warnings of this possibility by scientists (**Q53**). The demise of the Aral Sea, which was postulated by Russian scientists, was ignored until the slow loss of water to the sea led to a tipping point, triggering a "surprise" desiccation of the waters (Glantz, Rubinstein & Zonn 1993). Such surprises are pervasive to human–environmental systems throughout history (Kates & Clark 1996). Many of these surprises were, in fact, predicted in laboratory experiments or projected through models but largely ignored by society at large, or could have been anticipated by considering alternative epistemologies (ways of knowing) (Streets & Glantz 2000).

For these and other reasons, sustainability science pays attention to the precautionary principle. Although ill-defined by scientific standards (Aven 2011; Gardiner 2006;

Stefánsson 2019), this principle is promoted in government regulations throughout various parts of the world and advanced in several international declarations about sustainability. It states that if an innovation or activity holds the potential to harm humans or the environment, precautionary measures should be taken (i.e. preventing harm by reducing negative consequences). Such measures are especially significant if the uncertainties in the science and potential impacts on the social–environmental system are large. How uncertainty and harm are interpreted, however, is commonly contested, especially in those instances in which regulations are the proposed precautionary measures (van Asselt & Vos 2006). The significance of the precautionary principle for sustainability science at large resides in its warning that care must be taken by way of monitoring and improved understanding of the problem at hand, including enlarging the range of anticipatory possibilities, owing to the complex dynamics of the Earth system relative to human activities and behaviour.

References

Aven, T. (2011). "On different types of uncertainties in the context of the precautionary principle". *Risk Analysis: An International Journal* 31(10): 1515–25. doi.org/10.1111/j.1539-6924.2011.01612.x.

Ciuffo, B. *et al.* (2012). "Dealing with uncertainty in sustainability assessment". *Report on the Application of Different Sensitivity Analysis Techniques to Field Specific Simulation Models*. EUR 25166. JRC Technical Report, Institute for Environment and Sustainability. Luxembourg: European Commission.

Fawcett, W. *et al.* (2012). "Flexible strategies for long-term sustainability under uncertainty". *Building Research & Information* 40(5): 545–57. doi.org/10.1080/09613218.2012.702565.

Gardiner, S. (2006). "A core precautionary principle". *Journal of Political Philosophy* 14(1): 33–60. doi.org/10.1111/j.1467-9760.2006.00237.x.

Glantz, M., A. Rubinstein & I. Zonn (1993). "Tragedy in the Aral Sea basin: looking back to plan ahead?" *Global Environmental Change* 3: 174–98. doi.org/10.1016/0959-3780(93)90005-6.

Hsiang, S. *et al.* (2017). "Estimating economic damage from climate change in the United States". *Science* 356(6345): 1362–69. doi:10.1126/science.aal4369.

Kates, R. & W. Clark (1996). "Environmental surprise: expecting the unexpected?" *Environment* 38(2): 6–34. doi.org/10.1080/00139157.1996.9933458.

Quaas, M. F. *et al.* (2007). "Uncertainty and sustainability in the management of rangelands". *Ecological Economics* 62(2): 251–66. doi.org/10.1016/j.ecolecon.2006.03.028.

Soltani, S. *et al.* (2013). "Sustainability through uncertainty management in urban land suitability assessment". *Environmental Engineering Science* 30(4): 170–78. doi.org/10.1089/ees.2012.0193.

Streets, D. & M. Glantz (2000). "Exploring the concept of climate surprise". *Global Environmental Change* 10(2): 97–107. doi.org/10.1016/S0959-3780(00)00015-7.

Stefánsson, H. (2019). "On the limits of the precautionary principle". *Risk Analysis* 39(6): 1204–22. doi.org/10.1111/risa.13265.

van Asselt, M. & E. Vos (2006). "The precautionary principle and the uncertainty paradox". *Journal of Risk Research* 9(4): 313–36. doi.org/10.1080/13669870500175063.

IS THE EARTH SYSTEM APPROACHING ITS PLANETARY BOUNDARIES TO FUNCTION?

Short answer: perhaps, in some instances, the critical Earth system processes may not be sustained because the state of the controlling variables (by human activity) has increased the risk of changing the processes.

Carrying capacity estimates of the Earth project the maximum number of people that the biosphere can sustain (**Q75**). Embedded in these projections are assumptions about the delivery of environmental services (**Q7**) that the Earth system can sustain (Steffen *et al.* 2015; Rockström *et al.* 2009), given the processes that have prevailed during the Holocene. Planetary boundaries identify increases in the risk of non-linear, abrupt changes in the function of the Earth system and its capacity to deliver the services in question. Control variables (e.g. carbon dioxide for climate change) are used as the metrics to address the risk to various processes illustrated in Figure 88.1: stratospheric ozone depletion (**Q53**), atmospheric aerosol loading (**Q45**), ocean acidification (**Q41**), biochemical flows (phosphorus and nitrogen) (**Q4, Q43**), freshwater use (**Q33, Q34**), land-system change (Section II), biosphere integrity (functional diversity, genetic diversity) (Section V), climate change (Section IV) and novel entities (innovations affecting the Earth system).

The planetary boundary framework codes the risk that a boundary has been exceeded or not (Fig. 88.1). Green is below the boundary, or a safe condition. Yellow indicates increasing risk that the boundary has been or is about to be passed. Red denotes that the boundary has been surpassed. Grey, in contrast, means that the evidence is not yet in a quantifiable state to assess the risk level. The authors of the framework determine that the control variable, affected by human activity, has surpassed the Holocene boundaries for biochemical flows and biosphere integrity and has likely or will soon pass those for land-system change and climate change. Freshwater use and stratospheric ozone depletion are well within the boundary, inferring that innovations and regulations designed to reduce freshwater withdrawals and ozone-depleting emissions have succeeded. Ocean acidification remains safe but is approaching the boundary limit.

Conceptually, the framework is thought-provoking, formalizing the Cassandra (**Q73**) shift from questions about maximum planetary population to the condition of the biosphere. For this reason alone, the framework is useful. Its initial formulation, however, has

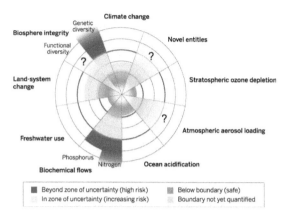

Figure 88.1 The planetary boundary framework
Source: Steffen *et al.* 2015.

been critiqued on a number of grounds, including the choice of control variables, such as the use of biota extinction rates for biosphere integrity (Montoya, Donohue & Pimm 2018; Mace *et al.* 2014) and the Holocene-set boundary for nitrogen (De Vries *et al.* 2013). In addition, debates continue about the existence of fixed threshold limits and the possibility of inappropriate policies based on such limits (Lewis 2012; Rockström *et al.* 2018). The framework continues to be updated to improve its rigour (Steffen *et al.* 2015). Perhaps an overarching question, akin to that posited about carrying capacity (**Q75**), involves the role of innovation, such as the "novel entities" in the framework. These entities reference new control variables affecting Earth system processes, but potentially they could also improve the risk condition, such as carbon capture reducing the risks of climate change.

References

De Vries, W. *et al.* (2013). "Assessing planetary and regional nitrogen boundaries related to food security and adverse environmental impacts". *Current Opinion in Environmental Sustainability* 5(3/4): 392–402. doi.org/10.1016/j.cosust.2013.07.004.

Lewis, S. (2012). "We must set planetary boundaries wisely". *Nature* 485(7399): 417–17. doi.org/10.1038/485417a.

Mace, G. *et al.* (2014). "Approaches to defining a planetary boundary for biodiversity". *Global Environmental Change* 28: 289–97. doi.org/10.1016/j.tree.2017.10.004.

Montoya, J., I. Donohue & S. Pimm (2018). "Planetary boundaries for biodiversity: implausible science, pernicious policies". *Trends in Ecology & Evolution* 33(2): 71–3. doi.org/10.1016/j.tree.2017.10.004.

Rockström, J. *et al.* (2009). "Planetary boundaries: exploring the safe operating space for humanity". *Ecology and Society* 14(2): 32. www.ecologyandsociety.org/vol14/iss2/art32/.

Rockström, J. *et al.* (2018). "Planetary boundaries: separating fact from fiction. A response to Montoya *et al.*" *Trends in Ecology & Evolution* 33(4): 233–4. doi.10.1016/j.tree.2018.01.010.

Steffen, W. *et al.* (2015). "Planetary boundaries: guiding human development on a changing planet". *Science* 347(6223). doi:10.1126/science.1259855.

ARE TIPPING ELEMENTS IN THE EARTH SYSTEM REACHING THEIR TIPPING POINTS?

Short answer: some large-scale components of the Earth system appear to be moving toward conditions that could flip them into new conditions owing to human activity endangering the biosphere.

Complex systems commonly have tipping points or thresholds in various parts of the system that, once passed, flip the system into a new state or condition that no longer operates as the earlier system did (Dakos *et al.* 2019). In sustainability, tipping points are addressed in terms of the Earth system's capacity to render the environmental services of the biosphere, such as the relatively steady climate parameters that have prevailed throughout the Holocene. These parameters (factors delimiting a system) could change – a tipping point – owing to Earth system feedbacks from climate warming.

Tipping elements, in contrast, are large-scale components of the Earth system with thresholds that can be surpassed by small perturbations or disturbances reaching a tipping point and shifting the components into a new state or condition. Various tipping elements have been identified that might trigger changes in the climate system of Earth (e.g. **Q40**) and which are affected by human activity that could be regulated, if difficult in many cases, such as reducing greenhouse gas emissions and tropical deforestation (Lenton *et al.* 2008; Lenton 2011). Examples are illustrated in Figure 89.1, which would generate significant melting of ice, biome shifts and circulation changes, all linked to global warming. Nine elements are identified and the consequences of them crossing tipping points are noted in the figure and expanded in Table 89.1.

Identification of tipping elements approaching a threshold or tipping point may be possible (Pearce & Prater 2020; Wassman & Lenton 2012). Significantly, modelling efforts indicate a domino effect among certain elements, such that crossing one threshold and the changes that follow create cascading threshold crossings in other elements, with special attention placed on the loss of Arctic ice sheets (Fig. 89.2; see **Q40** for weakening of the AMOC) (Wunderling *et al.* 2021). An argument can be made that self-reinforcing feedbacks in the Earth system, given tipping point crossings, could cause a "hothouse Earth", regardless of attempts to stabilize the climate, leading to significantly higher global average temperatures (**Q47**) and dramatic sea-level rises (**Q39**) (Steffen *et al.* 2018).

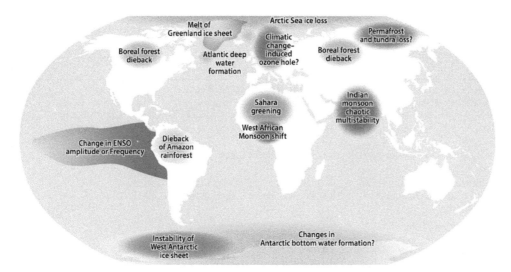

Figure 89.1 Tipping elements potentially affecting the Earth's climate system

The "Atlantic deep-water formation" is part of the thermohaline circulation (or Atlantic meridional overturning circulation; **Q40**) commonly associated with a slowdown in that circulation.

Source: Lenton *et al.* (2008); basemap: Esri.

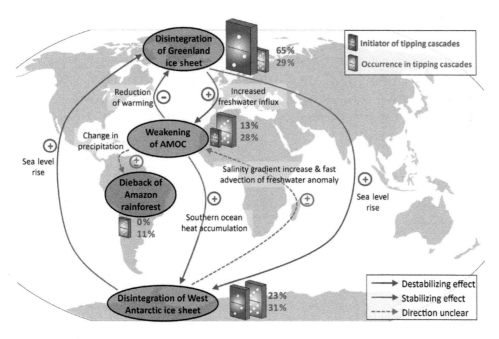

Figure 89.2 Tipping elements cascades

Four climate tipping elements (grey ovals) and their linkages examined via Monte Carlo ensembles (multiple runs of the models). For example, loss of the Greenland ice sheet increases freshwater influx to the North Atlantic, weakening the AMOC (Atlantic Meridional Overturning Circulation or thermohaline circulation **[Q 40]**), a destabilizing effect, whereas this weakening reduces the warming of the North Atlantic, stabilizing the Greenland ice sheet. The size of red dominoes indicates how proportion ensembles initiate tipping cascades and the blue dominoes indicate the proportion of cascades involved.

Source: Wunderling *et al.* (2021).

Table 89.1 Tipping elements, direction of change and major consequences

Tipping element	Direction of change	Major consequences
Arctic summer sea-ice	Area extent −	Amplified warming and ecosystem change
Greenland ice sheet	Ice volume −	Sea-level rise 2–7 m
West Antarctica ice sheet	Ice volume −	Sea-level rise 5 m
Atlantic thermohaline circulation	Overturning −	Regional cooling, sea-level change, ITCZ change*
El Niño–Southern Oscillation (ENSO)	Amplitude +	Drought in Southeast Asia and elsewhere
Indian summer monsoon (ISM)	Rainfall −	Drought and agriculture −
Sahara/Sahel and West African monsoon	Vegetation fraction +	Agriculture +
Amazon rainforest	Tree fraction −	Biodiversity − and rainfall −
Boreal forest	Tree fraction −	Biome Δ
Antarctic bottom water (AABW)	Formation −	Ocean circulation Δ and carbon storage −
Tundra	Tree fraction +	Amplified warming and biome Δ
Permafrost	Volume −	Carbon dioxide and methane emission +
Marine methane hydrates	Hydrate volume +	Amplified warming
Ocean anoxia	Anoxia +	Marine mass extinction
Arctic ozone	Column depth −	Surface ultraviolet radiation +

*ITCZ = intertropical convergence zone

Source: Lenton *et al.* (2008).

Tipping elements and tipping points are complementary to, but not the same as, planetary boundaries (**Q88**). The boundary concept refers to the risk that certain broad dimensions of the Earth system may cease to function and deliver their services. Tipping elements and points tend to address specific Earth system dynamics, such as the Atlantic meridional overturning circulation's or thermohaline circulation's capacity to operate or shut down.

References

Dakos, V. *et al.* (2019). "Ecosystem tipping points in an evolving world". *Nature Ecology & Evolution* 3(3): 355–62. doi.org/10.1038/s41559-019-0797-2.

Lenton, T. (2011). "Early warning of climate tipping points". *Nature Climate Change* 1(4): 201–09. doi.org/10.1038/nclimate1143.

Lenton, T. *et al.* (2008). "Tipping elements in the Earth's climate system". *Proceedings of the national Academy of Sciences* 105(6): 1786–93. doi.org/10.1073/pnas.0705414105.

Pearce, R. & T. Prater (2020). "Explainer: Nine 'tipping points' that could be triggered by climate change". www.carbonbrief.org/explainer-nine-tipping-points-that-could-be-triggered-by-climate-change.

Steffen, W. *et al.* (2018). "Trajectories of the Earth system in the Anthropocene". *Proceedings of the National Academy of Sciences* 115(33): 8252–59. doi.org/10.1073/pnas.1810141115.

Wassmann, P. & T. Lenton (2012). "Arctic tipping points in an Earth system perspective". *Ambio* 41(1). doi.org/10.1007/s13280-011-0230-9.

Wunderling, N. *et al.* (2021). "Interacting tipping elements increase risk of climate domino effects". *Earth System Dynamics* 12: 601–19. doi.org/10.5194/esd-12-601-2021.

WHAT IS THE GAIA HYPOTHESIS AND WHAT ARE ITS IMPLICATIONS FOR EARTH SYSTEM SUSTAINABILITY?

Short answer: the Gaia hypothesis proposes that the biotic world interacts with the abiotic world to maintain the biosphere, altering orthodox understanding of Earth system functioning.

The concept of Gaia – the Greek "mother of life" – emerged from a provocative query. Solar irradiance (**Q10**) has gradually increased over the existence of our solar system. The planet Earth, however, has not continually warmed with this increase in energy, but has oscillated between cooler and warmer conditions, largely within parameters that do not lead to icehouse or greenhouse conditions (Section V introduction). Why is this? The Gaia concept proposes that this apparent contradiction between increasing irradiance and Earth's temperate conditions is caused by life (carbon-based organisms) interacting with the abiotic world to create a self-regulating, complex system that maintains the biosphere – an Earth system favourable for life (Lovelock 2003). This proposition runs counter to historical and prevailing evolutionary thought in which life responds to abiotic changes, not vice versa.

The hypothetical Daisyworld illustrates the proposed relationship. The Daisyworld planet is similar to Earth in its dimensions, distance from its sun-star, and experiences gradually increasing solar irradiance. It differs from Earth in that life is reduced to black and white daisies (Fig. 90.1). As the average temperature rises from the increasing solar irradiance, the white daisies outperform the black ones because they reflect the short-wave radiation from its sun back into space. The cooling impact of the white daisies, however, gives an advantage to the black daisies, which absorb short-wave radiation and emit long-wave radiation, thereby warming the planet. The feedbacks of the flowers to the atmosphere cool and warm Daisyworld within a relatively stable temperature range suitable for the daisies' existence. The biotic and abiotic worlds interact to maintain a relatively constant temperature. Such processes have been proposed for Earth regarding global surface temperature, atmospheric oxygen and oceanic salinity, all of which have remained relatively constant over long periods of our planet's history.

The Gaia concept has long been critiqued for a variety of reasons (Free & Barton 2007; Kirchner 2002), including its teleological implications (biota purposively maintain the biosphere) and its failure to provide the mechanisms in natural selection leading to

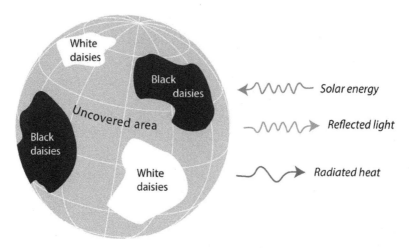

Figure 90.1 Daisyworld's stable temperature
Source: author.

self-regulations (Tyrell 2013). Its evolutionary impact, however, has opened consideration of how the biotic world affects the abiotic world (Kirchner 2002), with attention to linking evolution to Gaia ideas (Doolittle 2017).

While the Gaia hypothesis is explored and debated in evolutionary discourse, it has not been employed, surprisingly, among worldviews that champion non-regulatory approaches toward the environment. One outcome of the Gaia hypothesis – a relatively stable average Earth temperature, for example – could be expanded to support the view that an emergent property of the Earth system is a rather stable Earth average temperature, requiring minimal societal regulation. Interestingly, some opponents of societal regulation mistakenly associate the concept as asserting intentionality in the biotic–abiotic relationship – rather than that of an emergent property of the interactions – and thus disregard the concept rather than draw upon it to support their view (Briggs 2015).

References

Briggs, W. (2015). "The Gaia hypothesis is either trivial and useless or false and ridiculous". 8 July. wmbriggs.com/post/16400/#comments.

Doolittle, W. (2017). "Darwinizing Gaia". *Journal of Theoretical Biology 434*: 11–19. doi.org/10.1016/j.jtbi.2017.02.015.

Free, A. & N. Barton (2007). "Do evolution and ecology need the Gaia hypothesis?" *Trends in Ecology & Evolution* 22(11): 611–19. doi.org/10.1016/j.tree.2007.07.007.

Kirchner, J. (2002). "The Gaia hypothesis: fact, theory, and wishful thinking". *Climatic Change* 52(4): 391–408. doi.org/10.1023/A:1014237331082.

Lovelock, J. (2003). "Gaia: the living Earth". *Nature* 426(6968): 769–70. doi.org/10.1038/426769a.

Tyrell, T. (2013). "The Gaia hypothesis: the verdict is in". *New Scientist* 220(2940): 30–31. doi.org/10.1016/S0262-4079(13)62532-4.

WHAT IS THE ENVIRONMENTALIST'S PARADOX AND WHAT DOES IT MEAN FOR SUSTAINABILITY?

Short answer: historically, the degradation of environmental goods and services to provision humankind has increased global aggregate human well-being, owing to technological innovation and substitution; will this relationship continue as pressures mount on the Earth system at large?

The degradation of environments and the Earth system combined with their negative implications for humankind constitutes a major concern for the Anthropocene and global environmental change (Q88, Q89). Estimates that humankind now has an ecological footprint 1.7 times that which can be supported sustainably by the Earth system drive home this concern (Lin *et al.* 2018). Such pressures, and the environmental degradation that commonly follows in ecosystems, lead to declines in the carrying capacity and negative impacts on biota (Q75). At the global aggregate level, however, human well-being has increased, while environmental services and Earth system functions degrade. This last observation constitutes the environmentalist's paradox (Raudsepp-Hearne *et al.* 2010), a re-articulation of the conditions long noted in the narratives and stock-taking of the human–environmental sciences (Q77) and undergirding the Cassandra–Cornucopian debates (Q73). At least one argument holds that the paradox ceases to exist if addressed between human well-being and ecosystem services for a defined area or ecosystem (Delgado & Marín 2017), illustrated by various cases of societal collapse and area depopulation (Q93). The global aggregate scale of assessment is consistent with the paradox, however, as the average material well-being of people globally and Earth system degradation have never been higher in the history of our species.

Three propositions provide partial explanations for the paradox (Raudsepp-Hearne *et al.* 2010): (1) food provisioning services per se are increasing, whereas many of their supporting services (Q22), such as soil quality, are declining (Q21); (2) technology (or innovation substitutability) has decoupled human well-being from nature (Q14); and (3) a lag time may exist between environmental declines and those in human well-being. In reality, the first two propositions are strongly linked because various synthetic substitutes – nitrogen fertilizer and crop genomics (Q14) – have increased food production on less land (Q22). Water availability may prove to be a more limiting factor than land (**Q34**), however. Environmental degradation and enhancements will exist in different locations worldwide in the near future, whereas the aggregate material well-being of the global population will be likely to continue to increase. In these circumstances, the paradox may be resolved if those facets of the Earth system approaching planetary boundaries are reached (**Q86**). In these cases, the Earth system may cease to provision humankind at past levels of services.

References

Delgado, L. & V. Marín (2017). "Human well-being and historical ecosystems: the environmentalist's paradox revisited". *Bioscience* 76(1): 5–6. doi.org/10.1093/biosci/biw132.

Lin, D. *et al.* (2018). "Ecological footprint accounting for countries: updates and results of the national footprint accounts, 2012–18". *Resources* 7(3): 58. doi.org/10.3390/resources7030058.

Raudsepp-Hearne, C. *et al.* (2010). "Untangling the environmentalist's paradox: why is human well-being increasing as ecosystem services degrade?" *BioScience* 60(8): 576–89. doi.org/10.1525/bio.2010.60.8.4.

IN WHAT WAYS ARE THE CONCEPTS OF VULNERABILITY AND RESILIENCE APPLICABLE TO SUSTAINABILITY INTERESTS?

Short answer: they address the impacts of disturbances on social–environmental systems as well as the mitigation and adaptation options for dealing with such disturbances.

Social–environmental systems (SESs) are variously vulnerable and resilient to disturbances (aka, hazards, perturbations or stressors). The vulnerability concept, derived from social and health risk research, refers to the impact of a disturbance acting on a system. This impact is a product of the exposure (i.e. type and frequency), sensitivity (i.e. damage incurred) and adaptive capacity (i.e. response) of the system to a disturbance (Fig. 92.1) (Adger 2006). In this framing, adaptive capacity constitutes the coping capacity (or resilience) of the system to recover from the disturbance, including anticipatory actions that reduce exposure and sensitivity. Other SESs or elements operating beyond the SES in question affect and are affected by these dynamics (Turner 2003). The resilience concept, originally derived from ecology, refers to the capacity of the SES to retain its structure, function and identity while absorbing and reorganizing after a disturbance. System resilience is determined by its latitude (i.e. the extent to which the system can be changed and recover), resistance (i.e. ease or difficulty of its change), precariousness (i.e. proximity of the system to a threshold of change) and panarchy (influence of subsystems at scales above or below that system in question) (Fig. 92.1) (Walker *et al.* 2004).

The framing of vulnerability identifies the elements of an SES and indicates their interactions, suggestive of an explanatory model. That for resilience appears more definitional in kind, identifying the elements of an SES that equate to resilience. These differences notwithstanding, the two concepts are commonly described as the flip sides of sustainability, focused on either damage (vulnerability) or recovery (resilience) outcomes of disturbances on SESs, while including the other dimension internally. In other words, resilience is identified in the adaptive capacity of vulnerability, and vulnerability is implied in the combined elements of resilience (Nelson, Adger & Brown 2007; Cutter *et al.* 2008; Miller *et al.* 2010; Turner 2010).

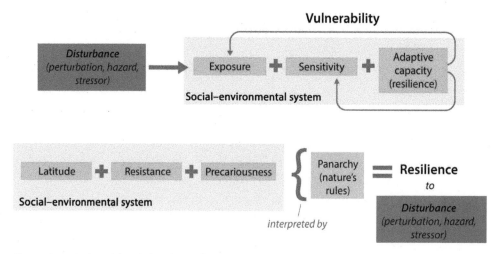

Figure 92.1 Vulnerability and resilience framing
Source: author, adapted from Turner *et al.* (2003) and Walker *et al.* (2004).

Perhaps the most significant differences in the use of the two concepts are the key interests of their practitioners. Resilience research has focused on the dynamics and organizing principles of SESs in general, providing broad insights about changes in SESs and hence their management (Carpenter *et al.* 2012). Vulnerability research, in contrast, tends to employ data-rich narratives or general metrics to inform decision-makers dealing with societal protection against disturbances, such as the environmental vulnerability index (Barnett, Lambert & Fry 2008; Cutter, Burton & Emrich 2010).

References

Adger, W. (2006). "Vulnerability". *Global Environmental Change* 16(3): 268–81. doi.org/10.1016/j.gloenvcha.2006.02.006.

Barnett, J., S. Lambert & I. Fry (2008). "The hazards of indicators: insights from the environmental vulnerability index". *Annals of the Association of American Geographers* 98(1): 102–19. doi.org/10.1080/00045600701734315.

Carpenter, S. *et al.* (2012). "General resilience to cope with extreme events". *Sustainability* 4(12): 3248–59. doi.org/10.3390/su4123248.

Cutter, S., C. Burton & C. Emrich (2010). "Disaster resilience indicators for benchmarking baseline conditions". *Journal of Homeland Security and Emergency Management* 7(1): 51. doi.org/10.2202/1547-7355.1732.

Cutter, S. *et al.* (2008). "A place-based model for understanding community resilience to natural disasters". *Global Environmental Change* 18(4): 598–606. doi.org/10.1016/j.gloenvcha.2008.07.013.

Miller, F. *et al.* (2010). "Resilience and vulnerability: complementary or conflicting concepts?" *Ecology and Society* 15(3): 11. www.ecologyandsociety.org/vol15/iss3/art11/.

Nelson, D., W. Adger & K. Brown (2007). "Adaptation to environmental change: contributions of a resilience framework". *Annual Review of Environment and Resources* 32: 395–419. doi.org/10.1146/annual.energy.32.051807.090348.

Turner II, B. (2010). "Vulnerability and resilience: coalescing or paralleling approaches for sustainability science?" *Global Environmental Change* 20(4): 570–6. doi.org/10.1016/j.gloenvcha.2010.07.003.

Turner II, B. *et al.* (2003). "A framework for vulnerability analysis in sustainability science". *Proceedings of the National Academy of Sciences* 100(14): 8074–9. doi.org/10.1073/pnas.1231335100.

Walker, B. *et al.* (2004). "Resilience, adaptability and transformability in social-ecological systems". *Ecology and Society* 9(2): 5. www.ecologyandsociety.org/vol9/iss2/art5/.

Q93

DO THE VULNERABILITY AND RESILIENCE OF PAST HUMAN–ENVIRONMENT RELATIONSHIPS PROVIDE INSIGHTS ABOUT CURRENT SUSTAINABILITY?

Short answer: not as direct analogues. Past events have led to occupational collapses of some regions/peoples and to recoveries of others, providing insights about the pace and surprises associated with changes in human–environment relationships.

History is replete with cases of societal vulnerabilities to human-induced environmental changes at large, almost all of them local to regional in spatial scale, such as cultivation practices on the Great Plains of the United States synergizing with drought to create the Dust Bowl (**Q21**) and massive, inappropriate irrigation systems assisting in the desiccation of the Aral Sea (**Q33**). In many cases, the regional environment recovered, and even the economy, such as agriculture on the Great Plains; in others, the environment and the pre-collapse economy has not, as in the case of the fishing industries and coastal towns of the Aral Sea.

Various examples exist in the distant past in which climate change was involved in occupational collapse, such as the Viking abandonment of Greenland (Dugmore *et al.* 2012; Zhao *et al.* 2022), or over-exploitation of the ecosystem as proposed for Easter Island (Rapa Nui) (Diamond 2007), although this interpretation is challenged by other evidence (DiNapoli *et al.* 2020). One of the most extensive depopulation and regional-level societal collapse examples is that of Classic Period Maya peoples in the interior uplands of the Yucatán, Mexico, northern Petén, Guatemala, and adjacent parts of Belize (Fig. 93.1). This tenth-century depopulation and abandonment was associated with extended periods of severe droughts cast over a dramatically engineered landscape that created a series of feedbacks with the changing climate. Among these was the sheer scale of deforestation for agriculture and timber for construction and fuel. The extent of deforestation reduced transpiration and amplified climatic drought (Turner & Sabloff 2012). The engineered landscapes were precisely that which made the Maya system resilient to previous extensive periods of drought. The failed response leading to the collapse took place at a time when the regional economy was significantly reduced. This economic decline appears to have been associated with shifts in north–south trading routes crossing the Yucatán Peninsula to seaborne transport around

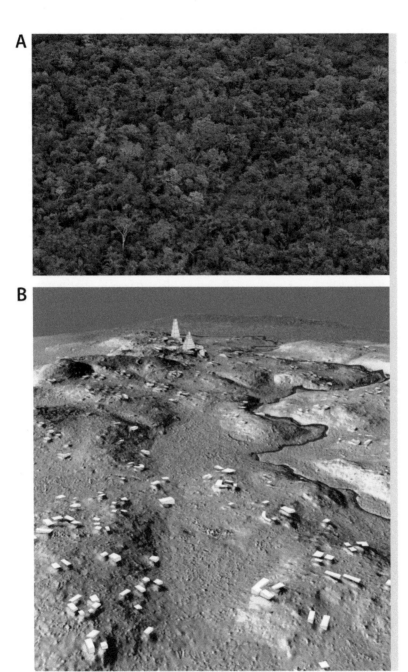

Figure 93.1 Long-term abandonment and Central Maya lowlands

Tropical forests cover the large majority of the (ca. 950—1000 CE to present) central and southern interior of the lowland Maya realm, a once densely settled and significantly open landscape abandoned about 950—1000 CE and subsequently sparsely occupied, even to this day. The little-known site of Gran Cacao, northwest Belize, is illustrated as seen in aerial photography (A), the forest hiding the past density of occupation revealed by lidar-enhanced imagery (B). Cacao Creek is a tributary of the Rio Bravo.

Source: photo by and base lidar image from T. Beach and S. Luzzadder-Beach; lidar-derived digital elevation model with enhanced buildings by B. A. Smith, Beach Labs in Geography and Environment, and the Programme for Belize Archaeology Project, University of Texas at Austin, 2022.

it, perhaps weakening the capacity of the Maya to confront the environmental feedbacks on their social–environmental system. It is not established, however, if the shift in trade routes took place during or after the significant depopulation of the area in question.

A central lesson from the Maya example and others is that what may appear to be environmental collapses tend to involve complex conditions and processes operating between the social and environmental subsystems, changing vulnerability and resilience of the combined system (Butzer 2012; Haldon *et al.* 2020). This complexity reduces the value of past collapse and depopulation examples as direct analogues for today and in the future because the social–environmental systems have changed so drastically. In addition, the past cases do not represent the totality of the Earth system changes under way now, nor the contemporary global socio-economic connections that reside in society today and increase and decrease risks to sustainability (Zscheischler *et al.* 2018). Instead, clues from the past that may inform current human–environmental dynamics include system-wide characteristics, such as trends and trajectories, frequencies, thresholds and alternative steady states, slow- and fast-acting variables, and legacies and contingencies (Dearing *et al.* 2010).

References

Butzer, K. (2012). "Collapse, environment, and society". *Proceedings of the National Academy of Sciences* 109(10): 3632–9. doi.org/10.1073/pnas.1114845109.

Dearing, J. *et al.* (2010). "Complex land systems: the need for long time perspectives in order to assess their future". *Ecology and Society* 15: 21. www.ecologyandsociety.org/vol15/iss4/art21/.

Diamond, J. (2007). "Easter Island revisited". *Science* 317(5845): 1692–4. doi:10.1126/science.1138442.

DiNapoli, R. *et al.* (2020). "A model-based approach to the tempo of 'collapse': the case of Rapa Nui (Easter Island)". *Journal of Archaeological Science* 116: 105094. doi.org/10.1016/j.jas.2020.105094.

Dugmore, A. *et al.* (2012). "Cultural adaptation, compounding vulnerabilities and conjunctures in Norse Greenland". *Proceedings of the National Academy of Sciences* 109(10): 3658–63.

Haldon, J. *et al.* (2020). "Demystifying collapse: climate, environment, and social agency in premodern societies". *Millennium* 17(1): 1–33. doi.org/10.1515/mill-2020-0002.

Turner II, B. & J. Sabloff (2012). "Classic Period collapse of the Central Maya Lowlands: insights about human–environment relationships for sustainability". *Proceedings of the National Academy of Sciences* 109(35): 13908–14. doi.org/10.1073/pnas.1210106109.

Zhao, B. *et al.* (2022). "Prolonged drying trend coincident with the demise of Norse settlement in southern Greenland". *Science Advances* 8(12): eabm4346. doi:10.1126/sciadv.abm4346.

Zscheischler, J. *et al.* (2018). "Future climate risk from compound events". *Nature Climate Change* 8(6): 469–77. doi.org/10.1038/s41558-018-0156-3.

CAN THE ECONOMIC VALUE OF GLOBAL ENVIRONMENTAL SERVICES – THE BIOSPHERE AND EARTH SYSTEM – BE CALCULATED?

Short answer: efforts to calculate the value of the biosphere and Earth system are highly contested, although the value is understood to be large.

Beyond certain natural capital (i.e. natural resources) such as timber or oil, a large array of environmental services, such as air quality and biodiversity, have been taken for granted in the history of human consciousness and actions. These services are not part of classical markets and, therefore, have no formal economic value. The environmental (ecosystem) service concept **(Q7)**, from individual services to their aggregation as an ecosystem or the biosphere, was developed and promoted as a means to amplify awareness of the value of nature (Daily *et al.* 2000), and where possible providing the price or economic cost of the service. Shadow prices can be used to estimate economic value registered by price, following such methods as contingent valuation, in which people are asked about their willingness to pay, or hedonic pricing, in which the price of a related market good with similar characteristics to the good in question can be used as a proxy for the value of that good (Spangenberg & Settele 2010). These methods of estimation, however, involve numerous assumptions and commonly result in significant price differences calculated by different practitioners, owing to the variations used to create the estimate. Nevertheless, efforts to create methods and models to examine various kinds of services continue to be developed. InVEST (Integrated Valuation of Economic Services and Trade-offs), for example, produces values of, and maps change in, ecosystem services as landscapes change (Nelson *et al.* 2009).

Given the difficulties in question, aggregating environmental services to the global level (i.e. valuing a large proportion of everything that supports life [Pimm 1997]) is a daunting task. The first attempt to do so for the year 1997 examined 17 environmental services, estimating an average price of $33 trillion yr^{-1}, based on estimates ranging from $16 to $54 trillion yr^{-1} (Constanza *et al.* 1998). This effort was roundly criticized on a large number of grounds, from the variances in valuation methods to a fundamental misunderstanding of what economic value means and how it is derived (i.e. trade-offs between well-defined alternatives) (Bockstael *et al.* 2000; Spangenberg & Settele 2010). For example, the multiplication of individual services by their unit value as undertaken in the global estimate errs by the rules practised in economics (Bockstael *et al.* 2000). Such critiques, however, did not impede another effort undertaken in 2011, which produced a value of $125 to $145 trillion yr^{-1}, based on unit value assumptions (Costanza *et al.* 2014). This second estimate was accompanied by the argument that the trade-off approach of economic valuation is not appropriate because most environmental services are public goods for which market trade-offs do not serve well.

Regardless of the robustness of global aggregate valuations, it is increasingly recognized that the life-support services of the biosphere are essential and that the Anthropocene challenges the condition and operation of the functioning of the biosphere as well as the Earth system at large. As such, attention to the costs of substituting for nature's services is undoubtedly large and continues to draw attention across a wide range of environmental services (Guerry *et al.* 2015).

References

Bockstael, N. *et al.* (2000). "On measuring economic values for nature". *Environment, Science and Technology* 34: 1384-9. doi.org/10.1021/es990673l.

Costanza, R. *et al.* (1998). "The value of the world's ecosystem services and natural capital". *Ecological Economics* 25(1): 3-15. doi.org/10.1038/387253a0.

Costanza, R. *et al.* (2014). "Changes in the global value of ecosystem services". *Global Environmental Change* 26: 152-8. doi.org/10.1016/j.gloenvcha.2014.04.002.

Daily, G. *et al.* (2000). "The value of nature and the nature of value". *Science* 289(5478): 395-6. doi:10.1126/science.289.5478.395.

Guerry, A. *et al.* (2015). "Natural capital and ecosystem services informing decisions: from promise to practice". *Proceedings of the National Academy of Sciences* 112(24): 7348-55. doi.org/10.1073/pnas.1503751112

Nelson, E. *et al.* (2009). "Modeling multiple ecosystem services, biodiversity conservation, commodity production, and tradeoffs at landscape scales". *Frontiers in Ecology and the Environment* 7(1): 4-11. doi.org/10.1890/080023.

Pimm, S. (1997). "The value of everything". *Nature* 387(6630): 231-2. Doi:10.1038/387231a0.

Spangenberg, J. & J. Settele (2010). "Precisely incorrect? Monetising the value of ecosystem services". *Ecological Complexity* 7(3): 327-37. doi.org/10.1016/j.ecocom.2010.04.007.

Q95

WHAT IS INCLUSIVE WEALTH AND ITS APPLICABILITY TO SUSTAINABILITY?

Short answer: inclusive wealth is an in-development measure of sustainable development that accounts for all capital assets, including natural capital, affecting human well-being intergenerationally.

As noted previously (**Q84**), the goal of sustainability is sustainable development. To assess this goal requires a means to measure the array of capitals that sustain humankind. Inclusive – or comprehensive – wealth constitutes a method and potential metric denoting the composite of these capitals: human (e.g. education), manufactured (e.g. goods and infrastructure), social (e.g. strength of shared values) and natural capital (e.g. environmental services, **Q7**) (Polasky *et al.* 2015). In addition, financial wealth (e.g. credit and liability) may be included. Inclusive wealth seeks to provide the value of these capitals to determine both current and projected future human well-being (Arrow *et al.* 2004). Sustainable development takes place if inclusive wealth does not decrease across generations (Text box 95.1).

Inclusive wealth differs from classic measures of economic performance, such as gross domestic product (GDP), which emphasizes goods and services traded in the marketplace, to provide a more complete measure of human–environment performance than other economic measures. In addition, it is built on economic theory, as opposed to some other measures (**Q96**) and, in principle, can be projected for future conditions (Polasky *et al.* 2015), unlike such metrics as the ecological footprint and the human development index. Recognizing distinctions in the data and methods that may be employed to calculate inclusive wealth (e.g. Arrow *et al.* 2012), the United Nations publishes inclusive wealth reports.

Text box 95.1 Simplistic calculation of inclusive wealth (IW)

Calculate the change (time$_1$ to time$_2$) in the various capitals: human (e.g. education); manufactured (e.g. goods), social (e.g. shared values) and natural (e.g. environmental services). Sum of the change in capitals equals positive or negative changes in IW. IW growth rate minus population growth rate = growth rate per capita + productivity over time (TFP or total factor productivity) = per capita inclusive growth rate accounting for TFP (source: Arrow *et al.* 2012).

That for 2018 indicates that, from 1990 to 2014, inclusive wealth increased on average 1.8 per cent yr^{-1} for 135 countries, although natural capital decreased at 0.78 per cent yr^{-1} (Managi & Kumar 2018). Such findings are largely consistent with other calculations, indicating that non-declining inclusive wealth is spurred by the various capitals other than natural capital, which constitutes a small fraction of the calculation (Arrow *et al.* 2012). Accounting for population, however, inclusive wealth calculations per capital do not appear to be as favourable as the base calculations that do not account for this factor. Also, inclusive wealth calculations yield results less than those based on GDP (Polasky *et al.* 2015),

Although conceptually sound, inclusive wealth confronts a series of challenges to become operationally useful (Roman & Thiry 2016). Various capitals are intangible, such as elements of social capital and various environmental services (**Q7**), generating controversy over the estimation of their value (**Q94**). These difficulties have led to proposals that would integrate monetary value, where appropriate, with biophysical metrics (Polasky *et al.* 2015; Roman & Thiry 2016). Other challenges include the suitability of monetary value to address human well-being, and the treatment of substitutability (e.g. replacement of natural capital through technology, such as industrial nitrogen fertilizers), and uncertainty in projecting future inclusive wealth.

Text box 95.2 The Gini coefficient

This coefficient (also index or ratio) measures the dispersion of something among entities, such as among populations or countries. In economics this dispersion may be income, wealth or assets. Equality–inequality is measured by a frequency distribution of 0 to 1 (or 0 to 100 per cent). Zero is perfect equality in which, for example, everyone has the same wealth. One is complete inequality in which one individual has all the wealth. Other such measures exist, such as the mean log deviation, common to economic analyses.

Equity poses another major challenge, recalling that sustainable development includes improved equity in human well-being. Inclusive wealth measures the aggregate changes in the capitals noted, not equality at any moment in time, although some efforts to do so have been entertained (Polasky *et al.* 2015). In terms of equity, global average human well-being has increased based on various measures, such as life expectancy or global-level income and wealth equality, as measured by the Gini coefficient (Text box 95.2), owing in large part to major advances in economic performances among such countries as China, India and Chile (Niño-Zarazúa, Roope & Tarp 2016). At the same time, large inequalities in income and wealth exist in many countries (Figs 95.1 and 95.2), including well developed ones, with trend lines increasing these inequalities.

Efforts to overcome the various challenges to inclusive wealth are under way, such as accounting for total factor productivity (the proportion of output not explained by labour and capital used in production) (Jumbri & Managi 2020), as well as treating equality, in which different measures create different, even contradictory, results (Niño-Zarazúa, Roope & Tarp 2016). Interestingly, explorations of inclusive wealth, which prove difficult given the

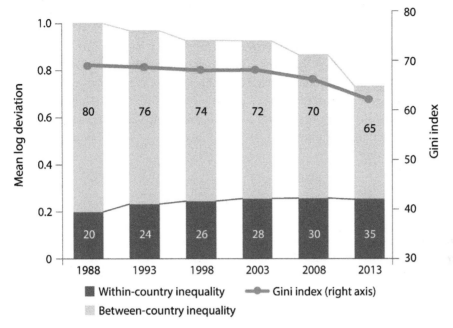

Figure 95.1 Changes in world Gini coefficients of income within and between countries, 1988-2013
The mean log deviation (left *y*-axis) is a measure of inequality as is the Gini coefficient (or Gini index, right *y*-axis). Bar height references the mean log deviation; the blue line is the Gini index; yellow bars are the level of between-country inequality; orange bars are the population weighted inequalities within countries. The numbers in the bars are the relative contributions (%) of the within- and between-countries inequalities contributing to total global inequality.
Source: World Bank (2016).

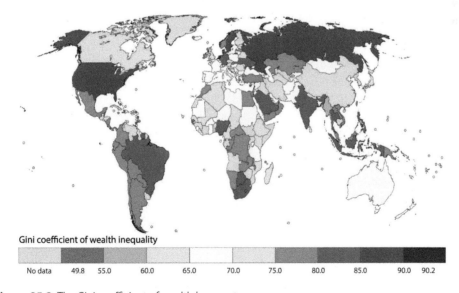

Figure 95.2 The Gini coefficient of wealth by country
The higher the Gini metric, the more wealth is held by a smaller percentage of individuals or families.
Source: DennisWikipediaWiki (2020).

data required, suggest that few countries are on a sustainable development pathway (Sato, Tanaka & Managi 2018).

A final note. A number of methods have been proposed or employed to calculate sustainable human welfare. Welfare, however, is in many senses a more subjective concept than is wealth and, thus, the attention to which economics enters the assessment. This attention notwithstanding, several assessments, such as that by Stiglitz, Sen & Fitoussi (2010), conclude that a single, composite measure of sustainability, such as an inclusive wealth calculation, prove wanting in various ways as noted above, and a dashboard of indicators may prove more useful.

References

Arrow, K. *et al.* (2004). "Are we consuming too much?" *Journal of Economic Perspectives* 3: 147–72. doi.10.1257/0895330042162377.

Arrow, K. *et al.* (2012). "Sustainability and the measurement of wealth". *Environment and Development Economics* 17(3): 317–53. doi.org/10.1017/S1355770X12000137

DennisWikipediaWiki (2020). "Gini coefficient of wealth inequality source". commons.wikimedia. org/wiki/File:Gini_Coefficient_of_Wealth_Inequality_source.png.

Jumbri, I. & S. Managi (2020). "Inclusive wealth with total factor productivity: global sustainability measurement". *Global Sustainability* 3: e5. doi.org/10.1017/sus.2020.1.

Managi, S. & P. Kumar (eds) (2018). *Inclusive Wealth Report 2018: Measuring Progress towards Sustainability*. Abingdon: Routledge.

Niño-Zarazúa, M., L. Roope & F. Tarp (2016). "Global inequality: relatively lower, absolutely higher". *Review of Income and Wealth* 63(4): 661–84. doi.org/10.1111/roiw.12240.

Polasky, S. *et al.* (2015). "Inclusive wealth as a metric of sustainable development". ***Annual Review of Environment and Resources* 40: 445–66. doi.org/10.1146/ annual-environ-101813-013253.**

Roman, P. & G. Thiry (2016). "The inclusive wealth index: a critical appraisal". *Ecological Economics* 124: 185–92. doi.org/10.1016/j.ecolecon.2015.12.008.

Sato, M., K. Tanaka & S. Managi (2018). "Inclusive wealth, total factor productivity, and sustainability: an empirical analysis". *Environmental Economics and Policy Studies* 20(4): 741–57. doi. org/10.1007/s10018-018-0213-1.

Stiglitz, J., A. Sen & J.-P. Fitoussi (2010). *Report by the Commission on Measurement of Economic Performance and Social Progress*. European Commission. ec.europa.eu/eurostat/documents/ 118025/118123/Fitoussi+Commission+report.

World Bank (2016). *Poverty and Shared Prosperity 2016: Taking on Inequality*. Washington, DC: World Bank. doi.org/10.1596/978-1-4648-0958-3.

WHAT OTHER MEASURES
OF SUSTAINABILITY ARE USED?

Short answer: several other measures are used to assess sustainability, such as the ecological footprint or renewable resources consumed per unit and the human development index, a measure of material well-being.

The "ecological footprint" (EF) and the "human development index" (HDI) are commonly applied in various sustainability assessment contexts. The EF refers to the total amount of renewable resources of the Earth (its biocapacity) that an entity (i.e. individual, community, city, country or planet) consumes, including disposal of waste (Wackernagel *et al.* 2019). In 2014, the EF of humanity was calculated at 1.7 Earths (Lin *et al.* 2018), meaning that our consumption is well beyond the capacity of the Earth system to sustain. Most measures, however, are presented in terms of global hectares per person or the land required to produce the resources consumed. The 2013 global average EF was 2.8 ha/capita, generated by a large range among countries (<1 ha/capita to >9 ha/capita) (Fig. 96.1) based on their level of economic development (Lin *et al.* 2018). Various critiques of EF exist, foremost that, as an aggregate measure, it is not sufficient for in-depth policy assessment (Wiedmann & Barrett 2010).

The HDI is a composite of three indicators – life expectancy, education and per capita income – reported by the United Nations Development Programme (UNDP) per country (Conceição 2019). Based on various critiques (Stanton 2007), a revised indicator to account for inequality – the inequality-adjusted HDI (or IHDI) – has been employed since 2010. Significant differences by country follow from the HDI shift to the IHDI, especially among the G7 countries. For example, based on 2018 data the United States ranked 15 by the HDI but dropped to 28 on the IHDI, whereas Japan's position at 19 by the HDI rose to 4 on the IHDI. For the most part, countries with high HDI scores also have large EF scores (Fig. 96.2). Only a few countries have high HDI scores and reside within the global sustainable development quadrant based on EF-biocapacity metrics; no country with a very high HDI approaches this quadrant (Lin *et al.* 2018). Various critiques of HDI and IHDI exist, including that regarding the faulty data used to calculate the scores (Ranis, Stewart & Samman 2006; Wolff, Chong & Auffhammer 2011).

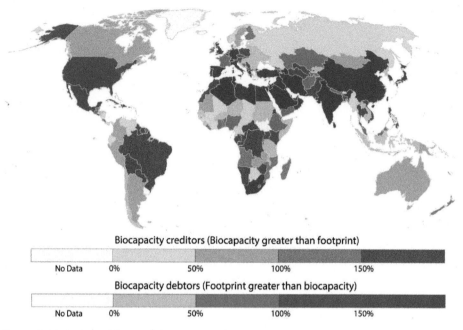

Figure 96.1 Ecological footprint–biocapacity relationship by country
Source: Lin *et al*. (2018) after GFN (2018).

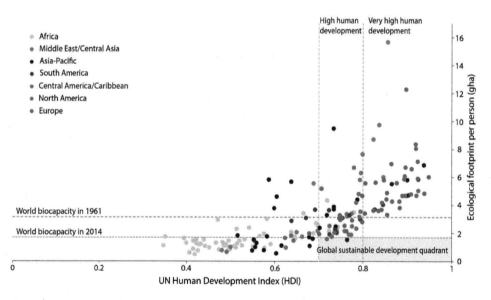

Figure 96.2 Ecological footprint–human development index relationship
In general, the higher the HDI the larger the EF; gha = global hectares.
Source: Lin *et al*. (2018) after UNDP (2016).

Other metrics exist as well, such as the "environmental performance index" (**Q67**) (EPI 2020) and the "energy, emergy and sustainability index" (Brown & Ulgiati 1999), among others. An even larger number of indicators exist covering both the development of the social subsystem and the maintenance of the environmental subsystem (**Q95, Q99**).

References

Brown, M. & S. Ulgiati (1999). "Emergy evaluation of natural capital and biosphere services". *Ambio* 28(6): 31–42. www.jstor.org/stable/4314939.

Conceição, P. (2019). *Human Development Report 2019: Beyond Income, Beyond Averages, Beyond Today: Inequalities in Human Development in the 21st Century.* United Nations Development Programme. hdr.undp.org/sites/default/files/hdr2019.pdf.

EPI (Environmental Performance Index) (2020). Yale Center for Environmental Law and Policy & the Earth Institute, Columbia University. epi.yale.edu/epi-results/2020/component/epi.

GFN (Global Footprint Network) (2018). *National Footprint Accounts.* Global Footprint Network. Oakland, CA.

Lin, D. *et al.* (2018). "Ecological footprint accounting for countries: updates and results of the national footprint accounts, 2012–2018". *Resources* 7(3): 58. doi.org/10.3390/ resources7030058.

Ranis, G., F. Stewart & E. Samman (2006). "Human development: beyond the human development index". *Journal of Human Development* 7(3): 323–58. doi.org/10.1080/14649880600815917.

Stanton, E. (2007). "The human development index: a history". PERI Working Papers, Political Economy Research Institute, University of Massachusetts Amherst. scholarworks.umass.edu/ cgi/viewcontent.cgi?article=1101&context=peri_workingpapers.

UNDP (United Nations Development Programme) (2016). *Human Development Report.* UNDP: New York.

Wackernagel, M. *et al.* (2019). "Defying the Footprint Oracle: implications of country resource trends". *Sustainability* 11(7): 2164. doi.org/10.3390/su11072164.

Wiedmann, T. & J. Barrett (2010). "A review of the ecological footprint indicator – perceptions and methods". *Sustainability* 2(6): 1645–93. doi.org/10.3390/su2061645.

Wolff, H., H. Chong & M. Auffhammer (2011). "Classification, detection and consequences of data error: evidence from the human development index". *Economic Journal* 121(553): 843–70. doi. org/10.1111/j.1468-0297.2010.02408.x.

WHAT ARE THE DISTINCTIONS AND IMPLICATIONS OF MITIGATION AND ADAPTATION FOR CLIMATE CHANGE?

Short answer: mitigation and adaptation are, respectively, strategies to reduce climate warming and improve societal responses to that warming. There are uncertainties in estimating future costs associated with either type of action.

Mitigation and adaptation capture two broad approaches that society may take in response to environmental change, foremost climate change. Mitigation is an action that reduces the trajectory of unwanted change in order to reduce its impacts. For climate change, mitigating actions following the Paris Agreement (Appendix) involve worldwide uses of technologies and practices that reduce greenhouse gas emissions in order to stabilize global average temperature increases below 2°C, preferably 1.5°C, over pre-industrial levels. Adaptation, in contrast, refers to societal adjustments to cope with increasing impacts of change, in this case rising temperatures. These adjustments might include revising working hours, migration to high latitudes and elevations, or willingness to move toward a "green" economy. Of course, both mitigation and adaptation may operate in tandem (Fig. 97.1, intersecting circles).

Historically, humankind has adapted to new or changing environmental conditions and mitigated against the consequences of those conditions. Two simple examples illustrate these strategies. To inhabit certain coastlines, many communities worldwide have adapted to high tides and storm surges by building residences on stilts (Fig. 97.2). Alternatively, the construction of dykes and surface drainage features are used to mitigate the impacts of standing water (drainage) and saline tidal water intrusions on croplands and homesteads, as in the case of the polders of the Netherlands or the dykelands of the Acadians in Nova Scotia (Fig. 97.3). Interestingly, while such undertakings are referenced as mitigation in their planning and development, historical assessment may refer to these wetland strategies as adaptations.

Which strategy or combination to undertake is determined by any number of societal conditions – for example, worldviews about human–environment relationships or concern about future generations. Commonly, however, such strategies are influenced by their projected short-term costs, and it is here that the contestation over climate warming has escalated. Are the risks of climate warming worth the cost of mitigation or would adaptation provide a less costly strategy? The economics involved are contentious, foremost because of the assumptions employed to estimate the costs. These issues were illustrated by a 2006

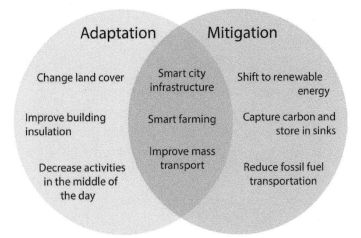

Figure 97.1 Examples of mitigation and adaptation to climate change
Source: author.

report (*The Stern Review* on the economics of climate change; Stern 2007) for the United Kingdom which concluded that the economic risks for society were sufficiently large to warrant mitigation. The report determined that to keep atmospheric CO_2 within the 450–550 ppm range would incur the significant cost of about 1 per cent per annum of the global GDP, but failing to do so would cost about 5 per cent of that GDP every year into the long-term future. Political and ecological reactions to the report notwithstanding, its major controversy involves the methods and assumptions applied to produce the cost estimates.

Figure 97.2 Stilt housing, southern Chile coast
Source: photo by author.

Figure 97.3 "The Acadians and the Creation of the Dykeland 1680–1755" by L. Parker
Source: www.landscapeofgrandpre.ca/the-acadians-and-the-creation-of-the-dykeland-1680ndash1755.html.

These controversies are complex and are simplified here to the question of the discount rate (trade-off between present and future benefits) applied to estimate the future value (costs) relative to the current value. Such estimates require assumptions about the marginal utility of consumption in the future, technological innovations creating substitutes, new means of dealing with the problem, and the uncertainty of future benefits. Using a high or low discount rate invariably provides significant changes in the estimates of the costs of mitigation versus adaptation. Much of the controversy over the Stern report involves different opinions about the appropriate discount rate to employ is such assessments (Nordhaus 2007; Weitzman 2007), along with a large range of other issues, such as intergenerational equity (Dasgupta 2007), the need to focus on non-substitutable natural capital (Neumayer 2007), the under or over cost estimates (Helm 2008) and so on. Such issues will not be resolved in the near-term, if ever, leading to claims, counterclaims and disagreement among the expert community, in part because estimating the future costs of climate change is a wicked problem (**Q86, Q87, Q100**) (Hsiang *et al.* 2017). This problem includes the unpredictable variations that will surely follow climate change, which are likely to escalate the economic and social costs to humanity (Calel *et al.* 2020) and predictable variations in which the habitability of some places and livelihoods will be enhanced and others not. Such consideration confounds the robustness of economic approaches that have proven useful for other environmental problems. As a result, support or not of mitigation strategies among economists and other expert researchers appears to be linked to their worldviews about human–environment relationships more so than the analytics (Arrow 2007).

Transformational adaptations – the reorganization of human–environment systems (Kates, Travis & Wilbanks 2012) to reduce vulnerabilities or increase resilience – and large-scale, novel mitigation efforts are not easy for society at large to undertake voluntarily, especially because such adaptations are likely to involve significant behavioural changes and the mitigations tend to be costly and uncertain. Assessing broad sweeps of history, societal adjustments to environmental changes appear to be more involuntary in kind – choices must be made and the options are few. Today, society may favour technological solutions to mitigate or adapt, especially if they stave off behavioural changes. For example, given the difficulties in reducing global carbon emissions as proposed in the Paris Agreement, attention is given to geoengineering responses to climate change, such as carbon capture and storage technology to reduce CO_2 emission (Q44) or injecting aerosols (Q45) into the atmosphere to reflect solar radiation into space (Caldeira, Bala & Cao 2013). Large-scale carbon capture is not yet developed (Schneider, Haul & Pressel 2020), however, and aerosol injections involves atmospheric complexities that generate other problems, possibly affecting food production by disrupting seasonal weather patterns (MacMartin & Kravitz 2019).

References

Arrow, K. (2007). "Global climate change: a challenge to policy". *The Economists' Voice* 4(3). doi.org/10.2202/1553-3832.1270.

Caldeira, K., G. Bala & L. Cao (2013). "The science of geoengineering". *Annual Review of Earth and Planetary Sciences* 41: 231–56. doi.org/10.1146/annurev-earth-042711-105548.

Calel, R. *et al.* (2020). "Temperature variability implies greater economic damages from climate change". *Nature Communications* 11: 5028. doi.org/10.1038/s41467-020-18797-8.

Dasgupta, P. (2007). "Commentary: the Stern Review's economics of climate change. *National Institute Economic Review* 199. doi.org/10.1177/0027950107077111.

Helm, D. (2008). "Climate-change policy: why has so little been achieved?" *Oxford Review of Economic Policy* 24(2): 211–38. doi.org/10.1093/oxrep/grn014.

Hsiang, S. *et al.* (2017). "Estimating economic damage from climate change in the United States". *Science* 356(6345): 1362–9. doi:10.1126/science.aal4369.

Kates, R., W. Travis & T. Wilbanks (2012). "Transformational adaptation when incremental adaptations to climate change are insufficient". *Proceedings of the National Academy of Sciences* 109(19): 7156–61. doi.org/10.1073/pnas.1115521109.

MacMartin, D. & B. Kravitz (2019). "Mission-driven research for stratospheric aerosol geoengineering". *Proceedings of the National Academy of Sciences* 116(4): 1089–94. doi.org/10.1073/pnas.1811022116.

Neumayer, E. (2007). "A missed opportunity: the Stern Review on climate change fails to tackle the issue of non-substitutable loss of natural capital". *Global Environmental Change* 17(3/4): 297–301. doi.org/10.1016/j.gloenvcha.2007.04.001.

Nordhaus, W. (2007). "A review of the Stern Review on the economics of climate change". *Journal of Economic Literature* 45(3): 686–702. doi:10.1257/jel.45.3.686.

Schneider, T., C. Haul & K. Pressel (2020). "Solar geoengineering may not prevent strong warming from direct effects of CO_2 on stratocumulus cloud cover". *Proceedings of the National Academy of Sciences* 117(48) 30179–85. doi.10.1073/pnas.2003730117.

Stern, N. (2007). *The Economics of Climate Change: The Stern Review 2006*. London: HM Treasury.

Weitzman, M. (2007). "A review of the Stern Review on the economics of climate change". *Journal of Economic Literature* 45(3): 703–24. doi:10.1257/jel.45.3.703.

WHAT CAN WE LEARN FROM THE CFC-OZONE AND CO₂-CLIMATE CHANGE EXPERIENCES ABOUT THE PROS AND CONS OF INTERNATIONAL ENVIRONMENTAL AGREEMENTS TO REDUCE DEGRADATION OF THE EARTH SYSTEM?

Short answer: the means by which the two agreements were developed, the incentives and enforcement mechanisms, and substitutions for the sources creating the emissions and the subsequent environmental damage differ significantly.

The stratospheric ozone layer that protects the Earth's surface from harmful ultraviolent (UV) radiation (**Q53**) is recovering from human-made ozone-depleting substances (ODSs). The layer is expected to recover fully by 2060 owing to the Montreal Protocol (MP) of 1987, despite recent increases in CFC-11 emissions (**Q53**). In contrast, greenhouse gas emissions (**Q44**) to the atmosphere, especially CO_2, have not yet declined to prevent the warming of the planet by 1.5–2°C, despite the 2015 Paris Agreement (also known as the Paris Climate Accords [PCA]) (see the UN Framework Convention on Climate Change, Appendix). Why has one international environmental agreement succeeded while the other is projected not to do so? Both problems involve global commons or global public goods and services (i.e. ozone layer protection and maintenance of a less warming climate). The PCA have proven inadequate, in part, because incentives to defect from the agreement exist (Nordhaus 2019). It confronts the prisoner's dilemma; countries' incentives to benefit themselves exacerbates the problems for others (Barrett & Dannenberg 2016).

A more complete, if simplistic, interpretation holds that the MP was successful because ultimately a science consensus of the cause emerged, the global public expressed concerns about the immediate dangers of UV radiation, and rapid innovations of substitutes for CFCs and HCFCs (**Q53**) – the agents of ozone depletion – were developed. In contrast, and despite a science consensus that CO_2 emissions are a principal cause of global warming (**Q44**), the global public has not shown high levels of concern, until recently perhaps, despite estimates that 80 per cent of the world's land area on which 85 per cent of the global population resides has already incurred changes in temperature and precipitation (Callaghan *et al.* 2021). Significantly, there are no relatively simple means to rapidly wean economies off fossil fuel use and attempts to do so by way of emission reductions are interpreted by some countries and many industries as endangering their economic well-being.

Such simplifications fail to highlight the role of the development and design of the MP in its success relative to the PCA (Albrecht & Parker 2019; Daniel *et al.* 2012). The MP development process included the following: negotiations in small and informal groups to build trust among countries; funds to assist developing countries to comply with ODS depletions; mechanisms to add more ODSs to the initial emphasis on CFCs; and enforcement through trade limitations on those parties not participating. In contrast, the PCA and their predecessor, the 1997 Kyoto Protocol, involved disagreements over who should burden the economic impact of reduced CO$_2$ emissions, and pledges to reduce them by participating countries were voluntary, with no means of enforcement other than public shaming by way of an international accountability mechanism (Falkner 2016). Essentially, the PCA seek collective action among countries with minimal incentives and penalties for those failing to reduce emissions (Barrett *et al.* 2017).

The Anthropocene cries out for successful international (or multilateral) environmental agreements (IEAs) to reduce the negative impacts of human activities of the Earth system and to improve sustainable development **(Q84, Q99)**. Some IEAs have been stable and successful, while many have not or appear to be problematic, as illustrated in the MP and PCA examples, respectively (e.g. Kellenberg & Levinson 2014; Marrouch & Ray Chaudhuri 2016). Significantly, the expert community has yet to crystallize a simple range of attributes that lead to success or not, in part because systematic, comparative data on IEAs has only recently become available (Mitchell *et al.* 2020), and assuredly because so many IEAs dealing with the Earth system have large economic implications for countries and economic sectors.

References

Albrecht, F. & C. Parker (2019). "Healing the ozone layer: the Montreal Protocol and the lessons and limits of a global governance success story". In P. 't Hart & M. Compton (eds), *Great Policy Successes*, 304–22. Oxford: Oxford University Press.

Barrett, S. & A. Dannenberg (2016). "An experimental investigation into 'pledge and review' in climate negotiations". *Climatic Change* 138(1): 339–51. doi.org/10.1007/s10584-016-1711-4.

Barrett, S. *et al.* (2017). "Rethinking approaches to climate change policy". *Report No. 2, Anxieties of Democracy Program*, WG-CC, Subgroup 2, Social Science Research Council, New York. s3.amazonaws.com/ssrc-cdn1/crmuploads/new_publication_3/rethinking-approaches-to-climate-change-policy.pdf.

Callaghan, M. *et al.* (2021). "Machine-learning-based evidence and attribution mapping of 100,000 climate impact studies". *Nature Climate Change* 11: 966–72. doi.org/10.1038/s41558-021-01168-6.

Daniel, J. *et al.* (2012). "Limitations of single-basket trading: lessons from the Montreal Protocol for climate policy". *Climatic Change* 111(2): 241–8. doi:10.1007/s10584-011-0136-3.

Falkner, R. (2016). "The Paris Agreement and the new logic of international climate politics". *International Affairs* 92(5): 1107–25. doi.org/10.1111/1468-2346.12708.

Kellenberg, D. & A. Levinson (2014). "Waste of effort? International environmental agreements". *Journal of the Association of Environmental and Resource Economists* 1(1/2): 135–69. dx.doi.org/10.1086/676037.

Marrouch, W. & A. Ray Chaudhuri (2016). "International environmental agreements: doomed to fail or destined to succeed? A review of the literature". *International Review of Environmental and Resource Economics* 9(3/4): 245–319. doi.org/10.1561/101.00000078.

Mitchell, R. *et al.* (2020). "What we know (and could know) about international environmental agreements". *Global Environmental Politics* 20(1): 103–21. doi.org/10.1162/glep_a_00544.

Nordhaus, W. (2019). "Climate change: the ultimate challenge for economics". *American Economic Review* 109(6): 1991–2014. doi:10.1257/aer.109.6.1991.

WHAT ARE THE UNITED NATIONS' SUSTAINABLE DEVELOPMENT GOALS AND WHAT ARE THEIR IMPLICATIONS FOR SUSTAINABILITY?

Brief answer: the goals and associated targets, if achieved, are proposed to lead to improved quality of life more equitably shared among the world's population and compatible with the maintenance of the biosphere.

The United Nations adopted 17 Sustainable Development Goals (SDGs) in 2015 (Fig. 99.1) as part of an agenda to achieve sustainable development by 2030 (Griggs *et al.* 2013; UN 2015b). These goals address human well-being, intermediate means of achievement, and maintenance of the biosphere (Text box 99.1). They replace the eight Millennium Development Goals set by the United Nations in 2000 and include a large number of specific targets and indicators for each goal. Achievement of these goals by the end date of 2030, it is proposed, will promote sustainable development – an improved quality of life that is more equitably shared among the peoples of the world and compatible with maintaining the biosphere. The goals are aspirational and fluid in that each country sets its own targets within the SDG framework based on national circumstances (Text box 99.1).

A number of critiques have been raised about individual or missing goals (Vasseur *et al.* 2017), the absence of an action plan (Fu *et al.* 2019), and the sheer number of targets (169) and indicators (231) applied to the 17 goals. Moreover, the SDGs skirt the issue that currently the higher scores a country has in regard to the human development index, the larger their ecological footprint **(Q96)**, raising the issue of increased pressures on environmental services and the Earth system (Wackernagel, Hanscom & Lin 2017). Notably, as well, the goals are less aggregated in their social dimensions and highly aggregated in their environmental dimensions, essentially collapsing all Earth system dynamics and their associated environmental services into three goals about climate change (13), the oceans (14) and the land-surface (15) (Fig. 95.1). While targets for the environmental goals increase specifications, the maintenance of the dynamics of the Earth system to support life are less emphasized than those focused on human well-being.

Such observations notwithstanding, as a unit the SDGs represent international recognition that human well-being and Earth system well-being are connected. This connection

Figure 99.1 United Nations' Sustainable Development Goals
Source: UN (2015a).

Text box 99.1 Categories of Sustainable Development Goals

Human well-being end goal
No poverty (Goal 1)
Zero hunger (Goal 2)
Good health and well-being (Goal 3)
Quality education (Goal 4)
Gender equality (Goal 5)
Clean water and sanitation (Goal 6)
Affordable and clean energy (Goal 7)
Decent work and economic growth (Goal 8)

Intermediate means of achievement
Industrial innovation and infrastructure (Goal 9)
Reduced inequalities (Goal 10)
Sustainable cities and communities (Goal 11)
Responsible consumption and production (Goal 12)
Peace, justice and strong institutions (Goal 16)
Partnerships for the goals (Goal 17)

Maintenance of the biosphere
Climate action (Goal 13)
Life below water (Goal 14)
Life on land (Goal 15)

Source: Costanza *et al.* (2014).

is illustrated by the tripartite linkage of the goals: the end goal of human well-being; the intermediate goal of economic, technological, political and ethical means of achieving that well-being; and a base goal of the maintenance of the biosphere (Text box 99.1) (Costanza *et al.* 2014). The immediate impacts of the SDGs have been the development of an SDG index indicating the "sustainability" level of the countries of the world (Sachs *et al.* 2021), although the results of the index raise a series of questions relative to the economic development of countries.

References

Costanza, R. *et al.* (2014). "An overarching goal for the UN sustainable development goals". *Solutions* 5(4): 13–16. www.researchgate.net/publication/271647074_An_Overarching_Goal_for_the_ UN_Sustainable_Development_Goals

Fu, B. *et al.* (2019). "Unravelling the complexity in achieving the 17 sustainable-development goals". *National Science Review* 6(3): 386–8. doi.org/10.1093/nsr/nwz038.

Griggs, D. *et al.* (2013). "Sustainable development goals for people and planet". *Nature* 495(7441): 305–07. doi.org/10.1038/495305a.

Sachs, J. *et al.* (2021). *Sustainable Development Report 2021. The Decade of Action for Sustainable Development Goals.* Cambridge: Cambridge University Press.

United Nations (2015a). "Sustainable development goals kick off with start of new year". UN News Centre. www.un.org/sustainabledevelopment/blog/2015/12/sustainable-development-goals-kick-off-with-start-of-new-year/.

United Nations (2015b). *Transforming our World: The 2030 Agenda for Sustainable Development.* New York: United Nations.

Vasseur, L. *et al.* (2017). "Complex problems and unchallenged solutions: bringing ecosystem governance to the forefront of the UN sustainable development goals". *Ambio* 46(7): 731–42. doi. org/10.1007/s13280-017-0918-6.

Wackernagel, M., L. Hanscom & D. Lin (2017). "Making the sustainable development goals consistent with sustainability". *Frontiers in Energy Research* 5: 18. doi.org/10.3389/fenrg.2017.00018.

HOW MIGHT THE KNOWLEDGE OF SUSTAINABILITY SCIENCE LEAD TO IMPROVED ACTIONS TOWARD SUSTAINABLE DEVELOPMENT?

Short answer: knowledge to action improves with strong salience, credibility and legitimacy elements, although new challenges exist for this framework.

The Sustainable Development Goals **(Q99)** for human and environmental well-being require the co-development and translation of knowledge gained from science to implementable actions that lead to a more sustainable human–environment relationship. This co-development, translation and implementation is referred to as a "knowledge to action" framework (Field *et al.* 2014), one in which knowledge assumes understanding (Kirchoff, Lemos & Dessai 2013; van Kerkhoff & Lebel 2006). The framework for environmental and sustainability science focuses on three critical elements that are effective for potential implemental action: salience, credibility and legitimacy (Cash *et al.* 2003; Cash & Belloy 2020).

Salience refers to the research or assessment relevancy to the needs of decision makers. Credibility deals with the scientific adequacy – the robustness of the evidence and arguments. Legitimacy involves the perception that the science in question respects the divergent value and beliefs of the stakeholders involved and is unbiased in its treatment of opposing views and interests. The interactions among these elements, changing perceptions of them, and the actions of non-trusted partners are challenges to this framework. Nevertheless, the evidence grows that the science about the environment and sustainability is more trusted and willing to be acted on if the three elements are present (Bremer & Meisch 2017; Cash & Belloy 2020).

Recent research, however, identifies four emerging stressors affecting the tripartite framework (Cash & Belloy 2020). Firstly, cross-scale dynamics of complex social–environmental systems affect solutions, indicating that decisions to address global sustainability problems have strong local dimensions in the solutions undertaken. Secondly, emphasis on equity encounters diversity of interpretations among stakeholders (i.e. what constitutes equity). The science community itself may view the engagement of equity as political, such that taking account of it may threaten the credibility of science. Thirdly, the digital world creates pros and cons for the framework. In one vein it can increase stakeholder involvement and improve analytical capabilities. In another vein, it reduces human contact and advantages those with the technology and skills to access and use the digital world. The fourth stressor is, perhaps, the most important and difficult to counteract. It is the "post-truth" world in which the objectivity and credibility of science dissipates and its confirmed results are contested. In this world, science may threaten the power of the status quo or challenge strongly held political positions, either of which may lead to disinformation campaigns against or outright denials of the science (**Q79**), commonly transmitted through the digital world of social media. These stressors to the knowledge to action framework require sustained research regarding their impacts on sustainability goals and on the possibility of the reformulation of the framework.

References

Bremer, S. & S. Meisch (2017). "Co-production in climate change research: reviewing different perspectives". *Wiley Interdisciplinary Reviews Climate Change* 8(6): e482. doi.org/10.1002/wcc.482.

Cash, D. *et al.* (2003). "Knowledge systems for sustainable development". *Proceedings of the National Academy of Sciences* 100(14): 8086–91. doi.org/10.1073/pnas.1231332100.

Cash, D. & P. Belloy (2020). "Salience, credibility and legitimacy in a rapidly shifting world of knowledge and action". ***Sustainability*** **12(18): 7376. doi.org/10.3390/su12187376.**

Field, B. *et al.* (2014). "Using the knowledge to action framework in practice: a citation analysis and systematic review". *Implementation Science* 9: 172. doi.org/10.1186/s13012-014-0172-2.

Kirchhoff, C., M. C. Lemos & S. Dessai (2013). "Actionable knowledge for environmental decision making: broadening the usability of climate science". *Annual Review of Environment and Resources* 38: 393–414. doi.org/10.1146/annurev-environ-022112-112828.

van Kerkhoff, L. & L. Lebel (2006). "Linking knowledge and action for sustainable development". *Annual Review of Environment and Resources* 31: 445–77. doi.org/10.1146/annurev-energy.31.102405.170850.

Q101

IS A SUSTAINABLE ANTHROPOCENE POSSIBLE?

Short answer: perhaps, but with increasing changes in the Earth system and variations in global development status.

The answer largely resides in the worldview of the interpreter, either the Cassandra or the Cornucopian **(Q73)**. The safest interpretation, perhaps, resides between the two polar views. The pressure of humanity on the Earth system will not cease, indeed it will undoubtedly increase in the near-term future, and is likely to provide surprises as new innovations become available and deployed to address the declines in and maintenance of environmental services **(Q7)**. In this sense, the states and flows of the Earth system **(Q3)** will become increasingly altered by the built environment we will create. Whether the planetary boundaries **(Q88)** or the tipping elements **(Q89)** will be surpassed sufficiently to change the Earth system functions radically is not clear. Proposed tipping points, such as the thermohaline circulation shutdown and its current slowing **(Q40)**, and the almost witnessed one, as in the case of the thinning of the ozone layer **(Q53)**, demonstrate how closely the Anthropocene has confronted significant, if not radical, changes in the Earth system.

Sustainability requires sustainable development **(Q81, Q84)**, foremost improving human well-being in a more globally equitable way. While various metrics indicate that such development is taking place among average global aggregate conditions, the distinctions between the material-have world and the material-have-not world remain large, and within-country Gini coefficients appear to be increasing **(Q95)**. A pivotal question is the consumption level that a more equitable global society can maintain without critical damages to the Earth system. This query, in turn, depends on innovations in the global political economy, increasing societal emphasis on the long-term human–environmental dynamics, global advances and technology innovations under way, such as those of synthetic and biosynthetic innovations **(Q15)**, and the environmental surprises that will surely follow from them.

Given the above, sustainable development at the global level is a wicked problem, and one that surely will entail numerous and significant surprises. It is one in which we must not only be cognizant of the vast changes that we have made to the Earth system and those looming, such as global climate change (**Q46**), but the magnitude of changes we continue to make. New methods to handle "big data" indicate that in a recent 25-year period (1990–2015) 1.6 Mkm^2 of native lands were converted for human uses, about 178 km^2 daily or over 12 ha min^{-1} (Theobold *et al.* 2020). Such state changes combined with those on flows (biogeochemical cycles; **Q4**) and the recognition of inertia built into our political and technological systems (Matthews & Wynes 2022) surely warrant attention to caution regarding the Earth system in the Anthropocene, no matter how opaque the precautionary principle may be (**Q87**).

References

Matthews, H. & Wynes, S. (2022). "Current global efforts are insufficient to limit warming to 1.5°C". *Science*, 376(6000): 1404–9. doi.org/10.1126/science.abo3378.

Theobald, D. *et al.* (2020). "Earth transformed: detailed mapping of global human modification from 1990 to 2017". *Earth System Science Data* 12(3): 1953–72. doi.org/10.5194/essd-12-1953-2020.

GLOSSARY

abiotic physical as in non-living parts of the environment (e.g. solar energy, wind, mantle).

acidification decreased pH levels creating a more acidic water source; in oceans by the uptake of carbon dioxide.

aerosols aero-solutions or fine-grain (micro-scale) solid particles or liquid droplets suspended in the atmosphere.

aerosols precursor gas creating an aerosol owing to condensation.

afforestation planting trees, including non-native species, in an area or in a greater number in an area than existed through natural land cover.

albedo surface (e.g. land, clouds, buildings) reflectivity of solar (shortwave) radiation; light- and dark-coloured surfaces have high and low reflectivity, respectively.

ammonia a gaseous compound of nitrogen and hydrogen.

anomaly (in this text) the difference from a longer-term average, such as an annual mean temperature being higher or lower than a 50-year average.

Anthropocene recognition of our current period in which humankind has become equivalent to nature as a driver of changes in the Earth system; debated that the period constitutes a geological epoch.

aquaculture farming of marine (fresh- or saltwater) organisms, fish to algae; some apply the term to freshwater farming only.

archaea single cell microorganisms; live in low oxygen environments.

Asian monsoon or East Asian monsoon is an air flow from the Indian and Pacific oceans that brings moisture to eastern Asia and the Indian subcontinent (see monsoon).

Atlantic meridional overturning circulation (AMOC) shift of warmer, northward flowing surface waters to cold, dense subsurface southward flow in the North Atlantic; part of the thermohaline circulation.

atmosphere the layers of gases that surround our planet.

backcasting see hindcasting.

biogenic produced by life (organisms).

biogeochemical cycles movement (flow) of elements and compounds through the different spheres (parts) of the Earth system, supporting the biosphere.

biogeophysical see biophysical.

biome plant and animal communities sharing common characteristics as a response to the physical climate.

biophysical referring to biotic and abiotic factors and their interactions that create, for example, landscapes or spheres of the Earth system; sometimes termed biogeophysical.

biosphere that portion of the Earth system supporting life.

biospheric values judging phenomena based on cost or benefits to the environment or biosphere.

biota plant and animal life.

birth rate crude birth rate is total number of live births per 1,000 population proportional to a specified period in years.

black carbon particulate matter ≤ 2.5 μm of carbon, such as soot.

bleaching or coral bleaching; corals expel algae in their tissues if water is too warm, turning coral white.

blue carbon carbon stored in marine ecosystems.

BP before present, meaning years before 1950.

carrying capacity maximum size of a species that can be sustained in a given environment.

Cassandras those holding a view that nature (environment or Earth system) is threatened and actions to maintain it are required.

catch fishing capturing wild marine life, such as through nets, traps or on lines.

chemical compound a substance composed of two or more different chemically bonded chemical elements, such as water (H_2O) or carbon dioxide (CO_2).

chemical element a pure substance that cannot be broken down chemically, such as oxygen (O) or carbon (C).

climate change change in the average climatic conditions, such as temperature and precipitation, over long periods of time, typically decades or longer.

co-adaptation (in this text) the mutual adaptation between human activities and the ecosystem.

co-design research or sustainability activities orchestrated and/or undertaken that involve researchers and stakeholders or those invested in the outcome of the research or activity.

continental volcanic arc multiple volcanoes typically aligned along the edges of continents where oceanic plates subduct below continental plates.

Coriolis force effect of the Earth rotation that moves air and water anticlockwise and clockwise in the Northern and Southern hemisphere, respectively.

Cornucopians those holding a view that human inventiveness overcomes threats to the environment, especially in regard to resources.

cryosphere those parts of the Earth system composed of solid water (e.g. snow or ice).

cultural theory (in this text) linked to cultural theory of risk in which social organizations imbue and reinforce the individual's view of the environment.

decarbonization reducing carbon emissions involved in production and consumption.

deep explanation a social science term signifying an outcome that can be traced through cascading interactions to a root phenomenon or process.

dematerialization reducing the materials and energy involved in production and consumption.

denitrification reduced nitrate or nitrite to gaseous forms of nitrogen.

desertification human actions that degrade arid lands, making their land-cover more xeric (drier) and less productive.

digital technological era also "Third Industrial Revolution"; the proliferation in the latter half of the twentieth century of digital electronic technology, such as the internet, changing the speed and amount of information globally.

disservice as in environmental or ecosystem, it refers to functions or properties of environmental systems viewed as harmful to people.

distal cause/driver a cause or linked factor that underlies but is remote from the immediate phenomenon or event to be understood.

DNA deoxyribonucleic acid holding the genetic instructions for living organisms.

drivers (or forcings) factors that influence human or environmental phenomena, common in environmental science, especially ecology, and in some social science research.

Earth's average temperature the aggregation and averaging of temperatures worldwide from land and water surfaces.

Earth system the spheres or components (e.g. atmosphere) that comprise our planet and its functioning.

eccentricity degree of variation from a circular-shaped orbit, in reference to the orbit of the Earth around the Sun; it is one element of the Milankovitch cycle.

eco-evolutionary unidirectional effect of an ecological process on evolutionary processes or vice versa.

ecological footprint the quantity of nature (i.e. environmental services) to sustain humankind (e.g. an individual, community, global population).

economic externalities third-party impacts created by production or consumption of other parties.

ecosystem interacting organisms creating a biological community with the physical environment, such as the species creating a prairie or a coral reef.

El Niño or ENSO the El Niño–Southern Oscillation, a periodic variation in winds and sea surface temperatures in the eastern Pacific Ocean tropics that has large climatic impacts beyond the tropics.

emergence an entity has properties that its parts do not (e.g. sand dune patterns versus sand grain).

emergent property the outcome of emergence (e.g. sand dune).

environment (in this text) the biophysical world and its operation independent of people.

environmental performance index (EPI) measure of environmental performance (i.e. outcomes) of a country's policies.

environmental services (or ecosystem services) the benefits from the environment (nature or Earth system) for humans.

evaporation liquid water transformation to a gaseous state, delivered from land and oceans to the atmosphere.

evapotranspiration the transfer of water from land (e.g. lakes, soil) by evaporation or through the "breathing" or transpiration of vegetation to the atmosphere.

exclusive economic zones an area of the sea in which countries have special rights regarding marine resources as prescribed in the 1982 UN Convention on the Law of the Sea.

explanation an articulation of the cause of a phenomenon or event, as opposed to the identification of an association or relationship with them; usually answers "why" the phenomenon or event exists.

externalities (see economic externalities).

flora plants of a particular landscape.

fodder feed for livestock.

forcings (or drivers) factors, commonly abiotic, that influence human or environmental phenomena, widely used in climate research.

fossil fuels the use of any buried carbon-based substance for fuel, such as natural gas, coal or oil; this use releases greenhouse gases to the atmosphere.

fossil water or paleowater is water collected and stored over millennia, common among aquifers; contemporary use tends to withdraw more of this water than nature recharges.

general systems (or general systems theory) organizing principles can be found that cross all system processes.

genomics study of genomes or the set of DNA of organisms; in this text, the creation and use of products from this study.

geoengineering intervention, usually infrastructure of some kind, to alleviate an environmental problem; commonly used in reference to combat climate change, such as withdrawing carbon dioxide from the atmosphere and storing it in the mantle of the Earth.

geological epoch geological time unit below, in order, eons, eras and periods; typically, tens of millions of years in length. The current epoch (Holocene), however, only began about 12,000 years ago.

geosphere used in two ways: the collection of all spheres comprising the Earth system or the lithosphere per se.

Gini coefficient (also Gini index or ratio) a statistical dispersion measure of the inequality among values of frequency distributions commonly used in economics; 0 = maximum equality, as in all people have the same capital assets; 1 = maximum inequality, as in all capital assets are held by one person.

glacier dense body of ice that moves slowly under its own weight created by the accumulation of snow.

global boundary stratotype section and points (GSSP) reference point of a stratigraphic section that defines the lower boundary of a geological timescale.

global environmental change (GEC) planetary-level change in the environment or Earth system.

global mean sea level (GMSL) weighted average of the elevation of oceans between high and low tides.

global mean surface temperature (GMST) weighted average of near (land) surface air and sea surface temperatures.

Great Acceleration the technological advances and the magnitude of their uses, beginning about 1950 and continuing today, creating significant consequences for the Earth system.

green economy an economy aimed at reducing environmental and Earth system impacts by the efficient use of natural capital and environmental services.

green revolution development and global use of high yielding varieties (hybrid) of food crops during the 1950s and 60s.

green spaces description of urban spaces dominated by vegetation – gardens, parks, golf courses, green roofs, arboretums – and absent of buildings, pavements and other impervious surfaces.

greenhouse Earth see hothouse Earth.

greenhouse gases (GHGs) trace gases and water vapour in the atmosphere that capture longwave radiation from the Earth system, warming the Earth's average temperature.

Gulf Stream warm, near-surface ocean waters flowing from the tropics through the Gulf of Mexico northward roughly following the North America coastline to about 40° latitude where it splits.

gyres large-scale circulating ocean currents created by winds and the Coriolis force.

Haber–Bosch process artificial nitrogen fixation process to produce ammonia.

halogens five non-metallic elements, compounds of which are GHGs and thin the ozone layer.

HANPP human appropriation of net primary production (NPP), metric of the percentage of global NPP that humans use.

herbivores animals (also insects) that only eat vegetation.

hindcasting (aka backcasting) method of testing the robustness of a model by using past empirical data to determine whether the model generates the known results.

Holocene current interglacial geological epoch beginning about 11,650 years before present.

hothouse Earth or greenhouse Earth, million-year periods in which the atmosphere has high levels of greenhouse gases and water vapour and no continental glaciers are present.

human–environmental relationships all interactions between humankind and the biophysical environment/Earth system.

hybrid crops offspring of cross-pollinating two different varieties of plants.

hydrological cycle the movement of water in all of its states (liquid, solid, gas) among land, atmosphere and oceans.

hydrosphere all portions of the Earth system maintaining solid, liquid and gaseous water, such as glacier, oceans and clouds, respectively.

hypoxia depletion of oxygen concentration in water such that aquatic life cannot exist.

ice age a long period in which the temperature of the Earth's surface cools, permitting an expansion of polar and continental ice cover.

ice sheet essentially a gigantic glacier covering at least 50,000 km² or 19,000 mi².

icehouse Earth million-year periods in which the Earth's atmosphere has low levels of greenhouse gases and water vapour and the polar regions are ice covered; without human-induced greenhouse gases the Earth has been in an icehouse stage that began 34 Mya.

impervious surface any surface resistant to water infiltration, usually signifying artificial structures such as pavements or buildings.

ICS International Commission on Stratigraphy, which formalizes geological time units, such as the existence or not of the Anthropocene as an epoch.

induced innovation change in relative price of production spurs innovation; for agriculture, focused on labour or land innovations depending on the scarcity of either factor.

induced intensification the intensity of cultivation is associated with level of demands placed on the system.

Industrial Revolution (in this text) the late eighteenth–early nineteenth centuries switch of energy use to fossil fuels, foremost coal; subsequent second to fifth industrial revolutions dealing with new fuels, materials and technologies are recognized but labelled differently in this text.

involution an agrarian condition in which communities maintain intensive systems of cultivation in which one unit of input derives little more than one unit of output (i.e. marginal returns ~1).

intertropical convergence zone (ITCZ) a low-pressure belt circling the Earth near the equator but seasonally oscillating north and south, where the trade winds (east to west flow) from both hemispheres meet.

irradiance (solar) output energy from the Sun received by the Earth, measured as Watts per square metre.

jet stream fast-flowing, meandering air currents; on Earth these are west to east flows near the tropopause, divided into the polar jets and subtropical jets.

Jevons paradox the increase in the efficiency of a resource to reduce its use increases the demand for that resource; linked to the Khazzoom–Brookes postulate (see below).

Karma line the limit of the Earth's atmosphere and the beginning of space, designated at 100 km (62 miles) above the Earth's surface.

Khazzoom–Brookes postulate micro-level energy efficiency improvements, leads to higher levels of energy consumption at the macro-level; links to the Jevons paradox (see above).

K–Pg or K–T extinction a mass extinction 66 million years ago, eliminating 75 per cent of all plants and animal in existence at that time.

land cover the biophysical conditions (phenomena) on the surface of the Earth, such as tropical forest or grasslands.

land grab or grabbing large-scale land acquisitions by external agents/entities; e.g. companies from one country obtaining the use of large blocks of land in another country.

land use the human intent/activity or lack of activity of the surface of the Earth, such as agriculture, settlements or nature reserves.

latitude a measure of the distance north and south of the equator.

leakage resources lost to conservation or preservation activities in one location that are gained from another location, usually discussed in terms of countries; e.g. timber lost to forest conservation in one country or region is gained by logging obtained from another country or region.

leeward directionally downwind from a point of reference.

legacy effect environmental changes following from antecedent human actions; often applied to the constraints these changes maintain on future human action.

lidar (LiDAR) light detection and ranging method of laser or LED light used for fine-resolution mapping variations in distance.

lithosphere the crust and uppermost mantle of the Earth in which plate tectonics operate.

Little Ice Age a period between 1650 and 1850 in which the overall temperature anomaly was cooler than the periods before and after the time frame, especially in the Northern hemisphere.

longwave radiation (or outward longwave radiation (OLR)) electromagnetic radiation of 3–100 μm emitted from the Earth and atmosphere to space, much of it captured by greenhouse gases and warms the Earth.

macroscopic visible by eye.

magnetic polarity the orientation of magnetic poles (positive and negative), typically of mineral in rock.

marginal returns additional output following a one-unit increase in the input, holding other factors constant.

mariculture a term replacing aquaculture applied to open ocean farming of marine organisms, including their by-products, such as pearls.

mass extinction large-scale and rapid loss of biota globally.

Meghalayan a newly formalized stage of the Holocene epoch, starting 4,200 years ago.

meridional overturning circulation (MOC) surface and deep ocean movement of water globally, referred to as the great ocean conveyor belt.

mesosphere an atmospheric layer above the stratosphere, roughly 50 km to 85 km (50–85 mi) above the Earth's surface.

meteorite meteors (rock or iron) that ultimately reach the surface of the Earth.

micronutrients essential elements required by organisms; this text focuses on those in soils for plant growth.

microplastics any type of plastic <5 mm in length polluting the environment.

middle-range theory used in some social sciences to designate "theory" as understood in most sciences owing to broader types of theories postulated in other parts of the social sciences.

Milankovitch cycles the long-term, repeated changes in solar radiation owing to three movements of the Earth relative to the Sun that create warming and cooling climates.

million species years one million species going extinct at the rate of one per year.

Mohorovičić discontinuity or Moho is the boundary between the crust and mantle of the Earth.

monsoon seasonal reversing of winds owing to land–ocean temperature differences creating wet and dry seasons; the rainy period often referenced as "the monsoon".

mortality rate (or crude death rate) total number of deaths per 1,000 population proportional to some period in years.

natural capital stock of environmental services or nature's resources.

nature's contribution to people (NCP) a term proposed to replace ecosystem or environmental services.

neo-colonial one country's strong political or economic influence on another country without maintaining formal rule over the influenced country.

neo-Malthusian individual or concept linked to the notion that population growth is central to environmental problems.

net primary productivity (NPP) rate at which plants produce chemical energy versus rate of transpiration, the remainder of which is used for growth and reproduction.

nitrogen fixation conversion of nitrogen in air to ammonia or to nitrogen compounds in soil.

non-linear dynamics conditions in which a change in output is not proportional to the change in input and which may appear to be chaotic or unpredictable.

normalized difference vegetation index (NDVI) remote-sensing measure of plant health based on reflected light frequencies.

Northwest Passage a shipping route to cross the Arctic Ocean between the Atlantic and Pacific Oceans north of continental Canada and Alaska.

Ogallala aquifer fossil water below the Great Plains of the United States, stretching from North Dakota to Texas, and source for irrigated agriculture.

paddy or padi a flooded field used for rice cultivation; also the rice plant itself.

parameter used slightly differently among the sciences; in this text it refers to the limits of a system or system variables.

peatland accumulation of decayed vegetation in soils that builds up in wetlands.

pedosphere the soil on top of the Earth's crust created by the interactions of the various spheres.

phenotypic change in genetics, a change in observable characteristics of an organism created by genes and/or environment.

photosynthesis process by which plants convert light energy to chemical energy that sustains their life.

plate tectonics the continental and marine lithosphere is partitioned in segments (plates) that slowly move owing to the interior heat of the Earth, changing the position of continents and oceans over multiple millennia.

Pleistocene the geological epoch before the Holocene lasting from about 2,580,000 to 11,700 years ago.

plutonium a radioactive chemical element (Pu) and marker of the use of nuclear activity.

$PM_{2.5}$ tiny particles, $\leq$ 2.5 microns, that are air pollutants.

polar vortex a low-pressure region of cold air masses over the polar regions.

precession (or axial precession) gradual cyclical shift in orientation of the Earth's axis; part of the Milankovitch cycle.

pre-industrial times before the Industrial Revolution or the use of large-scale fossil fuel energy in Europe and North America, variously dated between 1760 to 1840.

primary forests are old and show virtually no human disturbance, typically with unique ecological features.

primary productivity the amount of organic compounds generated by plants through photosynthesis.

progressive contextualization used in some of the social sciences to refer to a step-by-step understanding of causal linkages, causal identification becomes an outcome for the next causal link.

prokaryotic organisms (or prokaryotes) unicellular organisms, such as bacteria, archaea and algae.

proximate cause/driver the immediate cause or linked factor to the phenomenon or event to be understood.

Quaternary the geological period encapsulating the Pleistocene and Holocene.

radiative forcings difference between insolation (solar radiation) absorbed by the Earth and the energy reradiated back into space.

radiocarbon dating use of radiocarbon (^{14}C), a radioactive isotope of carbon, to measure the age of entities containing organic material.

radionuclides atoms that emit radiation; indicator of the use of nuclear items.

rebound phenomenon/effect reduced expected gains from new, improved efficiency in resource use owing to other factors, often behavioural changes.

reflexive agent in reference to humans who reflect and anticipate and thus can change their responses to events and processes.

reforestation natural or intentional restocking of a former existing forest.

remote sensing information gained on phenomena without physical contact; for the Earth system this information is usually obtained by satellite or airborne sensors.

ruminants herbivorous mammals that ferment plant foods in specialized stomachs (e.g. cows, camels, water buffalo).

selective logging taking of only certain timber/trees focused on size or species.

shifting cultivation (also swidden) after several years of cultivation, the field is fallowed (rested) for multiple years and cultivation shifted to another field.

shortwave radiation radiant energy (0.1–0.5 μm), including near-ultraviolet, and part of solar radiation.

sinks locations in the Earth system in which greenhouse gases, foremost carbon dioxide and methane, are stored, commonly for multiple millennia, such as oil and coal deposits.

social–environmental (or ecological) system/SES a set of societal and biophysical phenomena and processes interacting with one another, common to studies in the human–environmental sciences, including sustainability.

solar constant the total radiation from the Sun received per unit area by the Earth, averaged for the mean distance of our planet's elliptical orbit, measured as 1.366 kilowatts per square metre.

solar flares emit X-rays and magnetic fields that can disrupt transmissions from satellites but do not affect the temperature of the Earth.

solar irradiance the radiant flux or power of energy received from the Sun at different surfaces of the Earth (e.g. stratosphere or Earth surface).

solar radiation electromagnetic radiation or insolation received from the Sun as infrared, visible and ultraviolet light.

Southern Ocean the Antarctic or Austral Ocean, surrounds the Antarctic continent.

speciation process by which species evolve into new species.

staple crops food crops dominating diets

stratigraphy rock layers.

stratosphere gaseous layer of the atmosphere above the troposphere residing at 8–15 km to 50–60 km depending on latitude.

subduction process by which one tectonic plate (oceanic plate) moves under another plate (continental plate).

subsidence sinking or downward settling of the surface of land.

sun spots temporary regions on the Sun's surface of high magnetic fields and reduced temperatures, typically associated with solar flares.

sustainable development improving human well-being in a more equitable fashion without endangering the Earth system to provide environmental services now and for future generations.

swidden see **shifting cultivation**.

synthetic compounds (in this text) human-created chemical compounds, either adding to those existing in nature, such as ammonia, or creating non-natural compounds, such as CFCs.

taxa any unit of classification in biology/taxonomy.

tectonics the processes that affect the Earth's crust such as the movement of tectonic plates and volcanism.

thermohaline circulation (THC) also the meridional overturning circulation (MOC); the worldwide conveyor belt of oceanic water movement near the surface and at depth generated by differences in temperature and salinity that change water density.

thermosphere the layer of atmosphere above the mesosphere, about 80 km above the Earth's surface.

thick description in social science refers to explanations of social actions in which the context of the behaviours observed are examined.

transdisciplinary crossing disciplines to create a holistic approach beyond interdisciplinary activity; or adding non-research stakeholders to the study.

tropical cyclone (also hurricane and typhoon) large, rapidly rotating storms emerging from heat build-up over the tropical portion of the oceans and moving poleward.

troposphere the lowest gaseous layer of the atmosphere, from the surface of the Earth to 6–18 km above depending on latitude.

ultraviolet radiation electromagnetic radiation (100–400 nanometres) and part of solar radiation that is damaging to life.

unchecked population growth in population without voluntary or involuntary controls to reduced fertility.

upwelling deep, cold ocean water rises towards the surface to replace surface waters pushed away by winds, typically bringing nutrients to the surface.

urban heat island (UHI) effect (in this text) increases in the ambient (regional) air temperature over urban areas owing to energy and longwave radiation generated by urban activity and the concentration of impervious surfaces.

valuation in economics, various methods to project monetary value on items not traded in the market place.

volcanism eruption of molten rock onto the Earth's surface.

water grab or grabbing water acquisition (e.g. river water or aquifers) that may be associated with land-grabbing (see above).

windward directionally upwind from a point of reference.

wiring diagram a representation (map) of the components and their connections in a system.

Younger Dryas a period from about 12,900 to 11,700 BP in which the interglacial warming of the Earth was disrupted by glacial-like conditions in the northern hemisphere.

APPENDIX

International programmes, conventions and policies dealing with the environment and sustainability

The number of international research and applied programmes, agreements, conventions and accords that have dealt or are dealing with the global environmental and sustainability are considerable in number. Those identified in this section relate to various parts of this book and in some cases are mentioned in the questions and answers. They are restricted to intergovernmental and United Nations entities; regional, national, business and non-governmental organization efforts are not listed.

Alliance of Small Island States – an intergovernmental organization providing a consolidated voice to small island nations that are highly susceptible to increasing sea level. www.aosis.org

Convention on Biological Diversity (CBD; or Biodiversity Convention) – a multilateral treaty on biodiversity conservation, sustainability and equitable sharing of genetic resources sourced by UNEP.

Convention on International Trade in Endangered Species of Wild Fauna and Flora (or CITES) – a multilateral treaty to ensure that international trade in wild biota does not threaten species in the wild.

Convention on the Conservation of Migratory Species of Wild Animals (or Convention on Migratory Species, CMS) – a multilateral agreement to conserve migratory species and ranges sourced by UNEP.

Convention on the Protection and Use of Transboundary Watercourses and International Lakes (aka the Water Convention) – a United Nations negotiated treaty creating cooperative programmes to address transboundary water issues. www.imo.org/en/OurWork/Envi ronment/LCLP/Pages/default.aspx

Convention on the Prevention of Marine Pollution by Dumping of Wastes and Other Matter (aka the London Convention) – a United Nations initiative to control waste disposal by any kind of vessel transporting or anchored at sea, subsequently extended to include certain pollutants delivered from land (e.g. river–sea delivery).

Convention on the Law of the Sea or UNCLOS – a United Nations initiative addressing the rights and responsibilities of countries regarding the use and management of oceans. www.un.org/Depts/los/index.htm

Convention to Combat Desertification or UNCCD – a United Nations initiative to address arid land degradation and mitigate the impacts of drought on such lands. www.unccd.int

Food and Agricultural Organization (FAO) – a United Nations agency focused on international efforts to overcome hunger and improve nutrition and food security. www.fao.org

Future Earth – a United Nations inspired international programme focused on interdisciplinary knowledge fusing research and application to facilitate sustainable development. www.futureearth.org

Global Environmental Facility – an independent financial organization partnering countries and the private sector to support sustainable development, including various conventions on this list. www.thegef.org

Global Green Growth Institute – an intergovernmental organization supporting green growth development, balancing economic and environmental concerns. www.gggi.org

International Carbon Action Partnership – an open forum for public authorities to assist in the coordination of cap-and-trade systems to reduce carbon emissions. icapcarbonaction.com/en/

International Geosphere-Biosphere Programme (IGBP) – an International Council of Scientific Unions programme that coordinated international research on the process maintaining the Earth system and its interactions with humankind, creating much of the science on which global environmental change is based [dissolved].

International Human Dimensions Programme (IHDP) – an International Social Science Council programme that coordinated international research on social science themes relevant to global environmental change, assisting in linking social, economic and political dimensions into global change research [dissolved].

Intergovernmental Oceanographic Commission of UNESCO (IOC/UNESCO) – an United Nations organization promoting marine science and assisting countries with ocean and coastal management. ioc.unesco.org/node/2

Intergovernmental Panel on Climate Change (IPCC) – a United Nations initiative creating an intergovernmental body providing state-of-the-art science on climate change, its risks to society and response options. www.futureearth.org

Intergovernmental Science-Policy Platform on Biodiversity and Ecosystem Services (IPBES) – a United Nations initiative to improve the interface between science and policy on biodiversity and ecosystem services akin in principle to the IPCC. www.ipbes.net

International Tropical Timber Organization – an intergovernmental organization promoting tropical forest conservation and sustainable management. www.itto.int/

International Union for Conservation of Nature and Nature Resources (IUCN) – an international organization assisting societies in efforts to conserve nature and sustainably manage natural resources. www.iucn.org

Minamata Convention on Mercury – an international treaty to protect human health from mercury and mercury compound emissions and releases. www.mercuryconvention.org/

Montreal Protocol – an international treaty following from the Vienna Convention for the Protection of the Ozone Layer to phase out the use of ozone-depleting substances, such as chlorofluorocarbons, to protect the stratospheric ozone layer protecting the Earth's surface from ultraviolet radiation.

Organization of Economic Cooperation and Development – the OECD maintains several programmes related to sustainable development.

Paris Agreement/Accords – part of the United Nations Framework Convention on Climate Change, this international, voluntary agreement among countries seeks to hold climate change to an increase of no more 2°C from pre-industrial levels, with attempts to hold that increase to 1.5°C.

Ramsar Convention on Wetlands of International Importance especially for Waterfowl (or Convention on Wetland) – parties involved promote means to improve the preservation of wetlands.

Reducing Emissions from Deforestation and Forest Degradation and the Role of Conservation, Sustainable Management of Forests and Enhancement of Forest Carbon Stocks in Developing Countries (REDD+) – a programme emerging from the Paris Accords seeking to mitigate climate change via reducing net emissions of greenhouse gases from forest loss and degradation in developing countries. This effort is different from the UN-REDD program.

Stockholm Convention on Persistent Organic Pollutants – an international treaty to eliminate and restrict the use of persistent organic pollutants. www.pops.int/TheConvention/Overview/

United Nations Environment Program (UNEP) – developing and administering environmental programmes, with special attention to assisting developing countries with environmental activities and programmes. www.unep.org

United Nations Educational, Scientific, and Cultural Organization (UNESCO) – an agency promoting world peace and security, linked to various environmental and sustainability initiatives of the parent organization, including designation of World Heritage Sites, many of which help preserve the environment. www.unesco.org

UNESCO Man and the Biosphere Programme – a programme focused on establishing the science for improvement of human–-environment relationships.

UNESCO World Network of Biosphere Reserves (WNBS) – part of the Man and the Biosphere programme, fosters sustainable development on 686 reserves in 122 countries (2019).

United Nations Framework Convention on Climate Change (UNFCCC) – an international treaty seeking to reduce greenhouse gas emissions affecting climate that involved various accords and agreements leading to the Paris Agreement. unfccc.int/

United Nations Global Compact – a non-binding pact encouraging businesses worldwide to adopt and report on sustainable and socially responsible policies. unglobalcompact.org

United Nations Programme on Reducing Emissions from Deforestation and Forest Degradation (UN-REDD) – a collaboration of the UN FAO, UNDP and UNEP that seeks to reduce carbon emissions from and retain carbon stocks within forests while maintaining sustainable development.

United Nations Sustainable Development Goals (SDGs) – 17 goals, each with targets, adopted by the UN General Assembly (UN Resolution) to be reached by 2030 that promote human well-being, equity and sustainability. sdgs.un.org/

United Nations World Meteorological Organization (WMO) – an agency devoted to international cooperation on atmospheric science, climatology, hydrology and geophysics. wmo.int

The World Bank – the bank maintains a series of programmes dealing with sustainable development. www.worldbank.org/en/programs/global-program-on-sustainability

World Climate Research Programme (WCRP) – an international programme assisting to coordinate global climate research. www.wcrp-climate.org/

World Commission on Environment and Development (WCED or Brundtland Commission) – an effort to unite the world to pursue sustainable development, culminating in the report *Our Common Future* staking out the meaning of sustainable development [dissolved].

PERMISSIONS AND ACKNOWLEDGEMENTS FOR FIGURES AND TABLES

Permission has been obtained to publish all figures that are not open source or created by the author, with credits provided as sources for the figures. Where specific wording has been requested, it is provided below. Every effort has been made to trace copyright holders and obtain their permission for the use of copyright material. The publisher apologizes for any errors or omissions and would be grateful to be notified of any corrections that should be incorporated in future reprints or editions of this book.

Fig. 2.1 Reprinted by permission from Springer Nature. © (2019); Fig. 2.2 Reproduced with permissions from American Association for the Advancement of Science; Fig. 4.1 Reproduced with permissions from the Dewwool Team; Fig. 4.2 Reproduced with permissions from Michael Pidwirny; Fig. 5.1 Reproduced with permissions from Elsevier Science Ltd.; Fig. 5.2 Reproduced under the CC-BY-SA 2.5 licence; Fig. II.1 Adapted under the CC-By-SA 3.0 licence; Fig. 10.1 Reproduced with permissions from True Nature Foundation; Fig. 11.1 Reproduced with permissions from American Geophysical Union; Fig. 11.2 Reproduced with permissions from SAGE Publications; Fig. 13.1 Reproduced under the terms of Creative Commons Attribution 2.0 International licence (creativecommons.org/licenses/by/2.0); Table 15.1 Reprinted by permission from Springer Nature. © (2019); Fig. 17.1 Reproduced with permissions from American Association for the Advancement of Science; Fig. 17.2 Reproduced with permissions from Mongabay; Fig. 18.1 With open permission to reuse from UNEP/GRID-Arendal; Fig. 20.1 Reproduced under the terms of CC-By-SA 4.0 licence; Table 20.1 Reproduced under the terms of the Creative Commons Attribution 4.0 licence; Fig. 20.2 Reproduced with permissions from John Wiley & Sons, Inc.; Fig. 21.2 Reprinted by permission of the publisher (Taylor & Francis Ltd, www.tandfonline.com); Fig. 22.1 Reproduced under the Creative Commons Attribution 4.0 International licence. To view a copy of the licence, visit creativecommons.org/licenses/by/4.0/; Fig. 22.2 Reproduced under the terms of FAO's Open Data Licensing Policy: www.fao.org/3/ca7570en/ca7570en.pdf; Fig. 23.1A Reproduced under the terms of the Creative Commons Attribution-NonCommercial-NoDerivs licence; Fig. 23.1B Reproduced with permissions from PNAS; Fig. 24.1 Reproduced under the Creative Commons Attribution 4.0 International licence. To view a copy of the licence, visit creativecommons.org/licenses/by/4.0/; Fig. 25.1 Reproduced with permissions from Rob Pringle; Fig. 26.1 Reproduced under the CC-BY 3.0 licence;

Fig. 27.1 Reproduced under the Creative Commons Attribution 3.0 licence; Fig. 28.1 Reproduced under the CC-BY-SA 3.0 licence; Fig. 29.1 Reproduced with permission of PNAS; Fig. 29.1 Reproduced under the CC-BY-SA 4.0 licence; Fig. 30.1 Reproduced with permissions from Elsevier; Fig. 31.1 Reproduced with permission from the UN; Fig. 32.1 Reproduced under the terms of CC BY-SA 4.0; Fig. 32.2 Reproduced with permissions from Author Peter Gleick; Fig. 34.1 Database accessed with permissions from Sustainable Water Future Programme and Griffith University; Fig. 35.1 Reprinted by permission from Springer Nature. © (2012); Fig. 37.1 Reproduced under open licence conditions of the original publication; Table 37.1 Reproduced under CC-BY-SA 4.0 licence of the original publication; Fig. 37.2 Reproduced under CC-BY-SA 4.0 licence of the original publication; Fig. 38.1 Reprinted by permission from Springer Nature. © (2019); Fig. 40.1 Reproduced under the terms of GNU Free Documentation licence; Fig. 41.2 Reprinted with permission of the American Association for the Advancement of Science; Fig. 41.3B Reproduced under the terms of CC-BY-SA 3.0; Fig. 42.1 Reproduced under the terms of CC-BY-NC 4.0; Fig. 42.2 Reproduced under the terms of CC-BY licence; Fig. 42.3 Reprinted by permission from Springer Nature. © (2019); Fig. V.3 Reproduced with permission from NAS; Fig. 43.1 Available in public domain on Wikimedia Commons; Fig. 43.2 Reproduced under use of limited number permissions of the IPCC Report (www.ipcc.ch/copyright/); Table 44.1 Reproduced under use of limited number permissions of the IPCC Report (www.ipcc.ch/copyright/); Fig. 44.3 Reproduced with permissions from GCP; Fig. 44.4 Reproduced with permissions from GCP; Fig. 44.5 Reproduced with permissions from GCP; Fig. 44.7 Reproduced following the terms of the Creative Commons Attribution 4.0 licence of the original source; Fig. 46.1 Reproduced under use of limited number permissions of the IPCC Report (www.ipcc.ch/copyright/); Fig. 47.2 Reproduced under the terms of CC-BY-SA licence of the original publication; Fig. 47.3 Reproduced with permission from Berkeley Earth; Fig. 48.2 Reproduced under use of limited number permissions of the IPCC Report (www.ipcc.ch/copyright/); Fig. 51.1 Reproduced with permissions from Kerry Emanuel, MIT; Fig 54.2 Reproduced with permission. © (2018) *Scientific American*, a division of Nature America, Inc. All rights reserved; Fig. 55.2 Reproduced under the terms of CC-BY-SA 4.0; Fig. 56.1 Reproduced under PNAS open access conditions; Table 56.2 Reproduced under the Creative Commons Attribution-NonCommercial-ShareAlike 3.0 IGO (CC BY-NC-SA 3.0 IGO) explained here: creativecommons.org/licenses/by-nc-sa/3.0/igo/; Fig. 56.2 Reproduced with permissions of the photographer; Fig. 57.1 Reproduced under the terms of CC-BY-SA 3.0 licence; Fig. 58.1 Reprinted with permission of the American Association for the Advancement of Science; Fig. 59.1 Reproduced under the terms of CC-BY-SA 4.0 licence; Fig. 60.1 Reproduced under the terms of CC BY 3.0 licence; Fig. 60.2 Reproduced with permissions from Science; Fig. 62.1 Reproduced under the terms of Creative Commons Attribution 3.0 licence; Fig. 63.1 Reproduced under the terms of Creative Commons BY licence; Fig. 63.3 Reproduced under the terms of CC-BY-SA 3.0 licence; Fig. 63.4 Reproduced under the terms of Creative Commons Attribution 4.0 licence; Fig. 64.1 Reproduced under the terms of Creative Commons Attribution 4.0 International licence; Fig. 64.2 Reproduced under the Creative Commons Attribution-ShareAlike 4.0 International licence; Fig. 65.1 Reproduced with permissions from PNAS; Fig. 65.2 Reproduced under the terms of Creative Commons Attribution-Share Alike 4.0 International licence; Fig. 67.1 Reproduced under the terms of CC-BY-SA 3.0 licence; Fig. 67.2 Reproduced under the terms of CC-BY-SA 3.0 licence; Fig. 77.1 Reproduced with Permissions from PNAS; Table 78.1 Reproduced with permissions from Paul Kedrosky; Table 78.2 Reproduced with permissions from Paul Kedrosky; Fig. 79.1 Reproduced with permissions from Pew research; Fig. 82.1 Reproduced with permissions from CIESIN under the Creative Commons 3.0 Attribution licence; Fig. 82.1B Reproduced with permissions from CIESIN under the Creative Commons 3.0 Attribution licence; Fig. 88.1 Reproduced with permissions from American Association for the Advancement of Science; Fig. 89.1 Reproduced with permissions from PNAS. © (2008) National Academy of Sciences, USA; Table 89.1 Reproduced with permissions from PNAS. © (2008) National Academy of Sciences, USA; Fig. 89.2 Reproduced under terms of Creative Commons Attribution 4.0 licence; Fig. 93.1A/B Reproduced with consent from the creators. Fig. 95.1 Reproduced under the terms of Creative Commons Attribution 3.0 IGO licence (CC BY 3.0 IGO); Fig. 95.2 Reproduced under the Creative Commons Attribution-Share Alike 4.0 International licence; Fig. 96.1 Reproduced under the Creative Commons Attribution-ShareAlike 4.0 International licence; Fig. 97.3 Reproduced with permissions from Landscape of Grand Pre; Fig. 99.1 Icons reproduced with permissions from United Nations www.un.org/sustainabledevelopment/. Disclaimer: The content of this publication has not been approved by the United Nations and does not reflect the views of the United Nations or its officials or Member States; Table 99.1 Reproduced under the Creative Commons Attribution 3.0 IGO licence (CC BY 3.0 IGO) creativecommons.org/licenses/by/3.0/igo. Under the Creative Commons Attribution licence.

INDEX

Note: Page numbers in *italics* and **bold** denote figures and tables, respectively.